D1067745

Lecture Notes in Physics

The Lecture Notes in Physics

The series Lecture Notes in Physics (LNP), founded in 1969, reports new developments in physics research and teaching – quickly and informally, but with a high quality and the explicit aim to summarize and communicate current knowledge in an accessible way. Books published in this series are conceived as bridging material between advanced graduate textbooks and the forefront of research and to serve three purposes:

- to be a compact and modern up-to-date source of reference on a well-defined topic
- to serve as an accessible introduction to the field to postgraduate students and nonspecialist researchers from related areas
- to be a source of advanced teaching material for specialized seminars, courses and schools

Both monographs and multi-author volumes will be considered for publication. Edited volumes should, however, consist of a very limited number of contributions only. Proceedings will not be considered for LNP.

Volumes published in LNP are disseminated both in print and in electronic formats, the electronic archive being available at springerlink.com. The series content is indexed, abstracted and referenced by many abstracting and information services, bibliographic networks, subscription agencies, library networks, and consortia.

Proposals should be sent to a member of the Editorial Board, or directly to the managing editor at Springer:

Christian Caron
Springer Heidelberg
Physics Editorial Department I
Tiergartenstrasse 17
69121 Heidelberg / Germany
christian.caron@springer.com

K. Sekimoto

Stochastic Energetics

Ken Sekimoto
Matières et Systèmes Complexes
CNRS-UMR7057
Université Paris 7
France

and

Gulliver
CNRS-UMR7083
ESPCI ParisTech
Paris
France
ken.sekimoto@espci.fr
http://www.pct.espci.fr/~sekimoto/

Sekimoto, K.: *Stochastic Energetics*, Lect. Notes Phys. 799 (Springer, Berlin Heidelberg 2010), DOI 10.1007/978-3-642-05411-2

Lecture Notes in Physics ISSN 0075-8450 e-ISSN 1616-6361
ISBN 978-3-642-05410-5 e-ISBN 978-3-642-05411-2
DOI 10.1007/978-3-642-05411-2
Springer Heidelberg Dordrecht London New York

Library of Congress Control Number: 2009943129

Cover design: Integra Software Services Pvt. Ltd., Pondicherry

Printed on acid-free paper

Springer is part of Springer Science+Business Media (www.springer.com)

To my dear readers.

Preface

What is the *Stochastic energetics*? In brief, this is the framework that connects the "missing link" between the stochastic dynamics and the thermodynamics. See Fig. 1. Since the nineteenth century, the thermodynamics has been established through the efforts of Carnot, Mayer, Clausius, and many others. Later on the statistical mechanics is developed by Maxwell, Boltzmann, Gibbs, Onsager, Einstein, and others. The latter discipline relates the thermodynamics to the microscopic mechanics. The study of the stochastic dynamics, or the phenomena of thermal fluctuations, has also a long history since Robert Brown reported the so-called Brownian motion of a fine particle of pollen (Brownian motion, 1827). The modern framework of stochastic process is established by Einstein [1], Smoluchowski [2], Perrin [3], Langevin [4], Fokker and Planck [5, 6], Kramers [7], Itô [8], and others. More recently this framework has been justified from micro mechanics with the aid of the so-called projection methods by Zwanzig [9], Mori [10], Kawasaki [11], and others. It is, therefore, natural that there is a link between the stochastic dynamics and the thermodynamics, as Fig. 1 suggests. This book is the first lecture notes in English on this linkage.[1]

The *Stochastic energetics* adapts to the modern development of nanotechnologies. The vast improvement of the spatiotemporal resolution in these technologies has enabled the access to the individual thermal random processes at sub-micron ($1\mu\mathrm{m} = 10^{-4}\mathrm{cm}$) and sub-millisecond scales. While the entropy, which characterizes ensemble of fluctuating states and processes, is a core concept of the thermodynamics and statistical mechanics, a complementary approach to deal with the individual realizations is now needed also. Those what were mere thought experiments (Gedankenexperiment) in the mid-twentieth century are now testable models such as so-called *Feynman ratchet and pawl* [13] and many models related to the *Maxwell's demon* [14].

This book is, therefore, for those people who study and work on the scale of thermal fluctuations, such as microfluidics, nano machines, nano sensing devices,

[1] This is a highly enlarged and revised version of the Japanese book published before by the author [12]. The present version has about 1.5 times of pages with many new figures and a new chapter (Chap. 6) with respect to the Japanese version.

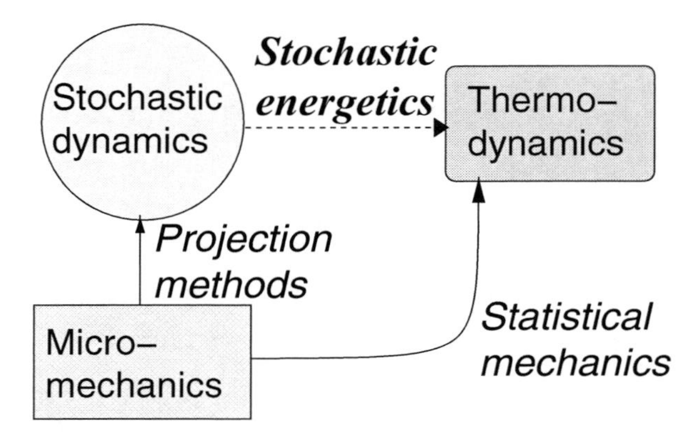

Fig. 1 The *Stochastic energetics* as completion of the "missing link"

nanobiology, nanoscopic chemical engineering, etc. In other words, those who are interested in the following questions will find the answer, or at least some clues, in this book:

- What is the heat associated to the thermal random (Brownian) motion of a mesoscopic particle?
- What work do we need for the operation and observation of small system?
- How much is the work to operate an ideal Carnot engine? Is it reversible?
- Can we cool a drop of water by agitating a nanoparticle immersed therein?
- How does the heat flow if a particle undergoing Brownian motion pulls a polymer chain?
- Is the energy conserved during an individual realization of Brownian motion?
- Is the projection methods, which eliminates rapid microscopic motions, compatible with reversible or quasiequilibrium process?
- Can we measure the free energy of the system by a single realization of stochastic process?
- Are there quantum mechanics-like uncertainty or irreversibility upon the measurement of thermal random process?
- Is the definition of the heat unique? Is the thermodynamics unique for any particular system ?
- Does a particle carry the chemical potential when it enters into an open system from the environment?
- Why does the chemical potential of a molecule depend on its density even if the molecule does not interact with other molecules?
- Do we need an irreversible work to make a copy of the information in a bit memory?
- Can we detect reversibly the arrival of a Brownian particle with 100% of sureness at finite temperature?
- Do molecular motors need to stock a large energy in order to do a large work?

The readers of this book are assumed to have the very basics of thermodynamics and Newtonian mechanics, as well as elementary analysis and the ordinary differential equation. No advanced knowledge at the level of physics graduate courses is required. Examples in the main text are all simple.

Limitation of the scope of this book: First, we do not deal with the quantum fluctuations. We assume that the temperature of the environment is high enough that the quantum interferences are negligible. The mesoscopic quantum systems are already an established field (see, for example, an inspiring book [15]). Second, we discuss very little about the subjects of nonequilibrium statistical mechanics around the fluctuation theorem (FT) and the (Jarzynski) nonequilibrium work relation. These subjects have been rapidly developed between the late 1990s and early 2000s, when *Stochastic energetics* has also been formulated. These subjects deal with the ensemble of stochastic processes, where the former framework has been used. I hope that some comprehensive books will be written by the people who initiated these vast subjects. Last, as the schema of Fig. 1 shows, the *Stochastic energetics* works where the mesoscopic scale is more or less well defined as distinguished both from micro level and from macro level. It, therefore, does not apply to the phenomena where the interference among these scales is strong [16].

Organization of the book: This book consists of three parts. See Fig. 2. Glance at the table of Contents will give you more detailed composition of each chapters. In the three chapters of *Part I* we will prepare the basics of the stochastic dynamics (mainly the Langevin equation), the thermodynamics, and the reaction dynamics (including the master equation). The following three chapters in *Part II* will introduce the basic concept of the heat on the mesoscopic scale and describe its consequences. In the last two chapters belonging to *Part III* we will see, through simple examples, various strategies and constraints in the fluctuating world. The asterisk * at the head of sections or subsections indicates the core part of the book. The technically advanced descriptions are given in the Appendix at the end of the book. There are already good textbooks of stochastic dynamics such as [17–19, 8], where the thermodynamic equilibrium and the fluctuation–response relations are well described in the context of stochastic dynamics. Complementarily to these textbooks, we consider in this book the stochastic dynamics whose parameter(s) are changed in time from outside or by other subsystems. This generalization opens a fertile field of physics and enables to fill up the "missing link" mentioned above. I hope that the readers enjoy the richness of physics in the *Stochastic energetics*.

About the references: The papers cited in the main text are those to which I owe directly their ideas, methods, and perspectives. Therefore, the references in this

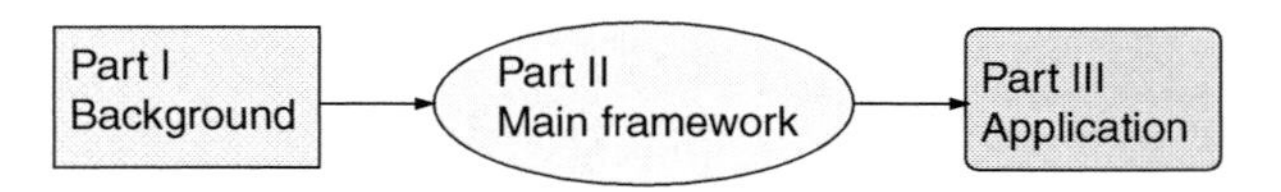

Fig. 2 Organization of the book

book do not cover the whole activities of the communities related to the stochastic dynamics. I apologize those people of whom I overlooked important papers either because of my limited capacity of understanding or bibliographical search.

Paris, December 2009 Ken Sekimoto

Acknowledgment

I would like to take this opportunity to express my deep thanks to my coauthors and the others to whom the contents of this book owes a lot:

(The coauthors on related subjects) T. Hondou, Y. Miyamoto, E. Muneyuki, E. Muto, T. Ooshida, C. Ribrault, S.I. Sasa, K. Sato, T. Shibata, F. Takagi, K. Tawada, and A. Triller. (The other reserchers) A. Allahverdyan, H. Asai, C. Bagshaw, M. Bazant, H. Hasegawa, S. Ishiwata, C. Jarzynski, T. Kodama, R. Kanada, R. Kamiya, K. Kawasaki, T. Komatsu, A. Libchaber, S. Leibler, N. Miyamoto, M. Mizuno, T. Munakata, N. Nakagawa, H. Noji, F. Oosawa, J. Parrondo, L. Peliti, K. Sasaki, F. Sasaki, T. Sasada, M. Schindler, U. Seifert, A. Shimizu, K. Sutoh, Y. Tanaka, H. Tasaki, T. Tsuzuki, M. Tokunaga, C. Van den Broeck, A. Yamada, T. Yanagida, R. Yasuda, T. Yuge, and S. Yukawa.

I also thank the colleagues of Groupe de Physico-Chimie Théorique of E.S.P.C.I. and the Laboratory of Matières et Systèmes Complexes of Paris 7th University, who provided me with intellectual and comfortable research environment.

Concerning the proposition of this book to Springer, I gratefully acknowledge those who have given invaluable encouragement and concrete help; M. Doi, R. Kawai, J. Kurchan, E. Muto, S. Nakamura, I. Ojima, and S. Ramaswamy.

I especially acknowledge R. Kawai, Y. Oono, S. Sasa, and the anonymous reviewers for reading through the draft and giving precious comments. Also R. Kawai kindly provided me with Fig. 4.10 related to their paper [20]. R. Adhikari helped me by pre-proofreading the beta version as a native English speaker and a physicist. Also I appreciate P. Marcq, who kindly prepared the translation of the first few chapters. Although all the comments by the above mentioned people have been taken into account as far as I can, the responsibility of all the eventual errors, ambiguities, or incompleteness is to the author.

I thank the senior editor of the Springer publishing, C. Caron, and the senior editor of Iwanami Books Ltd., N. Miyabe, for their professional editorship.

References

1. A. Einstein, *Investigation on the theory of the Brownian Movement*, (original edition, 1926) ed. (Dover Pub. Inc., New York, 1956), chap.V-1, pp. 68–75
2. M. von Smoluchowski, Ann. der Phys. **21**, 756 (1906)

3. J. Perrin, *Les Atomes* (Librairie Félix Alcan, 1913)
4. P. Langevin, Compt. Rend. **146**, 530 (1908)
5. A.D. Fokker, Ann. der Phys. **43**, 310 (1915)
6. M. Planck, Sitzungsber. Preuss. Akad. Will. Phys. Math. **K1**, 325 (1917)
7. H.A. Kramers, Physica **7**, 284 (1940)
8. K. Itô, *Stochastic Processes (Lecture Notes from Aarhus University 1969)* (Springer Verlag, New York, 2004)
9. R. Zwanzig, J. Chem. Phys. **33**, 1338 (1960)
10. H. Mori, Prog. Theor. Phys. **33**, 423 (1965)
11. K. Kawasaki, J. Phys. A **6**, 1289 (1973)
12. K. Sekimoto, *Stochastic Energetics* (Iwanami Book Ltd., 2004, in Japanese)
13. R.P. Feynman, R.B. Leighton, M. Sands, *The Feynman Lectures on Physics – vol.1* (Addison Wesley, Reading, Massachusetts, 1963), Sects. 46.1–46.9
14. H.S. Leff, A.F. Rex, *Maxwell's Demon: Information, Entropy, Computing* (A Hilger (Europe) and Princeton U.P. (USA), 1990)
15. Y. Imry, *Introduction to Mesoscopic Physics* (Oxford University Press, UK, 1997)
16. Y. Oono, *Introduction to Nonlinear World* (University of Tokyo Press, 2009, in Japanese)
17. N.G. van Kampen, *Stochastic Processes in Physics and Chemistry*, Revised ed. (Elsevier Science, 2001)
18. C.W. Gardiner, *Handbook of Stochastic Methods for Physics, Chemistry for Natural Sciences*, 3rd ed. (Springer, 2004)
19. H. Risken, *Fokker-Planck Equation*, 2nd edn. (Springer-Verlag, Berlin, 1989)
20. R. Benjamin, R. Kawai, Phys. Rev. E 77, 051132 (2008)

Contents

Part I
Background of the Energetics of Stochastic Processes

Chapter 1
Physics of Langevin Equation

As a broad introduction, we describe what we will discuss in this chapter.

Many different ways to describe fluctuations have been developed in different fields: physics, chemistry, mathematics, economics, or genetics, to mention a few. For physicists there are good textbooks that discuss fluctuations, for example, [1, 2]. Among those different ways of the description, we will describe in this book mainly the method of Langevin equation, because it describes individual processes of fluctuation. Throughout this book we regard the Langevin equation as an "equation of motion."

We first introduce the notion of fluctuations. The notion of fluctuation is related to statistics and uncertainty. We argue that the appearance and the consequence of fluctuations differ according to the space-time scales of observation or operation. A typical example of temporal fluctuation is the *Brownian motion*, the fluctuations of mechanical origin. The Brownian motion is related to diffusion. The Brownian motion is caused by the thermal random force. The thermal random force is often modeled as a special stochastic process called Gaussian white noise. This reflects the fact that the Brownian motion is caused by many microscopic impacts on the degree of freedom (variable) that undergoes Brownian motion. We assume that the environment is in thermal equilibrium.

The Langevin equation is a way to describe the Brownian motion under the influence of additional forces as well as the thermal random force. Even if the Brownian particle behaves in a nonequilibrium manner under an external forcing, we assume that the environment remains in equilibrium. We show a heuristic argument to introduce the Langevin equation. We describe also a simple and solvable example presented by Zwanzig [3].

We emphasis, however, that the Langevin equation is not an axiom or a mere hypothesis introduced empirically from the observation. Under several well-defined assumptions such as the Markov approximation, the Langevin equation is derived from the microscopic mechanics governing the whole system consisting of the entity undergoing the Brownian motion and the other fast varying microscopic degrees of freedom. The thermal environment for the resulting Langevin equation is constituted by the latter degrees of freedom. We leave the mathematical details of the derivation to the Appendix and the original literatures.

Sekimoto, K.: *Physics of Langevin Equation*. Lect. Notes Phys. **799**, 3–66 (2010)
DOI 10.1007/978-3-642-05411-2_1

In mathematics the temporal fluctuation is treated as the stochastic process. The powerful mathematical framework for the stochastic process is called stochastic calculus. The *Wiener process* is the mathematical counterpart of the free Brownian motion, and the stochastic differential equation (SDE) corresponds to the Langevin equation. But the stochastic calculus is not merely a formulation for mathematicians. Without it the Langevin equation is sometimes not unambiguously defined. The stochastic calculus deals with the product of variables or integral of functions, just as the usual analysis. But the rules of the stochastic calculus are different from the latter, the difference which often causes nonnegligible consequences. An example is the *Itô's lemma*. Because of this difference, the numerical schemes for solving the Langevin equation also require special care.

Roughly speaking the Langevin equation takes the thermal random force as input and generates the stochastic process of the variable, such as the position of Brownian particle, as an output. Although the thermal random force is supposed to contain no memory, i.e., no finite temporal correlations, the Langevin equation transforms this stochastic process into the output which has memory of the past. For example, the trajectory of a Brownian particle tells where it found itself sometime ago. The Langevin equation can take different forms depending on the time resolution of the description or of the observation, even if they describe the same physical process. This difference comes from the above mentioned memory time. Elimination of the inertia effect of the Brownian particle or elimination of fine spatial modulations of the potential energy causing the force on the Brownian particle can, therefore, change the form of the Langevin equation.

Suppose that a Brownian particle is subject under (only) a potential force, the force due to a fixed potential energy as function of the Brownian particle, on top of the thermal random force. The longtime observation of a particular stochastic process, i.e., a particular solution of Langevin equation for this particle, will then give rise to a stationary probability distribution as function of the position, and the momentum if the inertia is not ignored. This probability distribution is the residence time distribution of the position (and momentum). This stationary state is the equilibrium state by definition, while the equilibrium statistical mechanics gives the probability distribution as the canonical distribution [4]. This fact imposes an essential constraint on the form of the Langevin equation. This constraint is a form of the relation called *fluctuation–dissipation relation*.

The statistical mechanics discusses the *ensemble* of copies of the systems and the probability distribution on this ensemble. In the stochastic process, the related notion is the ensemble of the solutions of Langevin equation. The partial differential equation called the Fokker–Planck equation describes the evolution of the instantaneous probability distribution function. Using the Fokker–Planck equation, we can discuss the "first passage problem."

In most of the chapters we discuss the stochastic process of continuous variable, such as the position and momentum of a Brownian particle. For the stochastic process of discrete variable, or for the system undergoing the transitions among discrete states, we can construct the counterpart of the Langevin equation. The environment is still assumed to be in equilibrium.

1.1 Random Events or Fluctuations

*1.1.1 * Introduction: What Is Fluctuation?*

We are surrounded daily by the phenomena and the properties which we call fluctuations, such as the density of traffic on a highway, the variations within a biological species, the up-and-down of stock markets, the twinkling of stars, and so on. Let us characterize qualitatively the fluctuation. The fluctuation is related to the irregularity or stochasticity among the realizations, i.e., among the individual specimens of the data. The two main characteristics of the irregularity or stochasticity are the statistical nature and the uncertainty. By the statistical nature we mean that its characterization requires an ensemble of data. By the uncertainty we mean that, given a part of the data or the data up to a given moment of time, we cannot predict precisely the remaining data or the data in the future.

Because of this uncertainty, time series including fluctuations look as if their data were generated spontaneously. Time series containing fluctuations have often, but not as a rule, the property that the effect of an external stimulus on them disappears gradually in time.

How much fluctuations there are in a set of data depends on the resolution with which the data are analyzed and also on the information that we have on the data. A classical mechanical model for the motion of water molecule determines their velocities and positions at all time. The only source of fluctuation comes from the initial condition of which the chaotic dynamics is extremely sensitive. In the opposite limit, when we look at the water on the scale of hydrodynamics, the statistical properties of the microscopic motion of water molecules appear only through the viscosity coefficient, which is related to the temporal correlation of the stress field. In order for the underlying microscopic motions to manifest as "fluctuations" in the sense of the word, the characteristic scales of the fluctuations should be neither too small nor too large, as compared with the resolution of the observer who analyzes the fluctuations. In molecular dynamic description, everything is deterministic, while in hydrodynamic description, the macroscopic observable obeys deterministic equations of hydrodynamics. Neither of these describes directly the fluctuations. Therefore, we should be conscious about the space-time scales when we discuss the fluctuation.

As another example, let us take the winds, which include the atmospheric fluctuations of the air. The winds are neither completely predictable nor completely unpredictable. As long as we know that the wind direction does not change too quickly, and if we can detect it with a good enough time resolution, we can reorient wind turbines (windmill) along the temporary wind direction so that the turbines convert the energy of the winds into electric power. By analogy, we might ask if protein molecular motors have evolved under the selective pressure so that they can make profit of certain characteristics of the thermal fluctuations in the cytosol.

We should also note that whether a fluctuating quantity is regarded as a noise or a signal depends on the timescale with which we can respond to this fluctuation. Related to the last examples, suppose that there is a receptor on a lipid membrane of a cellular organelle waiting for the arrival of a molecular signal from the surrounding cytoplasm. Or, suppose that a protein motor, say, myosin is waiting for the arrival of an ATP molecule. The arrival of a signal molecule or an ATP molecule is at random in the sense that there is no regularity except for its average frequency, which serves for nothing as a signal. However, this random event can become a useful signal or a source of mechanical work because the receptor or the protein motor are designed so that they can respond to such event of arrival by changing their internal state, called the conformational change [5]. The unbinding of those molecules can be made difficult by this conformational change, and the receptor or the protein motor will start a cascade of functional process.[1]

1.1.2 Random Events Are Represented Through Random Variables

1.1.2.1 * Random Variables and Stochastic Processes

A particular realization vs. random variable: Now we will introduce several concepts and methods to deal with fluctuations at a more quantitative level. First we will define the variables that undergo fluctuations, called the random variables, and will introduce the related notations.

In order to refer generically to the number on a die which is cast properly, it is convenient to introduce a variable $\hat{n}$ that can take integer values from 1 to 6. We then use this variable to denote, for example, the square of the number on the die just by writing $\hat{n}^2$. If the number was eventually 3 in a realization, that is, upon a particular cast, we say that we had $\hat{n} = 3$. However, the value of $\hat{n}$ will not be fixed at 3 henceforth, but $\hat{n}$ remains a variable. We will call a variable such as $\hat{n}$ the random variable. When we need to distinguish a random variable among the other variables, we shall attach the hat (ˆ) to the former, like $\hat{n}$. If $f(x)$ is a function of x, the substitution of a random variable $\hat{n}$ defines a new random variable, $\hat{m} = f(\hat{n})$. For example, $\hat{m} = \hat{n}^2$ for $f(x) = x^2$.

Probability associated to each value of the random variable: We assume that an a priori probability is associated to each value that a random variable can take. In the case of a proper die, the probability 1/6 should be associated to each realizable value, i.e., the integers from $\hat{n} = 1$ to $\hat{n} = 6$. Once such association is done for a random variable, the probabilities associated for any function of this random variable are also known. For example, the probabilities for $\hat{m} \equiv \hat{n}^2$ are 1/6 to realize $\hat{m} = 9$ and 0 to realize $\hat{m} = 10$.

Average and variance: It is then possible to evaluate the average, or the mean value, of a given function of the random variable. Let us denote by $Prob[\hat{m} = i]$ the probability of the realization of $\hat{m} = i$. We denote by $\langle \hat{m} \rangle$ the average of a random variable, $\hat{m}$. Therefore,

[1] For the receptors or the motor proteins to function, we also need the global nonequilibrium of the environment, which promotes the process in one direction against the reversed one.

$$\langle \hat{m} \rangle = \sum_i i \; Prob[\hat{m} = i].$$

Also we define the variance of $\hat{n}$ by

$$\langle (\hat{m} - \langle \hat{m} \rangle)^2 \rangle = \sum_i (i - \langle \hat{m} \rangle)^2 Prob[\hat{m} = i].$$

In the case of the proper die, we have the average of $\hat{n}$ or $\hat{m} \equiv \hat{n}^2$, i.e., $\langle \hat{n} \rangle = 7/2$ and $\langle \hat{n}^2 \rangle = 91/6$, or the variance, $\langle [\hat{n} - \langle \hat{n} \rangle]^2 \rangle = 35/12$. The random variable bears both the statistical nature and the uncertainty, and therefore fits with our notion of the variable which undergoes fluctuations.

What if we want to know the average $f(\hat{n})$ in knowing $Prob[\hat{n} = i]$? The answer is

$$\langle f(\hat{n}) \rangle = \sum_i f(i) \, Prob[\hat{n} = i].$$

Two advanced examples of the discrete random variables (binomial distribution and Poisson distribution) are given in Appendix A.1.1.

Continuous random variable: We may consider a random variable $\hat{y}$ that can take any real value, for example, on the interval $[0, 1]$. If we chose completely at random a point y on this interval, every value of y is, by definition, equally likely. But the probability to realize a particular real value y is 0. Still we can discuss the probability of finding y within a segment $[a, b]$ with $0 \le a < b \le 1$. In this case $Prob[\hat{y} \in [a, b]] = b - a$. In general, if a random variable $\hat{y}$ takes a value on the real axis $\mathbb{R}$, we define the probability associated to $\hat{y}$ in terms of the probability density, $p(y)$, a nonnegative-valued real function such that

$$Prob\left[\hat{y} \in \left[y - \frac{dy}{2}, y + \frac{dy}{2}\right]\right] = p(y)dy.$$

If the probability is singularly concentrated on some points, y_α, we suppose that $p(y)$ contains the delta "functions" $\propto \delta(y - y_\alpha)$. Physicists symbolically use, or abuse, this useful function to write $p(y) = \langle \delta(y - \hat{y}) \rangle$, because this expression reproduces correctly the average of any function of the random variable $\hat{y}$, such as, $f(\hat{y})$ in that $\langle f(\hat{y}) \rangle \equiv \int f(y) p(y) dy$.[2]

Characteristic function for continuous random variable:[3] The random variable $f(\hat{y}) = e^{i\phi\hat{y}}$ (ϕ: real) has a special utility because its average $\Phi_y(\phi) \equiv \langle e^{i\phi\hat{y}} \rangle$ is related to the Fourier transformation of the probability density:

---

[2] Note that $\int f(y) \langle \delta(y - \hat{y}) \rangle dy = \langle \left[\int f(y) \delta(y - \hat{y}) dy\right] \rangle = \langle f(\hat{y}) \rangle$.

[3] Those who are not interested or familiar to the analysis may skip the details of the characteristic functions and also its applications to the Gaussian random variables which appear later on.

$$\Phi_y(\phi) = \int_{-\infty}^{+\infty} e^{i\phi y} p(y)dy. \tag{1.1}$$

Therefore, the characteristic function $\Phi_y(\phi)$ contains (almost) the same amount of information as $p(y)$, as long as this integral is convergent. The advantage of $\Phi_y(\phi)$ with respect to $p(y)$ is that the average of any other function $g(\hat{y})$ can be obtained simply by the differentiations of the former function:

$$\langle g\left(\hat{y}\right)\rangle = g\left(\frac{1}{i}\frac{d}{d\phi}\right) \Phi_y(\phi)\Big|_{\phi=0}. \tag{1.2}$$

For example, for $g(y) = y^2$, we have $(\frac{1}{i}\frac{d}{d\phi})^2 \langle e^{i\phi\hat{y}}\rangle\Big|_{\phi=0} = \langle \hat{y}^2\, e^{i\phi\hat{y}}\rangle|_{\phi=0} = \langle \hat{y}^2\rangle$.

Gaussian random variables: Among the continuous random variables, the those called the Gaussian random variables are of particular importance for the reasons which we will see below in different occasions. The probability density for this type of random variables (Gaussian distribution, for short) is of the following form:

$$p^{(G)}(y) = \frac{1}{\sqrt{2\pi\sigma^2}} e^{-\frac{(y-y_0)^2}{2\sigma^2}}. \tag{1.3}$$

Here y_0 is the average, $y_0 = \langle\hat{y}\rangle$, and σ^2 is the variance, $\sigma^2 = \langle[\hat{y} - \langle\hat{y}\rangle]^2\rangle$. The characteristic function $\Phi_y(\phi)$ for the Gaussian random variable takes especially simple form,

$$\Phi_y^{(G)}(\phi) = e^{i\phi y_0}\, e^{-\frac{\sigma^2\phi^2}{2}}.$$

In case of $y_0 = 0$, the average $\langle\hat{y}^{2n+1}\rangle$ ($n \geq 0$: integer) are 0 by symmetry, while

$$\frac{\langle\hat{y}^{2n}\rangle}{(2n)!} = \frac{\langle\hat{y}^2\rangle^n}{2^n\, n!}. \tag{1.4}$$

The reader might verify the identity $\langle y\, f(y)\rangle = \langle y^2\rangle\langle f'(y)\rangle$, which is a simple case of so-called Novikov's theorem.

A particular aspect of Gaussian distribution is explored in the context of the relation between fluctuation and response [6]. In Appendix A.1.2 we describe its outline.

Mathematician's view of random variables: The mathematician's view of probability is apparently somehow different from the physicists' view, though the essential notions are the same. The former view assumes a "fundamental" random variable, $\hat{\omega}$, with the probabilities (or "measure") P associated to each value of $\hat{\omega}$ in the domain Ω. With some technical precision for the admissible ensemble (set) of those values,[4] called *family of sets*, $\mathcal{F}$, they define the "probability space" $(\Omega, P, \mathcal{F})$. If Ω is a continuous set, one may imagine a monochrome canvas as Ω where the local brightness corresponds to the density of probability.

In this setup, any other random variable, say $\hat{X}$, should be the function of the fundamental one, i.e., $\hat{X} = f_X(\hat{\omega})$. Each random variable $\hat{X}$ has its own domain Ω_X and is associated by its own probabilities P_X as well as the family of sets $\mathcal{F}_X$. That is, a new probability space $(\Omega_X, P_X, \mathcal{F}_X)$ is produced. For the physicists' view, this change from $\hat{\omega}$ to $\hat{X}$ could be interpreted as a change of resolution of observation or measurement of a particular aspect of fluctuations. The ultimate fundamental random variable should describe everything, leaving no room for the fluctuation.

1.1.2.2 Stochastic Process

Random variable with more than one components: The random variable can have more than one component. Coming back to the example of die, we can define a random variable $\hat{n} \equiv \{\hat{n}_1, \hat{n}_2, \hat{n}_3\}$ such that its three components $\hat{n}_i$ $(i = 1, 2, 3)$ represent the numbers on a die upon three consecutive acts of casting. As another example, we can define $\hat{m} \equiv \{\hat{n}_1+\hat{n}_2, \hat{n}_2+\hat{n}_3\}$ with using the same $\hat{n}_i$ $(i = 1, 2, 3)$ as above. In the latter case, the two components of $\hat{m}$ are not independent. We can still associate a probability to each realization of $\hat{m}$. For example, we associate to $\hat{m} = \{4, 8\}$ the probability 1/108 since this is realizable by the two cases, $\{\hat{n}_1, \hat{n}_2, \hat{n}_3\} = \{1, 3, 5\}$ and $\{2, 2, 6\}$, each of which is realized with a probability 6^{-3}. If the continuous random variable has two components, say $(\hat{y}_1, \hat{y}_2)$, the probability density has also two variables, i.e., $p(y_1, y_2)$, and the average of a random variable $f(\hat{y}_1, \hat{y}_2)$ is the double integral, $\langle f(\hat{y}_1, \hat{y}_2)\rangle = \iint f(y_1, y_2) p(y_1, y_2) dy_1\, dy_2$.

A multicomponent Gaussian random variable $\boldsymbol{y} = (y_1, \ldots, y_d)$ is defined through its probability density

$$p^{(G)}(\boldsymbol{y}) = \frac{1}{(2\pi)^{d/2}\sqrt{\det(\mathsf{M})}} e^{-\frac{1}{2}(\boldsymbol{y}-\boldsymbol{y}_0)^t \mathsf{M} (\boldsymbol{y}-\boldsymbol{y}_0)}, \tag{1.5}$$

where $\boldsymbol{y}_0$ is the average, $\boldsymbol{y}_0 = \langle\hat{\boldsymbol{y}}\rangle$, M is positive definite symmetric $d \times d$ matrix. The characteristic function $\Phi_y(\boldsymbol{\phi}) \equiv \langle e^{i\boldsymbol{\phi}\cdot\hat{\boldsymbol{y}}}\rangle$ is calculated as

$$\Phi_y^{(G)}(\boldsymbol{\phi}) = e^{i\boldsymbol{\phi}\cdot\hat{\boldsymbol{y}}_0} \exp\left[-\frac{1}{2}\boldsymbol{\phi}^t \mathsf{M}^{-1}\boldsymbol{\phi}\right]. \tag{1.6}$$

[4] Though we do not go into the detail of the technical precision, this is an essential point concerning the continuous random variables. The situation is somewhat analogous and related to the fact that the rational numbers are dense in the real number, but is very rare compared with irrational numbers. See, p. 23 of [1].

The meaning of the matrix M is made clear from $\Phi_y^{(G)}(\boldsymbol{\phi})$, which tells immediately that $\langle (y_k - y_{k,0})(y_l - y_{l,0})\rangle = (\mathsf{M}^{-1})_{kl}$, or M is the inverse of the covariance matrix, $\langle (\boldsymbol{y} - \boldsymbol{y}_0)\,(\boldsymbol{y} - \boldsymbol{y}_0)^t\rangle$.

Stochastic process as a random variable: Especially important case of multicomponent random variable is the case where the number of components is infinite and the index to distinguish those components of the variable represents the "time."[5] We call this random variable the stochastic process. Each realization of a stochastic process is, therefore, a function of time. The time index, say t, of a stochastic process, say $\hat{\xi}$, is represented either as the suffix like $\hat{\xi}_t$ or as the argument of a function like $\hat{\xi}(t)$. In order to remind that a realization of stochastic process is a random variable, we will sometimes call it the "path" and denote as $\xi()$ where the generic index, t, is suppressed.

When we talk about a function of a stochastic process, say $f[\hat{\xi}]$, in fact $f[y]$ is the functional which takes a function $\xi()$ as variable. So is the probability density, $p[y]$. Then its average, $\langle f[\hat{\xi}]\rangle$, is obtained by performing infinitely multiple integrals called the functional integral of the product $f[y]\,p[y]$. Each integral is for each variable of integration, $\xi(t)$, with the index, t. In this book we will not go into details of the functional integral approach to the stochastic process. However, we should always keep in mind that any average $\langle f[\xi]\rangle$ is not the average over the value of $\xi(t)$ with particular index t, but over the whole paths $\xi()$. It is only when each component of $\hat{\xi}$ obeys the mutually uncorrelated fluctuations that the average $\langle \hat{\xi}(t)\rangle$ reduces to the average over a single random variable, $\xi(t)$.

Characteristic functional for the stochastic process: As an extension of the characteristic function $\Phi_y(\phi)$ for the single random variable, $\hat{y}$, we introduce the characteristic functional for the stochastic process, $\hat{\xi}$ [7]:

$$\Phi_\xi[\phi] \equiv \left\langle \exp\left[i \int_{-\infty}^{\infty} \phi(t)\hat{\xi}(t)dt \right]\right\rangle, \tag{1.7}$$

where $\phi(t)$ is any real function of time with proper smoothness. The brackets have the same meaning as before, that is, we should take the sum over all the possible realizations of $\hat{\xi}$.

The Gaussian stochastic process is a straightforward generalization of Gaussian random variable of multiple components. Later we will assert that the thermal random force is a special type of stochastic process. We define this process by its characteristic functional, $\Phi_\xi[\phi]$. For simplicity, we consider the case where $\langle \hat{\xi}(t)\rangle = 0$.[6] Instead of the double sum in the bilinear form $\phi^t \mathsf{M}^{-1}\,\phi$ in (1.6), we need the double integral with respect to the index of ϕ:

---

[5] The range of t is supposed to be fixed unless mentioned otherwise explicitly.

[6] We can simply redefine $\hat{\xi}(t) - \langle \hat{\xi}(t)\rangle$ as new $\hat{\xi}(t)$.

$$\Phi_{\xi}[\phi] = \exp\left[-\frac{1}{2}\int_{-\infty}^{\infty}\int_{-\infty}^{\infty}\phi(t_1)K(t_1,t_2)\phi(t_2)dt_1dt_2\right]. \tag{1.8}$$

As any order correlation $\langle \hat{y}^n \rangle$ was given in terms of the variance and the average of $\hat{y}$ for a single-component Gaussian random variable, any N-point correlation function for $\hat{\xi}(t_1), \cdots, \hat{\xi}(t_N)$ of a Gaussian stochastic process is represented in terms of the covariances, $\langle \hat{\xi}(t_j)\hat{\xi}(t_k)\rangle$. Taking twice the functional derivative of (1.8) with respect to $\phi(t)$ and $\phi(t')$, and then letting $\phi(t) = 0$ for all t, we see that the kernel $K(t,t')$ yields the correlation function of $\hat{\xi}$, that is,

$$K(t,t') = \langle \hat{\xi}(t)\hat{\xi}(t')\rangle. \tag{1.9}$$

As another example of the usage of the characteristic function, let us calculate the probability density $\langle \delta(\xi - \hat{\xi}(t))$ for $\hat{\xi}(t)$ at a time t. By choosing $\phi(t) = z\delta(t)$, (1.8) and (1.9) yield $\Phi_{\xi}[z\delta(t)] = \langle e^{iz\hat{\xi}(t)}\rangle = e^{-\langle \hat{\xi}(t)^2\rangle z^2/2}$. We then have[7]

$$\langle \delta(\xi - \hat{\xi}(t))\rangle = \int_{-\infty}^{+\infty} e^{-iz\xi}\left\langle e^{iz\hat{\xi}(t)}\right\rangle \frac{dz}{2\pi} = \frac{e^{-\frac{\xi^2}{2\langle \hat{\xi}(t)^2\rangle}}}{\sqrt{2\pi}\langle \hat{\xi}(t)^2\rangle^{\frac{1}{2}}}.$$

1.1.2.3 * Statistical Properties of Fluctuations as a Sum of a Large Number of Random Variables

Independence of random variables: As a preparation to the description of the thermal random force, we discuss here the statistical properties of the sum of a large number of random variables, $\hat{a}_i$ ($i = 1, \ldots, N$ with $N \gg 1$).[8] We suppose that these variables are mutually independent (uncorrelated). That is, the probability to realize $\hat{a}_i = a_i$ (where $i = 1, \ldots, \nu$ are mutually distinct index) writes

$$Prob[\{\hat{a}_i = a_i\}] = \prod_i Prob[\hat{a}_i = a_i], \quad \{\hat{a}_i\}\text{: independent.}$$

We assume further that these random variables share the same statistical properties.

For example, let us suppose that there is a huge container filled with a gas and that we observe, in this container, N nonoverlapping domains of volume Δv. We define the random variable $\hat{a}_i$ as the total kinetic energy of the gas particles found at a time within the ith domain. We assume that the sum of the volumes we observe, $N\Delta v$, is much smaller than the total volume of the container and that each domain is well separated from the other domains. Then, unless the gas is extremely dilute, we expect that the random variables $\{\hat{a}_i\}$ are mutually almost independent and that

[7] We use $\int_{-\infty}^{+\infty} e^{-iz\xi}dz = 2\pi\delta(\xi)$ as well as $\int_{\infty}^{\infty} e^{-ax^2}dx = \sqrt{\pi/a}$.

[8] We use the notation $\hat{a}_i$ instead of $\hat{y}_i$ simply to emphasize the aspect that $\{\hat{a}_i\}$ are different random variables, rather than the components of a multicomponent random variable $\boldsymbol{y}$.

their probability densities are almost identical. In terms of the probability density, we can express the independence as

$$p(a_1, \ldots a_\nu) = \prod_i p_i(a_i), \quad \{\hat{a}_i\}\text{: independent.}$$

As another example, let us suppose that we observe a temporary fluctuating quantity $\hat{\xi}(t)$ every Δt seconds and denote the result as a_i $(i = 1, \ldots, N)$, that is, $\hat{a}_i = \hat{\xi}((i-1)\Delta t)$. One could imagine that $\hat{\xi}(t)$ represent the local temperature at time t within a 1-hour period in a room with the interval $\Delta t = 10$ min, or the intensity of sound produced by the flow of a river, and that we are interested in the fluctuation between the result $\{a_i\}$ for a 1-hour period and that for another 1-hour period. We suppose that the interval Δt is sufficiently long so that we can assume that the random variables $\{\hat{a}_i\}$ $(i = 1, \ldots, 6)$ are statistically almost independent.

Independently and identically distributed (i.i.d.) random variables: In the above example, we can further assume that the probabilities $Prob[\hat{a}_i = a_i]$ or the probability densities $p_i(y)$ $(i = 1, \ldots, \nu)$ are identical. In this case we call these random variables as independently and identically distributed (i.i.d.) random variables. If the random variables $\hat{a}_i$ $(i = 1, \ldots, N)$ are i.i.d., the probability of the realization of $\{a_i \le \hat{a}_i \le a_i + da_i\}$ $(i = 1, \ldots, N)$ is $\prod_{i=1}^{N}(p(a_i)da_i)$ with a common probability density $p(a)$ for a single random variable $\hat{a}$.

Empirical average of i.i.d. random variables: We now define the *empirical average* (or *sample average*) of these i.i.d. random variables:

$$\hat{A}_N \equiv N^{-1}\sum_{i=1}^{N}\hat{a}_i.$$

Unlike the average of a random variable, the empirical average is still a random variable. It is easy to verify that $\langle\hat{A}_N\rangle = \langle\hat{a}_i\rangle(\equiv a^*)$. Our interest is how the statistical properties of $\hat{A}_N$ change with N.[9]

We may intuitively expect that the probability for $\hat{A}_N$ to take the value away from a^* decreases with N. In fact this is true but there are three different levels of expression for this result, which we summarize briefly and intuitively here and leave the detailed explanations in Appendix A.1.3.

Law of large numbers:

$$Prob[|\hat{A}_N - a^*| > \epsilon] \to 0 \quad \text{for } N \to \infty. \tag{1.10}$$

[9] In composing the part below the author has referred to the lecture note by Y. Oono at the University of Illinois at Urbana Champaign together with his private communications.

Roughly speaking, the error regarding the empirical average $\hat{A}_N$ as the true average a^* becomes 0.

Central limit theorem:

$$Prob\left[\frac{|\hat{A}_N - a^*|}{\frac{\sigma}{\sqrt{N}}} < c\right] \to \int_{-c}^{c} \frac{e^{-\frac{w^2}{2}}}{\sqrt{2\pi}} dw \quad \text{for } N \to \infty, \tag{1.11}$$

where σ^2 is the variance of the individual random variable $\hat{a}_i$ and $c(> 0)$ is a number independent of N. Roughly speaking, the error in $\hat{A}_N$ with respect to a^* is typically confined within the range of $\frac{\sigma}{\sqrt{N}}$

Large deviation property: Given a x independent of N,

$$Prob[\hat{A}_N - a^* = x] \sim e^{-N\,I(x)} \quad \text{for } N \to \infty, \tag{1.12}$$

where $I(x)$ is independent of N and satisfies $I(0) = 0$. Roughly speaking, the probability that $\hat{A}_N$ deviates from a^* by x, which is eventually large as compared with $\frac{\sigma}{\sqrt{N}}$, becomes exponentially small with N.

The result (1.11) first indicates that, usually, the fluctuations of the "macro-variable" $\hat{A}_N$ ($\sim N$) do not fluctuate visibly because its typical deviations, $\hat{A}_N - a^*$, are of order of $N^{-\frac{1}{2}}$. Second, the typical deviations, $\hat{A}_N - a^*$, for large N can be simulated by a Gaussian distribution to a good approximation.

The reason that the probability of large deviations x in (1.12) decreases exponentially could be understood qualitatively as follows: For the random variable $\hat{A}_N - a^*$ to realize a large value beyond $\sim N^{-1/2}$, it is necessary that a certain fraction of the components $\{\hat{a}_i\}$, say ϕ $(0 < \phi < 1)$, must take those values which are not close to its average value, $\langle \hat{a}_i \rangle$. If we denote by p_* the probability for an individual random variable $\hat{a}_i$ to realize such exceptional values, then we have a rough estimation: $P_N(A) \sim {p_*}^{\phi N} = e^{-N\phi|\ln p_*|}$.

1.1.2.4 * Thermal Random Force – Idealized Fluctuating Force Due to Micro-Mechanics

The law of large numbers, central limit theorem, and the large deviation property introduced in Sect. 1.1.2.3 are the properties of the random variables which are not limited to the Brownian movement. In this book, however, we will mainly consider the case where the fluctuations have their origin in the mechanical motions of a large number of microscopic molecules. Moreover, we will limit our scope to the fluctuations in the environment near thermal equilibrium. For the simplicity of description, most of the descriptions hereafter refer to the system in one spatial dimension.[10]

---

[10] Sections 1.1.3.4 and 1.2.1.5 deal with the more than one dimensions.

As an example, let us consider the simple Brownian motion, i.e., the random motion of a fine particle (Brownian particle) freely suspended in the fluid at rest with a homogeneous temperature. The fluid molecules undergo the collisions with the Brownian particle. We are interested in the character of the force from the fluid molecules onto the Brownian particle. It would, however, be too difficult to describe those forces with the temporal resolution of the microscopic characteristic time τ_{m}, which is related to the collisions of individual fluid molecules upon the Brownian particle. Only the full molecular dynamic simulations can do it. But we are not interested in the detailed information of the movement of the fluid molecules. We, therefore, focus on the following aspects:

Single process: We look for a model of stochastic process of the Brownian particle, i.e., the generator of individual realizations of the motion of the Brownian particle. It means that we do not model the ensemble average of the force at a given instant of time. If we do the average of some quantity, it should be related to the temporal accumulation or empirical average (see Sect. 1.1.2.3) in a single realization.

Mesoscipic object: We are not interested in the detailed motion of the water molecules either causing the motion of the Brownian particle or responding to the Brownian particle. We only look at the Brownian particle and need a good description on the mesoscopic level.

Mesoscopic timescale: We want to know the coarse-grained force along the time axis, i.e., the rate of momentum transfer, on the Brownian particle, being averaged over a timescale which is much larger than the microscopic time, τ_{m}.

Statistical reproducibility: It is sufficient to have a good dynamical model which reproduces the statistical characteristics of the actual force or actual motion, and we do not need to predict the random motion of the Brownian particle from a particular initial condition.[11] In other words, we look for a model of the stochastic process which generates the ensemble of paths approximately identical to the ensemble of observed processes.

We should immediately say that it is not evident to be able to find the closed description for the force that satisfies the above conditions. If the environment of the Brownian particle has a significant longtime memory, we may not separate the motion of the Brownian particle from the dynamics of the environment. In this book we will not study what microscopic conditions are necessary to realize such an environment. When the environment allows the descriptions under the above constraints, we call such environment the thermal environment.

[11] In fact, any high-resolution setup of the initial condition cannot control the chaotic molecular motion for the time $\gg \tau_{\mathrm{m}}$.

Next we shall consider the collision force due to the fluid molecules on the Brownian particle over the duration of observation Δt.[12]

(1) First we consider a hypothetical situation where the Brownian particle is fixed in space. For Δt much smaller compared with τ_m, the force of collision by the fluid molecules from the left or from the right will be at most a single spike (Fig. 1.1a, top).
(2) For $\Delta t \gg \tau_m$, the number of collisions by water molecules, $\hat{n}$, will be numerous with various amplitudes of force (Fig. 1.1a, bottom). The cumulated momentum transfer during Δt, or $\int_{\Delta t} \hat{\xi}_{tot}(t)dt$, may obey approximately the Gaussian distribution (Fig. 1.1b).[13]

    From the symmetry of the setup, the average of this distribution must be 0. Moreover, since the Brownian particle undergoes many random collisions with water molecules during Δt, the cumulated momentum transfer over this timescale should be sharply peaked around this average, according to the law of large number. In the same context, the fluctuation around the average, which is $\int_{\Delta t} \hat{\xi}_{tot}(t)dt$ itself, should behave as a Gaussian random variable, except for the very rare large deviations.

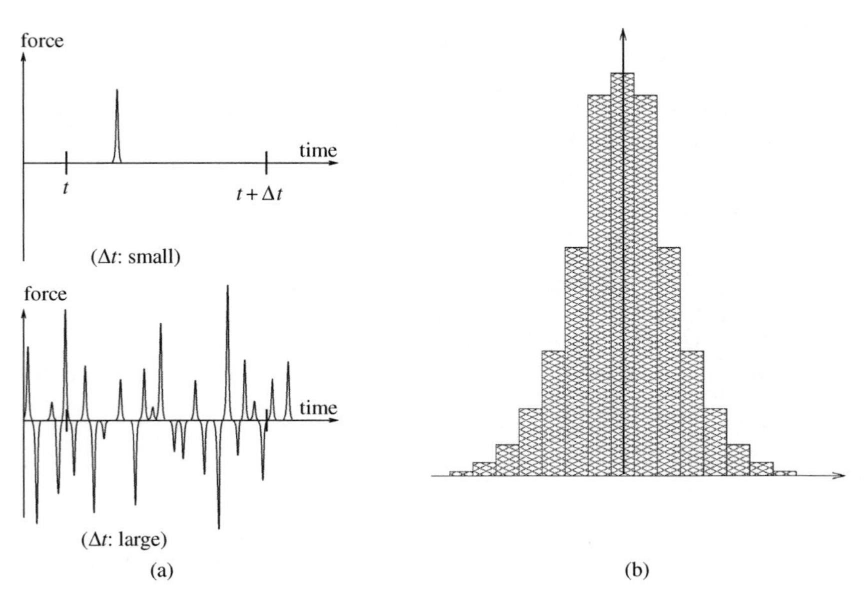

Fig. 1.1 (**a**) Schematic time series of the force on a Brownian particle during Δt, which is small (*top*) or large (*bottom*) as compared with microscopic time. (**b**) The distribution of the cumulated momentum transfer during a large Δt

[12] The argument below is suggested by R. Kawai, Arabama University.

[13] The force on the timescale Δt is $(\Delta t)^{-1}$ times the total momentum transfer cumulated over Δt. (When the Brownian particle is fixed by an apparatus, this momentum is further transmitted to the apparatus.)

If the water does not have a long-term memory as we suppose, the random variable $\int_{\Delta t} \hat{\xi}_{\text{tot}}(t)dt$ for a time interval Δt and that for another time interval Δt just next to the former should be independent. If Δt is small enough as compared with the characteristic timescale of the Brownian movement itself, we may find such an $N \gg 1$ that $N\Delta t$ is still small compared with the latter timescale. We then have a model of random force as i.i.d. random variables.

(3) Next we suppose that the Brownian particle moves forward with a constant velocity V relative to the fluid environment, which we define to be at rest. The dominant collisions then come from the front of the moving particle. For the velocity V not being very large, we may assume a linear relationship between $\int_{\Delta t} \hat{\xi}_{\text{tot}}(t)dt$ and V. We will characterize this relationship by a coefficient $\gamma(> 0)$ as follows:

$$\left\langle \int_{\Delta t} \hat{\xi}_{\text{tot}}(t)dt \right\rangle = -\gamma V \Delta t, \tag{1.13}$$

where the minus sign on the right-hand side indicates that the average force should act so as to resist the movement of the Brownian particle. By this physical interpretation, we call γ the viscous friction coefficient.

(4) What about the remaining fluctuations of $\int_{\Delta t} \hat{\xi}_{\text{rem}}(t)dt \equiv \int_{\Delta t} \hat{\xi}_{\text{tot}}(t)dt - (-\gamma V \Delta t)$? By definition the average of $\int_{\Delta t} \hat{\xi}_{\text{rem}}(t)dt$ is zero. Whether $V = 0$ or not, there underlie many independent degrees of freedom. By the same reason as in (2) above, the fluctuation must, therefore, be well approximated by a Gaussian random variable.[14] We further approximate that as Gaussian random variable $\int_{\Delta t} \hat{\xi}_{\text{rem}}(t)dt$ for $V \neq 0$ is of the same character as $\int_{\Delta t} \hat{\xi}_{\text{tot}}(t)dt$ at rest ($V = 0$).

We then follow the approach by Langevin [8]. We introduce a stochastic process $\hat{\xi}(t)$ (t being the time) as an idealized force acting on the Brownian particle at rest. This idealized force $\hat{\xi}(t)$ is aimed to be used on the time resolution Δt much larger than the microscopic time τ_{m}. Under this constraint, the time correlation function (i.e., the covariance) of $\hat{\xi}(t)$ can be made simple:

$$\langle \hat{\xi}(t)\hat{\xi}(t') \rangle = 2b\,\delta(t - t') \quad \text{(thermal random force).}$$

Moreover, by the above argument (2) we assume that $\hat{\xi}()$ is a Gaussian random variable or Gaussian (stochastic) process. That is, in (1.8) we choose $K(t, t')$ as

$$K(t, t') = \langle \hat{\xi}(t)\hat{\xi}(t') \rangle = 2b\,\delta(t - t') \quad \text{(thermal random force),} \tag{1.14}$$

where $b(> 0)$ is a constant to be determined later in relation to the coefficient γ and also to the temperature of the environment. We will call $\hat{\xi}()$ thus defined the thermal random force.

---

[14] The above argument is adopted from the lecture note by Y. Oono, UIUC.

Remark 1 Generally the Gaussian stochastic process with the property of (1.14) is called *white Gaussian (stochastic) process*, since its frequency spectrum does not depend on the frequency like white light.[15] As mathematical object, $\hat{\xi}$ is singular and cannot be formulated as it is. The Wiener process and the stochastic calculus which we will describe later have been developed to circumvent the difficulty.

Remark 2 As model of physical random force on the Brownian particle, however, we should remember that in the context of (1.14) $t \neq t'$ always implies $|t - t'| \gg \tau_{\mathrm{m}}$ and that the equality $t = t'$ has to be understood as $|t - t'| \leq \Delta t$, where $\Delta t \gg \tau_{\mathrm{m}}$. In this context, we may replace $\delta(t)$ by any sharply peaked but smooth function $\tilde{\delta} t$ that mimics $\delta(t)$, keeping the identity $\int_{-\infty}^{+\infty} \tilde{\delta} t dt = 1$. The formula about the δ-function such as $-t\delta'(t) = \delta(t)$ may no more hold for $\tilde{\delta}(t)$ since $-t\tilde{\delta}'(t)$ has double peak. However, as long as the peak is narrower than Δt the replacement, $\delta(t) \to \tilde{\delta}(t)$, makes no difference on the calculation of quantities of the time resolution (smoothness) coarser than τ_{m}.

Remark 3 With (1.8) and (1.14), the thermal random force as white Gaussian stochastic process is characterized by the characteristic functional:

$$\left\langle e^{i \int_{-\infty}^{\infty} \phi(t)\hat{\xi}(t)dt} \right\rangle = e^{-b \int_{-\infty}^{\infty} \phi^2(t)dt}. \tag{1.15}$$

1.1.3 Free Brownian Motion Is the Force-Free Motion of Mesoscopic Objects in a Thermal Environment

1.1.3.1 * Free Brownian Motion

We now relate the motion of a Brownian particle to the thermal random force. We continue to assume that the thermal environment of the Brownian particle is macroscopically homogeneous at a temperature T and that the environment is at rest as a whole. We will ignore the gravitational force and the buoyant force on the Brownian particle. Then, we define free Brownian motion as the random motion of the Brownian particle as stochastic process which is caused by the thermal random force characterized by the parameters γ and $k_{\mathrm{B}}T$ (k_{B} is the Boltzmann constant).

The free Brownian motion is fundamental for the dynamics of a particle in the fluctuating world just as that the force-free motion at a constant speed in vacuum was fundamental for the classical mechanics governed by the Galilean principle of relativity and the Newton's laws. In both cases the motions are realized under no distant external forces on the moving object, one in vacuum and the other in a thermal environment. We will see later that the general motions of a mesoscopic object in a thermal environment are described as modifications of the free Brownian motion in the presence of external forces, just like that the motions in classical mechanics are described as modifications of the force-free motion due to the extrenal forces.

---

[15] The remarks on the mathematical treatment of $\delta(t)$ will be given later in Sect. 1.2.3.1.

However, the microscopic origin of the free Brownian motion is the deterministic mechanics. Therefore, we seek to formulate the free Brownian motion from the viewpoint of Newtonian mechanics while we keep the timescale of interest much larger than τ_{m}. To this end, we take into account all the mechanical forces that the Brownian particle receives from the thermal environment. From the argument in the Sect.1.1.2.4, first of all there is a random thermal force, $\hat{\xi}(t)$, that is, a white Gaussian process (1.14) with the average 0, $\langle\hat{\xi}(t)\rangle = 0$. Another force is the viscous friction force, $-\gamma v(t)$, where $v(t)$ is the velocity of the Brownian particle at time t with respect to the thermal environment. (See (1.13) in the previous subsection.) The viscous friction force can be written either as $-\gamma p/m$ or as $-\gamma dx/dt$, where m, p, and x are, respectively, the mass, momentum, and position of the Brownian particle:[16] We, therefore, have the following Newtonian-like equations of motion of free Brownian motion:[17]

$$\frac{dp}{dt} = -\gamma \frac{p}{m} + \xi(t), \qquad \frac{dx}{dt} = \frac{p}{m}. \tag{1.16}$$

See Fig. 1.2a for the trajectory of $x(t)$ obtained by solving (1.16).[18]

Einstein introduced the basic concepts of the Brownian motion [9]. We refer the readers to Sect. 1.2.1 of [1] for his original reasoning. Einstein's paper finally convinced people that the heat is molecular motions [4]. The Brownian particle was the testimony of the thermal motion. In this book we will see how the heat in the thermal environment turns into the energy of the Brownian particle and vice versa.

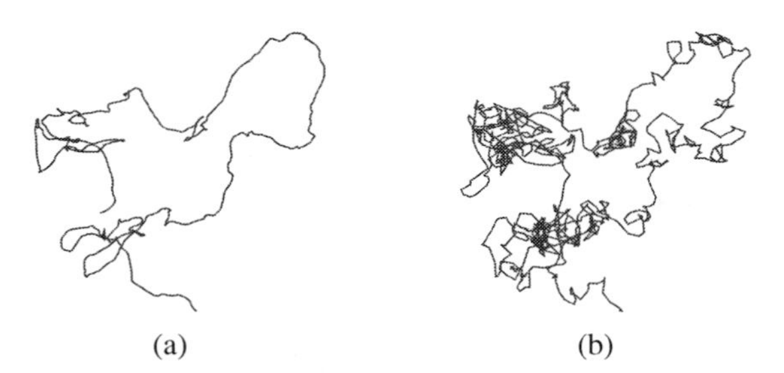

Fig. 1.2 Two-dimensional trajectories of a free Brownian particle. (**a**) A solution $x(t)$ of (1.16), (**b**) A solution $x(t)$ of (1.19). In (**a**), the velocity of the Brownian particle ($\sim v_{\mathrm{th}}$) is approximately maintained along the segment for $\sim v_{\mathrm{th}}(m/\gamma)$. In (**b**) the trajectory shows fictive fine structures even below the length $v_{\mathrm{th}}(m/\gamma)$

[16] We ignore the renormalization of the mass m due to the entrainment of the fluid molecules. See the remark after (1.16).

[17] Hereafter, we will often omit the hat, "^", which signifies the stochastic process. In fact $\xi()$, p, and x in (1.16) are stochastic processes.

[18] The figures are drawn by simple program: $(p_{n+1} - p_n)/\Delta t = -\Gamma(p_{n+1} + p_n)/(2m) + \Xi_n$, $(x_{n+1} - x_n)/\Delta t = (p_{n+1} + p_n)/(2m)$, where $m = \Delta t = 1$ and Ξ_n are i.i.d. Gaussian random variables with zero average. In (a) $\Gamma = 0.1$ and in (b) $\Gamma = 2$.

Remark About the Effect of Inertia

The above equations are not invariant under the Galilee transformation, $p \to p - mV$ and $x \to x - Vt$, with a constant velocity, V. It is understandable because Galilean invariance applies to the whole system consisting of the Brownian particle *plus* the thermal environment. If we also displace the latter, the term $\xi(t)$ should also undergo the Galilean transformation so that the whole equation is invariant.

On the microscopic scale, those microscopic molecules which once collided with the Brownian particle can be reflected by other molecules. Then the former molecules can eventually return the momentum to the Brownian particle, either directly or indirectly. This type of correlated collisions makes a network, and the whole effect may have a memory.

The universal point of view to see the effect of inertia, either macroscopic or microscopic scale, is the conservation of momentum. Equations (1.16) do not conserve the momentum by itself. In fact, the momentum conservation of the particle *plus* environment adds a memory effect to $\xi(t)$, called (hydrodynamic) longtime tail [10, 11]. In constructing (1.16) we assumed that the effect of inertia of the Brownian particle is more important than that of the surrounding fluid. When the mass density of the bead is comparable to that of the surrounding fluid, the neglect of the total momentum conservation is very crude approximation and the memory in $\xi(t)$ is not negligible at short time. Equation (1.16), therefore, requires attentions when we compare with experiments. The same remark also applies to the Langevin equation with inertia ((1.31) below). On the other hand, the equation that neglects the inertia ((1.19) below) or its generalization to Langevin equation ((1.32) below) is more generally valid for actual situations. The trade-off is that the latter equations have poorer time resolutions, as we shall see below. The memory effect will be discussed in Chap. 6.

Einstein Relation

The compatibility requirement between (1.16) and the canonical equilibrium distribution (Maxwell distribution) of the particle velocity leads to the relation named after Einstein:

$$b = \gamma k_{\mathrm{B}} T, \tag{1.17}$$

where b appeared in (1.14). Therefore, the thermal random force $\xi(t)$ as the Gaussian white noise is finally characterized by the following relations:

$$\langle \xi(t) \rangle = 0, \quad \langle \xi(t)\xi(t') \rangle = 2\gamma k_{\mathrm{B}} T \delta(t - t'). \tag{1.18}$$

The derivation is as follows: From (1.16) we calculate the longtime average of the kinetic energy of the Brownian particle,

$$\overline{\frac{p^2}{2m}} = \lim_{t \to \infty} \frac{1}{t} \int_0^t \frac{p(t')^2}{2m} dt'.$$

By substituting into this integral a particular solution of the first equation of (1.16), say, $p(t') = \int_0^{\infty} e^{-(\gamma/m)u}\, \xi(t'-u)du$, we have

$$\overline{\frac{p^2}{2m}} = \lim_{t\to\infty} \int_0^{\infty}\int_0^{\infty} \frac{e^{-\frac{\gamma}{m}(u_1+u_2)}}{2m}\left[\frac{1}{t}\int_0^t \xi(t'-u_1)\xi(t'-u_2)dt'\right]du_1du_2.$$

Since the thermal random force ξ is the stochastic process having effectively no time correlation (see (1.14)), the longtime average (i.e., the empirical average) in the angular brackets with large t is the average over many independent random variables. By the law of large numbers (see Sect. 1.1.2.3), such average should converge in the limit of $t \to \infty$ to the (ensemble) average, $\langle \xi(t'-u_1)\xi(t'-u_2)\rangle = 2b\delta(u_1-u_2)$ (see (1.14)).[19] By performing the remaining integrals with respect to u_1 and u_2, we find

$$\overline{\frac{p^2}{2m}} = \frac{b}{2\gamma}.$$

Now that we apply the same logic of the law of large numbers directly to the left-hand side, i.e., the kinetic energy of the Brownian particle. If the temporal fluctuation of the kinetic energy has only a finite correlation time, which we expect to be of order of m/γ, its longtime average (i.e., the empirical average) must converge to the (probabilistic) average. The latter can be calculated using the canonical partition function at temperature T. The result is $\overline{\frac{p^2}{2m}} = \frac{k_B T}{2}$, being known as the law of equipartition [4].[20] Equating the last two estimates of the kinetic energy, we obtain $b = \gamma k_B T$.

The Time Coarse Graining of the Equation of Free Brownian Motion

In case that the time resolution, Δt, is much larger than the characteristic time, $m/\gamma (\gg \tau_m)$, the two Equations (1.16) can be replaced by the following equation having no inertia term:

$$0 = -\gamma\frac{dx}{dt} + \xi(t). \tag{1.19}$$

See Fig. 1.2b for the trajectory of $x(t)$ obtained by solving (1.19).

The derivation is as follows: We substitute the solution of the first equation of (1.16), i.e., $p(t') = \int_0^{\infty} e^{-(\gamma/m)u}\, \xi(t'-u)du$, into the second equation of (1.16). The result writes

[19] Note that we have not used the ensemble average, but used the convergence of the empirical average, $\overline{\xi(t'-u_1)\xi(t'-u_2)}$, to the ensemble average, $\langle\xi(t'-u_1)\xi(t'-u_2)\rangle$.

[20] We did all the calculations in one dimension. The generalization to three dimensions does not change the result.

$$0 = -\gamma dx(t)/dt + \bar{\xi}(t), \qquad \bar{\xi}(t) \equiv \int_0^\infty \frac{\gamma}{m} e^{-\frac{\gamma}{m}u} \xi(t-u)du. \tag{1.20}$$

The newly defined thermal random force, $\bar{\xi}$, is also the Gaussian process with zero average, $\langle \bar{\xi}(t) \rangle = 0$. The temporal correlation of $\bar{\xi}$ is[21]

$$\langle \bar{\xi}(t)\bar{\xi}(t') \rangle = 2\gamma k_{\mathrm{B}} T \left[\frac{\gamma}{2m} e^{-\frac{\gamma}{m}|t-t'|} \right] \simeq 2\gamma k_{\mathrm{B}} T \, \delta(t-t').$$

The last equation is justified as long as the temporal resolution Δt satisfies $\Delta t \gg m/\gamma$. By redefining $\bar{\xi}$ as ξ we arrive at (1.19).

Remark 1 We have derived (1.19) without taking the statistical average of (1.16). That is, a particular realization of free Brownian motion looks differently under different time resolutions.[22] When an evolution equation bears a time resolution Δt, the solutions of the equation contain reliable information only in their frequency components with the frequencies smaller than $(\Delta t)^{-1}$. In other words, if we calculate the time integral of the solution with another function of time, the latter must vary on the timescale larger than Δt. Otherwise the result may lead to unphysical conclusion. For example, if we solve (1.19) $\xi(t)$ using a path of Gaussian white noise, the solution $x(t)$ behaves like that of (1.16) *only* on the timescales coarser than m/γ. This explains the difference in the trajectories in Fig. 1.2. If we were to calculate the velocity dx/dt using the (1.19), it is unboundedly large! [23] In short, (1.19) is useless for the study of the momentum, $p = m\, dx/dt$, while (1.16) describes a finite p.

Remark 2 The Eq. (1.19) contains no characteristic timescales. It implies that this result is universal for all the time resolutions well beyond m/γ. In other words, the further coarse graining of this equation leaves the form of equation invariant.

1.1.3.2 * Diffusion by the Free Brownian Motion and the Einstein Relation

The Eq. (1.16) implies that, due to the inertia effect, a Brownian particle keeps its velocity and orientation of movement during the time $\sim m/\gamma$, while the velocities at the times separated more than $\sim m/\gamma$ are no more correlated among each other. From (1.16) we can calculate the mean square displacement (MSD), which yields[24]

$$\langle [x(t) - x(0)]^2 \rangle = 2D|t|, \quad \text{for } |t| \gg m/\gamma, \tag{1.21}$$

[21] We already use the Einstein relation (1.17).

[22] It is like that the real function $\sqrt{x^2+1}$ behaves like $|x|$ for $x^2 \gg 1$ but not for $x^2 \sim 1$.

[23] In fact, if there were a bound M for the white Gaussian process such that $|\xi(t)| < M$, we would have $|\langle \bar{\xi}(t)\bar{\xi}(t') \rangle| < M^2$, which contradicts with the correlation condition (1.18).

[24] More generally, for the diffusion taking place in d-dimensional space ($d = 1, 2$, or 3), we should replace $2D$ in (1.21) by $2dD$.

where

$$D = \frac{k_B T}{\gamma}. \tag{1.22}$$

One can obtain the same formulas from (1.19), *formally* for any $|t|$, but $|t| \gg m/\gamma$ is always assumed (see the *Remark 1* of Sect. 1.1.3.1).

The result (1.21) implies that the net displacement of a Brownian particle during the time t is typically $\sim \sqrt{2Dt}$. This type of behavior is called diffusion with the diffusion coefficient, D. The formula (1.22) which relates D with the viscous friction coefficient γ is also called the Einstein relation. In Sect. 3.3.1.6 we will see that the Einstein relation is required for the *detailed balance condition.*

The relation (1.22) can be understood qualitatively by recalling the central limit theorem.[25] We approximate the trajectories of the Brownian particle as the piecewise linear steps, $x(t) - x(0) = \sum_{i=1}^{N} \hat{a}_i$. Each linear segment $\hat{a}_i$ corresponds to the time (m/γ) during which the Brownian particle moves at a constant velocity. This velocity is typically of thermal velocity, $v_{\text{th}} = \sqrt{k_B T/m}$, derived from the law of equipartition mentioned above. Then the distance of each segment $|\hat{a}_i|$ is $v_{\text{th}}(m/\gamma)$ and the number of segments N is therefore $N = t/(m/\gamma)$. Every steps are independent, and the MSD after the N steps is $\sqrt{N}\, v_{\text{th}}(m/\gamma)$. [26] Identifying the MSD with $2Dt$ we recover (1.22) up to a constant factor.

Let us consider the composition of the Einstein relation (1.22).

Effect of Temperature: We understand that the diffusion coefficient depends both on a kinetic factor γ and a thermodynamic factor $k_B T$. The former indicates how tightly a Brownian particle is coupled to its environment, while the latter represents how strong fluctuations are there in the environment. For liquid environments, γ is usually a decreasing function of temperature.[27] Therefore, D usually increases with temperature.

Effect of the size of Brownian particle: As to the dependence on the size of Brownian particle, γ is larger for a larger Brownian particle.[28] Thus, a larger Brownian particle receives stronger thermal random forces, since $\xi(t)$ scales with $\sqrt{\gamma k_B T/\tau_{\text{m}}}$

[25] Originally Einstein has derived this relation by considering a stationary density distribution of suspending particles in fluid under the gravitation [14]. He required the compatibility between the thermodynamic argument, "The stationary state is the balance between the gravitational force and the osmotic pressure," and the kinetic argument, "The stationary state is the balance between the sedimentation and the diffusion."

[26] $\left\langle \left(\sum_{i=1}^{N} \hat{a}_i\right)^2 \right\rangle = \sum_{i=1}^{N} \langle (\hat{a}_i)^2 \rangle = N \langle (\hat{a}_i)^2 \rangle$, where we used $\langle \hat{a}_i \hat{a}_j \rangle = 0$ for $i \neq j$ because of the independence.

[27] The fluid viscosity η is usually a decreasing function of temperature, and η and γ are proportional in many cases.

[28] For a spherical particle in an incompressible simple viscous fluid (Newtonian fluid), γ is linearly proportional to the radius of the particle [15].

(see (1.18)).[29] However, a larger particle receives also stronger frictional forces $\propto \gamma$. Since the latter effect is more important for large γ, the thermal random force has little influence on the motion of the macroscopic objects.

Which to measure, D or γ?: For some experiments the spontaneous movements are easier to measure than the response under external driving. For some theories the situation could be reversed. One can study either D or γ, thanks to the Einstein relation. For example, experimentalists [12] measured the 1D diffusion coefficient D of a rod-like filament (microtubule) on a substrate coated with inactivated molecular motors (dyneins). A theory [13] explained their result through the estimation of γ when the filament is driven at a constant velocity.[30]

Probability Density of the Displacement and the Diffusion Equation

With the time resolution of $t \gg m/\gamma$, we will evaluate the probability density for the displacement $\hat{x}(t) - \hat{x}(0)$.[31] It is written as $\mathcal{P}(X, t) = \langle \delta(X - [\hat{x}(t) - \hat{x}(0)]) \rangle$. Using (1.19), we can show that

$$\mathcal{P}(X, t) = \frac{1}{\sqrt{4\pi D t}} e^{-\frac{X^2}{4Dt}}. \tag{1.23}$$

That it takes the form of Gaussian distribution is the general consequence of the linear combination of Gaussian random variables. The derivation is given in Appendix A.1.4.

From (1.23) we can verify that the distribution function $\mathcal{P}(X, t)$ obeys the following partial differential equation called the diffusion equation:

$$\frac{\partial P}{\partial t} = D \frac{\partial^2 P}{\partial X^2}. \tag{1.24}$$

1.1.3.3 Diffusion Transport or Active Transport?

If we estimate the mean velocity of displacement $\sqrt{2Dt}/t$ from (1.21), it attains as large as the thermal velocity of the Brownian particle, $v_{\text{th}} = \sqrt{k_{\text{B}} T/m}$. Small biological organisms use this rapid mechanism of transport more efficient than, for example, the protein molecular motor. However, for a large distance or for a long time, the mean velocity $\sqrt{2Dt}/t$ becomes eventually very small, and the diffusion is no more an efficient way of transport.

What then is the characteristic length scale ℓ_D below which the transport by diffusion is efficient? In order to answer this question, let us compare the distance of diffusion $\sqrt{2Dt}$ with the distance by motor transport, $v_m t$, where v_m is the velocity

[29] The last factor τ_{m}^{-1} comes from $\delta(t - t')$ in (1.18).

[30] The cycles of stochastic binding and unbinding between the motors and the filament cause the dissipation called the "protein friction" [13].

[31] The hat ^ indicates again the random variables.

of a protein motor along a cytoskeletal filament. Assuming the typical values as $2D \sim 10^{-9}\mathrm{cm}^2/s$ and $v_m \sim 10^{-4}\mathrm{cm/s}$, the equality about the time $\frac{(\ell_D)^2}{2D} = \frac{\ell_D}{v_m}$ yields $\ell_D \simeq 10^{-5}\mathrm{cm}$ or 100 nm.

For example, the neurotransmitter diffuses the gap (~ 50 nm) between the presynaptic cell membrane and the postsynaptic cell membrane. On the other hand, the cellular organelle are transported by protein motors in the cells ($\sim$ μm). These two facts are consistent with the separation criterion by ℓ_D estimated above. By contrast, the diffusive signal transport of small "morphogen" molecules in embryos (~ 500 μm) looks surprising. In fact it was found recently that the morphogen transport is carried by an active intercellular transport (trans-cytocis) [16].

1.1.3.4 An Extension: Free Brownian Motion of an Anisotropic Object

A free Brownian motion in D-dimensional space can be generalized to the case where the diffusing object is anisotropic [17, 18]. One can imagine a rod-like object, for example. Such an object then undergoes the rotational random motion as well as spatial diffusion. Below we illustrate the 2D case [19, 21].[32]

For an object of any shape, there is a point what we call the "center of diffusion" [20], whose position we denote by $\boldsymbol{x}(t) = (x_1(t), x_2(t))$ in space-fixed coordinate, and two mutually orthogonal axes, which we distinguish by the symbol $\|$ and $\perp$. The orientation of the object at time t is defined by the instantaneous angle $\theta(t)$ made by the $\|$-axis of the object and the x-axis of an experimental frame. (Therefore, only θ modulo π has physical meaning.) The Brownian motion of the object can be described by the following equations [19]:

$$\frac{d\theta}{dt} = \sqrt{2D_\theta}\, \zeta_\theta(t), \tag{1.25}$$

$$\frac{d\boldsymbol{x}}{dt} = \left[\sqrt{2D_\|}\, \hat{\boldsymbol{u}}_\|\, \hat{\boldsymbol{u}}_\|^t + \sqrt{2D_\perp}\, \hat{\boldsymbol{u}}_\perp\, \hat{\boldsymbol{u}}_\perp^t\right] \cdot \boldsymbol{\zeta}(t), \tag{1.26}$$

where $D_\theta = k_{\mathrm{B}}T/\gamma_\theta$, $D_\| = k_{\mathrm{B}}T/\gamma_\|$, and $D_\perp = k_{\mathrm{B}}T/\gamma_\perp$ are the diffusion coefficients determined from the shape of the object, and $\hat{\boldsymbol{u}}_\|^t(t) \equiv (\cos\theta(t), \sin\theta(t))$ and $\hat{\boldsymbol{u}}_\perp^t(t) \equiv (-\sin\theta(t), \cos\theta(t))$ are the transpose of $\hat{\boldsymbol{u}}_\|(t)$ and $\hat{\boldsymbol{u}}_\perp(t)$, respectively. We have also introduced the (normalized) independent thermal random forces, $\zeta_\theta(t)$ and $\boldsymbol{\zeta}^t(t) = (\zeta_1(t), \zeta_2(t))$, such that

$$\begin{aligned}
\langle \zeta_\theta(t)\rangle &= \langle \zeta_1(t)\rangle = \langle \zeta_2(t)\rangle = 0,\\
\langle \zeta_\theta(t)\zeta_1(t')\rangle &= \langle \zeta_1(t)\zeta_2(t')\rangle = \langle \zeta_2(t)\zeta_\theta(t')\rangle = 0,\\
\langle \zeta_\theta(t)\zeta_\theta(t')\rangle &= \langle \zeta_1(t)\zeta_1(t')\rangle = \langle \zeta_2(t)\zeta_2(t')\rangle = \delta(t-t').
\end{aligned} \tag{1.27}$$

[32] Experimental realization of 2D Brownian motion requires attentions both by technical [21] and fundamental [22] reasons.

It takes a finite time to reorient the anisotropic object by the rotational diffusion. During this time, the object diffuses spatially in the direction for which the diffusion constant is large. By this reason, the spatial trajectories of the anisotropic object show some memory, even if we ignore the inertia. See Fig. 1.3 for the comparison between the diffusion of an isotropic molecule and that of an anisotropic one. If $D_{\parallel}$ is much larger than $D_{\perp}$, the trajectory looks to have the characteristic length scale $\sim \pi \left(D_{\parallel}/D_{\theta}\right)^{1/2}$ over which the preferential local axis of the diffusing object is kept approximately constant.

While Eq. (1.26) takes a simple form, the form is not similar to the previous Eq. (1.19), which resembles more to the Newtonian equation. In fact (1.26) can be rendered to the form more analogous to (1.19) [23]:

$$0 = -\Gamma \cdot d\boldsymbol{x}dt + \Xi(t) \tag{1.28}$$

with

$$\Gamma = \gamma_{\parallel}\, \hat{\boldsymbol{u}}_{\parallel}\, \hat{\boldsymbol{u}}^{t}_{\parallel} + \gamma_{\perp}\, \hat{\boldsymbol{u}}_{\perp}\, \hat{\boldsymbol{u}}^{t}_{\perp}, \qquad \Xi(t) = \hat{\boldsymbol{u}}_{\parallel}\, \xi_{\parallel}(t) + \hat{\boldsymbol{u}}_{\perp}\, \xi_{\perp}(t), \tag{1.29}$$

where the thermal random forces satisfy $\langle \xi_{\parallel}(t)\rangle = \langle \xi_{\perp}(t)\rangle = 0$, $\langle \xi_{\parallel}(t)\xi_{\perp}(t')\rangle = 0$ and

$$\langle \xi_{\parallel}(t)\xi_{\parallel}(t)\rangle = 2\gamma_{\parallel}k_{\mathrm{B}}T\delta(t-t'), \qquad \langle \xi_{\perp}(t)\xi_{\perp}(t)\rangle = 2\gamma_{\perp}k_{\mathrm{B}}T\delta(t-t'). \tag{1.30}$$

The latter form makes explicit that the actual motion of $\boldsymbol{x}$ is the superposition of (instantaneous) 1D Brownian motion along $\hat{\boldsymbol{u}}_{\parallel}$ and $\hat{\boldsymbol{u}}_{\perp}$. The transformation between the two representations is given by the formula:

$$\xi_{\parallel} = \sqrt{2\gamma_{\parallel}k_{\mathrm{B}}T}\, \hat{\boldsymbol{u}}_{\parallel} \cdot \boldsymbol{\zeta}(t), \qquad \xi_{\perp} = \sqrt{2\gamma_{\perp}k_{\mathrm{B}}T}\, \hat{\boldsymbol{u}}_{\perp} \cdot \boldsymbol{\zeta}(t).$$

Even if the orientation of the particle is not observable, the spatial trajectory presents some informations on the rotational Brownian motion. If $D_{\parallel} \neq D_{\perp}$, the data of $x(t)$ are in principle enough to find the values of these parameters *and* that

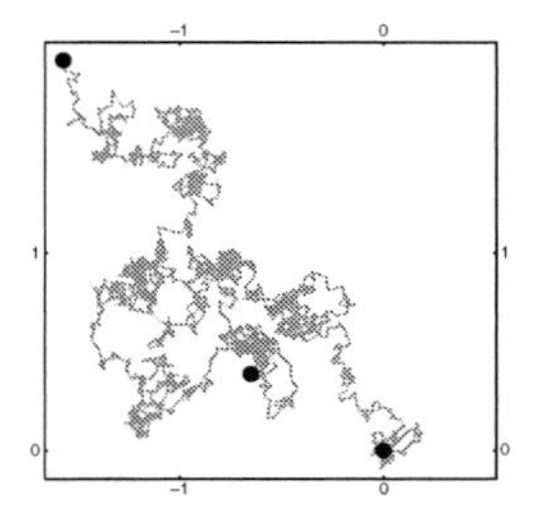

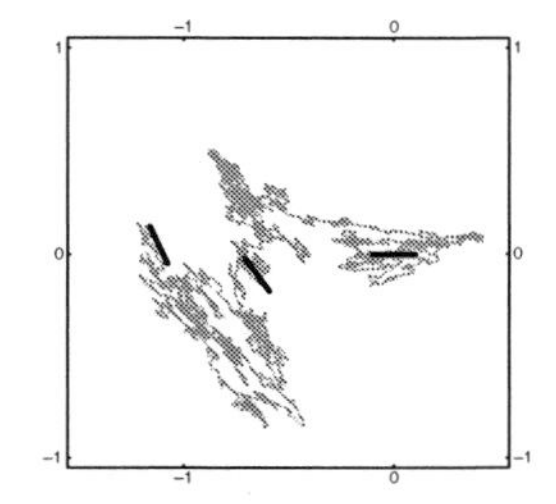

Fig. 1.3 Numerically generated trajectories of isotropic Brownian particle (*thick dot: left*) and of anisotropic Brownian particle (*thick bar: right*) (Adapted from Fig. 2 in [19])

of D_θ [21, 19].[33] The most important point is that the diffusion phenomena does not reflect the "head" and "tail" of the diffusing object. That is, even if the shape of the object does not have any mirror symmetry, the linear relation (1.26) between $\frac{d\boldsymbol{x}}{dt}$ and $\boldsymbol{\zeta}(t)$ preserves the invariance under $\hat{\boldsymbol{u}}_\parallel \leftrightarrow -\hat{\boldsymbol{u}}_\parallel$ and $\hat{\boldsymbol{u}}_\perp \leftrightarrow -\hat{\boldsymbol{u}}_\perp$. A consequence of this fact [24, 25] will be discussed in Sect. 1.3.4.2.

1.2 Construction of Langevin Equations

1.2.1 Langevin Equation Can Be Intuitively Constructed, But Is Derivable from Micro-Mechanics Through the Markov *Approximation*

The Langevin equation can be introduced as a natural generalization of the equation for the free Brownian motion. However, the Langevin equation can also be derived from microscopic equations of motion. A good and simple example has been presented by Zwanzig [3] (also by Ford, Kac, and Mazur [26]), where a Langevin equation is derived through direct integration of the variables of small "molecules." Then, by mid-1970s, Zwanzig, Mori, Kawasaki, and others have developed more general and systematic methods to derive the Langevin equation. They used the technique of projection operators.

1.2.1.1 * From Free Brownian Motion to Langevin Equation

The equation of motion of a particle in classical mechanics is based on the equation of force-free motion, $dp/dt = 0$, supplemented with the terms representing applied forces (Newton's second law). Similarly, the equation of motion of a particle in a thermal environment is generally described by supplementing the equation of free Brownian motion with terms representing applied forces.[34] Especially, if the applied force is caused by a potential energy, $U(x, a)$, the modified equation should take the following form, which we call the Langevin equation [8]:[35]

$$\frac{dp}{dt} = -\frac{\partial U(x, a)}{\partial x} - \gamma \frac{p}{m} + \xi(t), \qquad \frac{dx}{dt} = \frac{p}{m}, \tag{1.31}$$

[33] In two dimension, the equations are explicitly solvable by integration, since the angular part can first be solved separately. In three dimension, it is not the case because of the non-Abelian character of rotation [27], as well as the possible couplings between rotation and translation for chiral particles.

[34] What then corresponds to the Newton's third law? This is the subject the chapters from Chap. 4.

[35] English translation of the original paper of Langevin is found in [8].

where m, p, x, γ, and $\xi(t)$ are defined as before, and a is the parameter by which the potential energy U can be changed through an external operation.[36] For a moment we assume that the parameter a does not depend on the time, t.

If the time resolution of our interest, Δt, is such that $\Delta t \gg m/\gamma$, and if γ and T are constant in time, we can replace (1.31) approximately by the following equation without inertia term (cf. (b) in Sect. 1.1.3.1).

$$0 = -\frac{\partial U(x,a)}{\partial x} - \gamma \frac{dx}{dt} + \xi(t). \tag{1.32}$$

The derivation from (1.31) to (1.32) will be given later (Sect. 1.3.2). The Eqs. (1.31) or (1.32) are called the Langevin equations. In both cases we recall that $\xi(t)$ is the white and Gaussian stochastic process satisfying (1.18). The equations of motion for the free Brownian particle is the special cases of Langevin equation.

1.2.1.2 Macroscopically Mimicked Thermal Environment and Langevin Equation

It is possible to simulate mechanically the thermal random force. There, both $k_{\mathrm{B}}T$ and γ are the adjustable parameters and can be chosen so that the relation $b = \gamma k_{\mathrm{B}}T$ be satisfied. In the experimental setup of [28], macroscopic charged metallic balls are confined in a circular channel and subject under mechanical agitations with the (quasi) white noise spectrum. In the presence a harmonic binding force, the motion of a ball obeys the Langevin equation $m\frac{d^2x}{dt^2} = -\gamma\frac{dx}{dt} - Kx + \xi(t)$, where m and x are the mass and displacement of the ball, respectively, and K is the force constant of the binding force. In the steady state x and the momentum $p = m(dx/dt)$ should, therefore, reproduce a Gaussian canonical distribution. In the overdamped limit ($m/\gamma \ll \gamma/K$) where the inertia term can be neglected, the probability density for $x(t)$ at time $t(\gg m/\gamma)$ is known and called the Uhlenbeck–Ornstein formula [29] (see also [30]):

$$P_{\mathrm{UO}}(x,t|x_0) = \frac{K^{1/2}}{(2\pi k_{\mathrm{B}}T(1-e^{-2t/\tau}))^{1/2}} \exp\left[-\frac{K(x-x_0e^{-2t/\tau})^2}{2k_{\mathrm{B}}T(1-e^{-2t/\tau})}\right], \tag{1.33}$$

where $\tau = \gamma/K$ and x_0 is the initial position of the mass at $t = 0$. The comparison of (1.33) with the experimental observation gives the values of τ, K, and $k_{\mathrm{B}}T$. The $k_{\mathrm{B}}T$ and γ are the fictive temperature and the fictive friction coefficient of this mechanically mimicked thermal environment and the coupling between the mass and this environment. The parameters should be consistent with the formulas of equipartition $\langle p^2\rangle/(2m) = k_{\mathrm{B}}T/2$ (measurable only for the time

[36] In the literatures, the control parameters are often denoted by the greek characters, $\alpha, \lambda, \lambda_t$, etc., and are distinguished from the roman characters for the fluctuating variables, x, etc. In this book, however, we will often use the roman characters also for the control parameters, like a. Although such choice makes no essential differences, our standpoint is that the roles of x and a are relative in certain cases, especially when we are interested in different scales of descriptions.

resolution, $\Delta t \not\gg m/\gamma$) and $K\langle x^2\rangle/2 = k_B T/2$, and also with the Einstein's relation, $\gamma = k_B T/D$.[37] Finally the artificial thermal random force is characterized by $\langle \xi(t)\rangle = 0$ and $\langle \xi(t)\xi(t')\rangle = 2\gamma k_B T \delta(t-t')$.

1.2.1.3 * Derivation of a Langevin Equation from Micro-Mechanics: An Example

In Sect. 1.1.3.1 we supposed that the thermal environment around a Brownian particle represents a large number of microscopic molecules that interact with the Brownian particle. In order to obtain the Langevin equation for the Brownian particle, we need to eliminate from pure mechanics equations all the rapidly varying degrees of freedom, that is the position and the momentum of the fluid molecules, $\{x_j, p_j\}$ ($j = 1, \ldots, N$ with $N \gg 1$). Through this operation, an isolated mechanical system containing a Brownian particle and a large number of microscopic molecules is turned into a Brownian particle and the thermal environment, with or without external forces.

Zwanzig [3] presented an example in which the above mentioned operation can be explicitly performed. What we will present here is a simplified version of his model.[38] Let us denote by $\{X, P\}$ the position and the momentum of the Brownian particle and suppose that there is no other slowly varying degrees of freedom. We introduce the following Hamiltonian:

$$H = \frac{P^2}{2} + \sum_{j=1}^{N} \frac{{p_j}^2}{2m} + U(X, \{x_k\})$$

$$U(X, \{x_k\}) = U_0(X) + \sum_{j=1}^{N} \frac{m{\omega_j}^2}{2}\left\{x_j - \frac{\gamma_j}{m{\omega_j}^2}X\right\}^2.$$

We have assumed that the mass of the Brownian particle is unity while the microscopic molecules have the mass m. The equations of motion (Hamilton equations) for the Brownian particle and for the microscopic molecules then write

$$\frac{dX}{dt} = P, \qquad \frac{dP}{dt} = -\frac{\partial U_0(X)}{\partial X} + \gamma_j \sum_{j=1}^{N}\left\{x_j - \frac{\gamma_j}{m{\omega_j}^2}X\right\} \tag{1.34}$$

$$\frac{dx_j}{dt} = \frac{p_j}{m}, \qquad \frac{dp_j}{dt} = -m{\omega_j}^2 x_j + \gamma_j X. \tag{1.35}$$

Schematically, the Brownian particle is under the influence of its proper potential energy, $U_0(X)$, together with the harmonic potential energies through which

[37] D can be measured from the MSD $\langle[x(t) - x(0)]^2\rangle = 2Dt$ of the free diffusion ($K = 0$).

[38] Historically, this solvable model has been proposed in the course of the development of the general framework described in Sect. 1.2.1.5.

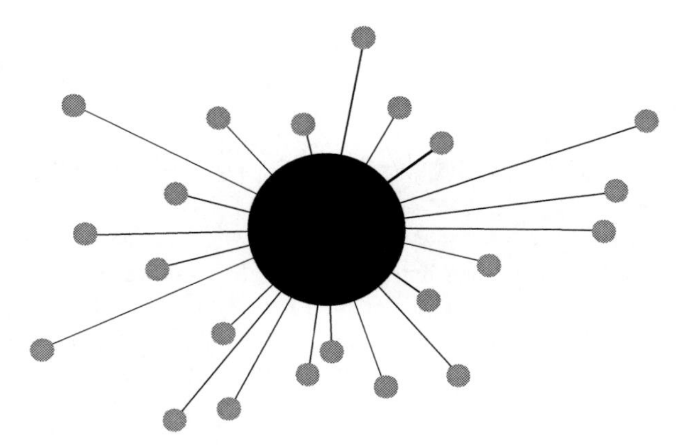

Fig. 1.4 Schematic representation of the solvable model by Zwanzig [3]. The huge disc and small ones represent, respectively, the Brownian particle and microscopic molecules. The lines represent the harmonic binding potential energies. In the text we discuss the 1D model

the surrounding microscopic molecules exert forces to the Brownian particle (see Fig. 1.4). The last equation for dp_j/dt indicates that the spring constant of the harmonic potential is $m\omega_j{}^2$ for the jth microscopic molecule and that γ_j represents the strength of the coupling between this molecule and the Brownian particle.[39] The object is to eliminate all of $\{x_j, p_j\}$ to find the equation of motion for X and P. Below we sketch the derivation. The points are the appearance of temperature and the change of the property of time-reversal symmetry.

Since the Equations (1.35) are linear in $\{x_j, p_j\}$, we can integrate these equations with respect to $\{x_j, p_j\}$ assuming that X is known as a function of time:

$$x_j(t) = \frac{\gamma_j}{m\omega_j{}^2}X(t) - \int_0^t \frac{\gamma_j}{m\omega_j{}^2}\cos\left[\omega_j(t-t')\right]\frac{dX(t')}{dt'}dt' + \left\{x_j(0) - \frac{\gamma_j}{m\omega_j{}^2}X(0)\right\}\cos(\omega_j t) + \dot{x}_j(0)\frac{\sin(\omega_j t)}{\omega_j}, \tag{1.36}$$

where $\dot{x}_j(0) = p_j(0)/m$ is the initial velocity of the jth microscopic molecule. Substitution of this solution into (1.34) yields the following concise form:

$$\frac{dX}{dt} = P$$

$$\frac{dP}{dt} = -\frac{\partial U_0(X)}{\partial X} - \int_0^t \zeta(t-t')\frac{dX(t')}{dt'}dt' + \xi(t). \tag{1.37}$$

Here the "friction coefficient" (with memory), $\zeta(t)$, and the "thermal random force" $\xi(t)$ are defined as follows:

[39] γ_j and the viscous friction coefficient, γ, have different dimensions.

$$\zeta(t) \equiv \sum_{j=1}^{N} \frac{\gamma_j{}^2}{m\omega_j{}^2} \cos(\omega_j t), \tag{1.38}$$

$$\xi(t) \equiv \sum_{j=1}^{N} \gamma_j \left[\left\{ x_j(0) - \frac{\gamma_j}{m\omega_j{}^2} X(0) \right\} \cos\omega_j t + \dot{x}(0) \frac{\sin\omega_j t}{\omega_j} \right]. \tag{1.39}$$

The memory term including ζ in (1.37) represents the dynamical response of the microscopic molecules at time t to the motion of the Brownian particle at the past time $t'(< t)$. Although Equations (1.37) take a form similar to (1.31), the former are purely mechanical in the following two aspects:

(i) There is no information on the temperature of the environment, that is, of the ensemble of the microscopic molecules.
(ii) The time-reversal symmetry is strictly observed.[40]

We now introduce nonmechanical aspects by modifying both (i) and (ii):

(i') We assume that the positions and velocities of the microscopic molecules at the initial time are distributed so that its probability density corresponds to the canonical equilibrium at temperature T of the statistical mechanics:

$$Prob\left[\{x_j(0), \dot{x}_j(0)\}\right] \propto \exp\left\{ -\beta \sum_{j=1}^{N} \left[\frac{m\dot{x}_j(0)^2}{2} + \frac{m\omega_j{}^2}{2} \left\{ x_j(0) - \frac{\gamma_j}{m\omega_j{}^2} X(0) \right\}^2 \right] \right\},$$

where $\beta \equiv (k_{\mathrm{B}}T)^{-1}$ is the inverse temperature. With this probability density, $\xi(t)$ becomes the Gaussian stochastic process[41] and moreover satisfies the following relations:

$$\langle \xi(t) \rangle = 0, \qquad \langle \xi(t)\xi(t') \rangle = k_{\mathrm{B}} T \zeta(t - t'), \tag{1.40}$$

where the average is taken with respect to the probability density, $Prob\left[\{x_j(0), \dot{x}_j(0)\}\right]$.

(ii') We introduce an approximation of replacing $\zeta(t)$ by $2\gamma\delta(t)$. It is justified if $\zeta(t)$ decays in time and the characteristic time of the decay is short compared with the time resolution of our interest to describe the motion of Brownian particle. In order for $\zeta(t)$ to decay in time, there are constraints on ω_j and γ_j, which we do not enter into detail. The assumption of a short decay time

[40] The time-reversal symmetry requires that the equations of motion are unchanged upon this operation. This operation is the replacement of $P \mapsto -P$, $p_j \mapsto -p_j$, $t \mapsto -t$, and $d/dt \mapsto -d/dt$, therefore, $\dot{x}_j \mapsto -\dot{x}_j$. Intuitively, this operation is like the playing back of movies. Concomitantly, the initial conditions are regarded as the terminal conditions.

[41] Any linear combination of random variables obeying Gaussian probability distributions also obeys Gaussian probability distribution.

comes from our assumption that only X and P are the slowly varying degrees of freedom. Once these assumptions and approximations are introduced, we do not use (1.39) anymore. Therefore, the characters of the microscopic molecules are not directly reflected anywhere in $\xi(t)$. Instead, we generate $\xi(t)$ as a white and Gaussian stochastic process satisfying (1.40) with $\zeta(t) \mapsto 2\gamma\delta(t)$.

With these two operations which are not the consequence of the mechanics, the equation of motion (1.37) for X and P is converted to the form of the Langevin equation:[42]

$$\frac{dX}{dt} = P,$$
$$\frac{dP}{dt} = -\frac{\partial U_0(X)}{\partial X} - \gamma \frac{dX(t)}{dt} + \xi(t). \tag{1.41}$$

This is the outline of the derivation of the Langevin equation from a simple and purely mechanical model.

What do the two steps (i') and (ii') change the physical properties of the equation of motion for X and P?

(a) Once the canonical distribution is adapted, the translational invariance and, therefore, the conservation relation of the total momentum are lost.
(b) Once the microscopic construction of $\xi(t)$ (see (1.39)) is abandoned, the operation of the time reversal, i.e., $P \mapsto -P$, $t \mapsto -t$ and $d/dt \mapsto -d/dt$, does not leave invariant the equation of motion. In order to bring the time-reversal symmetry into (1.41) we need to define the new time-reversal operation for the sum, $-\gamma dX(t)/dt + \xi(t)$.

1.2.1.4 * Markov Approximation, Markov Process, and the Notion of Irreversibility

The approximation to omit the memory effect of the environment, like $\zeta(t) \mapsto 2\gamma\delta(t)$ in the previous subsection, is called the Markov approximation.[43] By this approximation, the evolution of the variables of our interest, X and P, is now determined in the way that *if we know their values $X(t)$ and $P(t)$ at the present time t, their values in the past $X(t')$ and $P(t')$ $(t' < t)$ add no further information on their future values, $X(t'')$ and $P(t'')$ $(t'' > t)$.* Or, simply, the values of $X(t + dt)$ and $P(t + dt)$ $(dt > 0)$ depend only on $X(t)$ and $P(t)$ but not their past history. This property is called the Markovian property, and the process having this property is called the Markovian process. The time evolution through the Langevin equation

[42] We should evaluate the integral including the delta function with applying the rule, $\int_0^t \delta(t - t')dt' = \frac{1}{2}$, since its original form, $\zeta(t - t')$, is even function of its argument.

[43] In Sect. 1.1.3.1 we have already used the Markov approximation to derive (1.19) without mentioning it.

is not the only example of the Markovian processes. The master equation, which we will discuss in Chap. 3, also describes a Markovian process. The Hamiltonian equations of motion also describe Markovian processes without stochastic nature.[44]

Remark The loss of the mechanical time-reversal symmetry ((a) above) related to the Markovian approximation does not imply the incompatibility between the Langevin equation and the equilibrium thermodynamics. We should recall that the equation of free Brownian motion, (1.16), supplemented with the relation (1.18) can reproduce the property of equipartition in equilibrium. This implies that we need to distinguish the mechanical reversibility from the thermodynamic reversibility. In Chap. 5 we will see that not only the equilibrium states but the reversible thermodynamic processes can also be described on the basis of the Langevin equation.

1.2.1.5 Derivation of Langevin Equation from Mechanics by the Projection Method

In mechanics it is generally difficult to eliminate explicitly many degrees of freedom corresponding to the environment. A formal method to derive Langevin equations by eliminating these degrees of freedom has been established based on the concept of projection operator. In linear algebra or in functional analysis projection operators are linear operators in the space on which the scalar (inner) product is defined.

As the equations of motion in mechanics are generally nonlinear,[45] it is useless to apply projection operators to the phase space, i.e., the full set of position coordinates and momentum coordinates, $\{\boldsymbol{x}, \boldsymbol{p}\}$. The idea of the projection operator in mechanics is to study the evolution of the observable quantities which we expect to be slowly varying ("mesoscopic/gross variables"), instead of studying the motions of the state point in the phase space.[46] The projection operators then act in the (functional) space of these observable quantities.

However, this projection operator applied to a few gross variables gives rise only to a set of linear Langevin equation for these quantities (Mori formula [31]).

In order to go beyond the linear Langevin equation, we need to include the equation for all powers of the gross variables. This difficulty has been overcome by Kawasaki [32]. He considered the evolution equation for $\delta(\boldsymbol{a} - \boldsymbol{A})$, where $\boldsymbol{A}$ is the gross variable(s) and $\boldsymbol{a}$ is the parameter. Once the evolution for $\delta(\boldsymbol{a} - \boldsymbol{A})$ is known by the method of projection operator, the evolution of the observable $f(\boldsymbol{A})$ is known by the superposition, $f(\boldsymbol{A}) = \int f(\boldsymbol{a})\delta(\boldsymbol{A} - \boldsymbol{a})d\boldsymbol{a}$.

Leaving the outline of the derivation in Appendix A.1.5, the result writes in the following form of Langevin equation:

---

[44] The Schrödinger equation also describes deterministic Markovian process, where the wave functions represent the states. The classical or quantum Liouville equation describes also Marcovian process.

[45] That is, the interactions are not quadratic in their variables.

[46] The original idea of this type of projection operator is due to [33], where it was applied to the probability density, complementarily to the observable quantities.

$$\frac{d}{dt}A_i = v_i(\mathbf{A}) - \sum_j \frac{L^0_{ij}}{T}\frac{\partial U_{\text{eq}}(\mathbf{A})}{\partial A^*_j} + f_i(t). \tag{1.42}$$

Here $v_i = v_i(\boldsymbol{A})$ is called the convective term, which can be determined independent of the projection, such as $v_i(\boldsymbol{A}) = P/m$ for $\boldsymbol{A} = \{X, P\}$. The coefficient L^0_{ij} is related (under the Markovian approximation) to the term corresponding to the thermal random force, $f_i(t)$, by

$$\langle f_i(t) f^*_j(0)\rangle = 2L^0_{ij}\delta(t). \tag{1.43}$$

The last relation assures that the distribution of $\boldsymbol{A}$ in equilibrium obeys the canonical one:

$$\mathcal{P}^{\text{eq}}(\mathbf{A(t)}) \propto e^{-\frac{U_{\text{eq}}(\mathbf{A}(t))}{k_{\text{B}}T}}. \tag{1.44}$$

The "potential energy function" $U_{\text{eq}}(\boldsymbol{a})$ for the slow variable $\boldsymbol{A}$ is the Helmholtz free energy at the temperature T with the value of $\boldsymbol{A}$ being constrained at $\boldsymbol{a}$:

$$e^{-U_{\text{eq}}(\boldsymbol{a})/k_{\text{B}}T} = \text{Tr}\{e^{-H(\boldsymbol{x},\boldsymbol{p})/k_{\text{B}}T}\delta(\boldsymbol{A}(\boldsymbol{x},\boldsymbol{p}) - \boldsymbol{a})\}. \tag{1.45}$$

Remark 1 The fact that the "potential energy" $U_{\text{eq}}(\boldsymbol{a})$ is a (constrained) free energy is understandable since the projection treats many microscopic degrees of freedom as environment. This fact implies also the following thing: If one can find a thermodynamic relation, like the first law and so on, from the Langevin equation, then the thermodynamic variables thus derived should not necessarily be the same as those of the conventional thermodynamics. The conventional thermodynamic variables can be related, through the classical statistical mechanics, to the microscopic Hamiltonian with *bare* potential energy functions. In other words, the thermodynamic framework derived from the Langevin equation can describe or even find the thermodynamic relations at the levels of description different from the macroscopic one. From experimental point of view it is crucial to know the relation between the thermodynamics of different levels, the issue which will be addressed later in Chap. 6.

Remark 2 We can consider a system which interacts with two heat baths of different temperatures. Experimentally, it will be possible to realize such a system by using a fine charged particle which is immersed in fluid at a temperature T on the one hand and at the same time subject under an equilibrated electromagnetic field (i.e., a black body radiation) at another temperature T'. (We assume that the radiation field does not practically interact with the fluid.) In such setup the two thermal environments will interact via the motion of a Brownian particle. This is a mechanical model of heat conduction since the interaction transports the energy from the fluid to the radiation field and vice versa. We can model such a system as coupled Langevin equations. In a later chapter (Sect. 4.2.1.2) we will discuss this phenomenon on the basis of the energetics of the Langevin equation. Such equations, however, will not be derivable, at least straightforwardly, from the Hamiltonian mechanics since the projection method would then require two different definitions of the scalar product, each corresponding to one of the temperatures T and T'.

1.2.2 Stochastic Calculus is the Mathematical Framework that Removes Ambiguities in the Langevin Equation

The projection technique applied above eliminates the slowly varying component from the fluctuating force $\xi(t)$ in the Langevin equations of the form (1.31) or (1.32). If the correlation time in $\xi(t)$ is negligibly small as compared with the time resolution of our interest, the Markov approximation is justified and we replaced finally the time correlation of $\xi(t)$ by (1.14), i.e., $\langle\xi(t)\xi(t')\rangle = 2\gamma k_B T\delta(t-t')$. Such idealization involving the delta function has a merit of simplifying description and calculation of represented quantities. However, as $\xi(t)$ or $\delta(t)$ are not mathematically ordinary functions, a care must be taken when we do the calculations involving these objects. Otherwise, the results could be widely, even qualitatively, different from what one should have by the calculation without this idealization. (The same remark may apply also to the description of a "point charge" as a delta function, which can actually represent, for example, an ionized molecule of finite size.) For example, an integral $\int x(t)\xi(t)dt$ for the variable $x(t)$ obeying $\gamma dx/dt = \xi(t)$ does not have a finite average for the idealized white Gaussian process $\xi(t)$.

A naive solution to avoid the ambiguity related to the δ-function and other singular objects would be to come back to the description with a finite correlation time in $\xi(t)$ and to do all the integrals using such "smoothened" variables. But such option would require a case-by-case treatment of problem where complicated estimation of errors and justification of the (re-)limiting procedure are needed. It will, therefore, be more efficient if there are systematic rules to deal with these singular mathematical objects as they are. For this purpose, we need therefore (i) to know the situations where the rules of ordinary calculus of the analysis are not applicable to $\xi(t)$ or $\delta(t)$, (ii) to learn the rules of calculus specific to these singular objects, and (iii) to recognize the results of new calculus as a limit of ordinary calculus.

It is Itô [34] who formulated *as mathematical framework* the Langevin equation and any other dynamical processes involving $\xi(t)$ obeying (1.14). The formulation is principally summarized by what is called Itô's lemma (Sect. 1.2.2.3 below). Just as the axiomatization of Dirac delta function by Schwartz has opened the field of "distribution" in mathematics, the mathematical foundation of the (equivalent of) Langevin equation through Itô's lemma has opened the new field of analysis called the "stochastic differential equation" or the "stochastic calculus."

Below we describe briefly the main concepts and tools of the stochastic calculus which we will use later in this book.[47] We will give mathematical meaning to

[47] The author referred to the lecture note by Y. Oono at the University of Illinois at Urbana Champaign in composing the present section.

$\xi(t)$ (Sect. 1.2.2.1), then to the integral with $\xi(t)$ (Sect. 1.2.2.2), and to the functions containing $\xi(t)$ (Sect. 1.2.2.3), before reformulating the Langevin equation (Sect. 1.2.2.4). Those who are not interested in the mathematical details of the calculus may skip to Sect. 1.2.2.5.

1.2.2.1 Wiener Process

Wiener has introduced a stochastic process called Wiener process, B_t, as a basic object to interpret the white Gaussian stochastic processes such as $\xi(t)$. In the context of the thermal random force, B_t is defined as follows:[48]

$$\int_0^t \xi(s)ds = \sqrt{2\gamma k_B T}[B_t - B_0] \tag{1.46}$$

or

$$\int_t^{t+dt} \xi(s)ds = \sqrt{2\gamma k_B T} dB_t. \tag{1.47}$$

where $\xi(t)$ is as introduced in Sect. 1.1.2.4 and in Sect. 1.1.3.1, and we have defined

$$dB_t \equiv B_{t+dt} - B_t, \tag{1.48}$$

with $dt > 0$ infinitesimal positive increment of time. Intuitively, B_t is introduced as a normalized form of ξ so that dB_t/dt satisfies $\langle (dB_t/dt)(dB_t/dt')\rangle = \delta(t-t')$. With (1.19), we see that the position $x(t)$ of a free Brownian motion under negligible inertia effect is represented by $x(t) - x(0) = \sqrt{2Dt}(B_t - B_0)$. Mathematically, $\xi(t)$ or dB_t/dt cannot be treated as functions of time, but B_t does. One might ask why we did not write $\xi(t)dt = \sqrt{2\gamma k_B T} dB_t$ in (1.47). There are two reasons, which we will become clear later. In short, it is because dB_t is not of order $\mathcal{O}(dt)$ but "$\mathcal{O}(\sqrt{dt})$," and because dB_t should be distinguished from $B_{t+\theta dt} - B_{t-(1-\theta)dt}$ with any $\theta \neq 1$.

The Wiener process is one of the most elementary stochastic processes. This process has the following properties:

1. Equation (1.18), that is $\langle \xi(t)\rangle = 0$ for any t, implies $\langle B_{t'} - B_t\rangle = 0$ for any t and t'.[49] As the last equation holds also for infinitesimal time interval between t and $t + dt$, we can write

$$\langle dB_t\rangle = 0. \tag{1.49}$$

---

[48] We adopt the convention in the field of stochastic calculus which often represents the time argument t as a suffix of the stochastic process, e.g., B_t for $B(t)$.

[49] We recall that, $\langle\,\rangle$ means the average over the paths (realizations of stochastic process).

The "white" property of $\xi(t)$ is interpreted by B_t as

$$\langle dB_t\, dB_{t'}\rangle = 0, \quad \text{if } t \neq t',$$

where $t \neq t'$ implies that the two intervals, $[t, t+dt]$ and $[t', t'+dt']$, have no overlap.

2. Since $\xi(s)$ with s between t and t' obeys the Gaussian distribution, its linear functional, $B_{t'} - B_t$, does also.
3. The *total variation* of B_s between $s = 0$ and $s = t$ is defined by the $n \to \infty$ limit of $\sum_{k=1}^n |B_{\frac{k}{n}t} - B_{\frac{k-1}{n}t}|$. This limit is divergent with probability 1 for the Wiener process. It is then said that the Wiener process is not of bounded variation.[50]
4. The following limit holds with probability 1 ($t > 0$):

$$\sum_{k=1}^{n} \left| B_{\frac{k}{n}t} - B_{\frac{k-1}{n}t} \right|^2 \to t \quad (n \to \infty). \tag{1.50}$$

Since each segment of time, $\left[\frac{k}{n}t - \frac{k-1}{n}t\right]$, brings an independent contribution, the above result allows the following replacement:

$$(dB_t)^2 = dt, \tag{1.51}$$

almost everywhere (*a.e.*), that is, with probability 1. Notice that in (1.51) there is no average $\langle\, \rangle$ over the paths.

As for the last two properties, 3 and 4, we might have the following intuitive explanations:[51] Equations (1.47) and (1.18) yield $\langle |B_{\frac{k}{n}t} - B_{\frac{k-1}{n}t}|^2\rangle = t/n$, which then implies $|B_{\frac{k}{n}t} - B_{\frac{k-1}{n}t}| \sim \sqrt{t/n}$. If we add up such contribution for n intervals, we would have the sum $\sim \sqrt{nt}$. The last quantity diverges in the limit of infinitely fine division of the interval. Since $B_{\frac{k}{n}t} - B_{\frac{k-1}{n}t}$ with different values of k are mutually independent, $\sum_{k=1}^n |B_{\frac{k}{n}t} - B_{\frac{k-1}{n}t}|^2$ has the average, $\sum_{k=1}^n \langle |B_{\frac{k}{n}t} - B_{\frac{k-1}{n}t}|^2\rangle = t$, and the variance,[52] $\left\langle \left[\sum_{k=1}^n |B_{\frac{k}{n}t} - B_{\frac{k-1}{n}t}|^2 - t\right]^2 \right\rangle = 2t^2/n$. Therefore, the difference between $\sum_{k=1}^n |B_{\frac{k}{n}t} - B_{\frac{k-1}{n}t}|^2$ and t becomes effectively 0 in the limit of fine segmentation, $n \to \infty$. When several such calculus are combined, we should keep only

---

[50] A real-valued function $f(x)$ is said to be of *bounded variation* on the interval $[a, b]$ if, with any partitioning, $a = x_0 < x_1 < \ldots < x_N = b$, and with any natural number N, the sum $\sum_{k=0}^{N-1} |f(x_{k+1}) - f(x_k)|$ is less than a positive number, M.

[51] For more rigorous proofs the reader should consult a book on stochastic process, for example, Sect. 4.2.5 of [1].

[52] We recall the formula (1.4), i.e., $\langle y^{2p}\rangle/(2p)! = \langle y^2\rangle^p/(2^p p!)$ with integer $p \geq 0$ for y obeying a Gaussian distribution with the average 0. This formula can be derived by expanding the special case of (1.15), i.e., $\langle e^{i\phi\hat{y}}\rangle = e^{-\frac{1}{2}\langle \hat{y}^2\rangle\phi^2}$.

the terms up to the order of dt and neglect higher order ones, where dB_t is regarded as a quantity of the order of $\sqrt{dt}$. In summary, the following prescriptions for the infinitesimal quantities should be applied:

$$dt^2 \equiv 0, \quad dB_t dt \equiv 0, \quad (dB_t)^2 \equiv dt. \tag{1.52}$$

As mentioned at the end of Sect. 1.1.3.1, the solution of $dx/dt = \xi(t)$ or the Wiener process has no intrinsic length or timescale (scale invariance or similarity law). Other aspects of Wiener process are described in [35].

1.2.2.2 Itô Integral and Stratonovich Integral

Now we define the time integral of the form $\int f(t)\,dB_t$. Intuitively, this is related to the type of calculations like $\int f(t)\xi(t)dt$. The latter, however, is not unambiguously defined since $\xi(t)dt$ does not precisely represent dB_t.

The integral of the type $\int f(s)dg(s)$ is called the Stieltjes integral. If $g(s)$ is not of bounded variation, we cannot define the Stieltjes integral unambiguously within the standard analysis. For example, if we were to evaluate the integral $\int_0^t [B_s - B_0]dB_s$ as a Riemannian integral, we might have different results depending on the different ways of taking limit:

$$\alpha_1 \equiv \sum_{\ell=1}^{n} \left[\sum_{k=1}^{\ell} (B_{\frac{k}{n}t} - B_{\frac{k-1}{n}t}) \right] (B_{\frac{\ell}{n}t} - B_{\frac{\ell-1}{n}t})$$

$$\alpha_2 \equiv \sum_{\ell=1}^{n} \left[\sum_{k=1}^{\ell} (B_{\frac{k}{n}t} - B_{\frac{k-1}{n}t}) \right] (B_{\frac{\ell+1}{n}t} - B_{\frac{\ell}{n}t}).$$

Other interpretations are also possible. If B_t were of bounded variation, these two would have the same limit when $n \to \infty$. However, with B_t being the Wiener process, the difference between the above two, $\alpha_2 - \alpha_1$, is $\sum_{k=1}^{n} |B_{\frac{k}{n}t} - B_{\frac{k-1}{n}t}|^2 + \mathcal{O}(t/n)$. As we have seen above, the sum on the right-hand side tends to t in the limit of $n \to \infty$.

To avoid such ambiguity of the integral, we introduce detailed definitions of the integral of the type, $\int f(s)dB_s$, together with new notations. Below, Δs denotes a *positive* finite interval of time. *Calculus of Itô type*:

$$f(s)[B_{s+\Delta s} - B_s] \to f(s) \cdot dB_s, \tag{1.53}$$

Calculus of Stratonovich type:

$$\frac{f(s+\Delta s) + f(s)}{2}[B_{s+\Delta s} - B_s] \to f(s) \circ dB_s. \tag{1.54}$$

Notice the different symbols between $f(s)$ and dB_s. As the essential difference is only of the lowest order in Δs, the latter definition can also be regarded as the limit of $f(s+(\Delta s/2))[B_{s+\Delta s}-B_s]$. By these definitions we have, for example, the formulas as follows:

$$\int_{s=0}^{s=t}(B_s-B_0)\cdot dB_s=\frac{(B_t-B_0)^2}{2}-\frac{t}{2},$$
$$\int_{s=0}^{s=t}(B_s-B_0)\circ dB_s=\frac{(B_t-B_0)^2}{2}. \tag{1.55}$$

Nonanticipating property of the Itô-type calculus: As a tool of calculus, a merit of the calculus of Itô type is that $\langle f(t)\cdot dB_t\rangle=0$ holds always *if* $f(t)$ depends on nothing that occurs after the time t. This property of $f(t)$ is called nonanticipating. It is so because dB_t by definition (see (1.48)) concerns only what is after the time t, and therefore

$$\langle f(t)\cdot dB_t\rangle=\langle f(t)\rangle\langle dB_t\rangle=0 \qquad (f(t):\ \text{nonanticipating}).$$

Sometimes it is not apparent which type of calculus is used. In Appendix A.1.6, we give two examples and show the way to render the problems in the above described form.

1.2.2.3 Stochastic Differential Equation (SDE) and Itô's Lemma

In order to evaluate integrals including dB_t, it is not practical to go back each time to the definitions (1.53) or (1.54) and to use (1.50). We describe the prescription that Itô has introduced for such calculation.

Suppose that a stochastic process x_t is generated by the Wiener process B_t through the following general equation for the increment, $dx_t\equiv x_{t+dt}-x_t$, called the stochastic differential equation (SDE):

$$dx_t=a(x_t,t)dt+b(x_t,t)\cdot dB_t. \tag{1.56}$$

To make clear the implication of this equation, let us temporarily introduce the notation $\langle\,\rangle_{x_t}$ to denote the average over the processes for $[t,t+dt]$ with a given "initial condition" x_t. Such average can be defined because x_t is a Markovian process. Using $\langle dB_t\rangle=0$ and the nonanticipating property of the Itô-type product, we have

$$\langle dx_t\rangle_{x_t}=a(x_t,t)dt, \qquad \left\langle(dx_t)^2\right\rangle_{x_t}=[b(x_t,t)]^2dt. \tag{1.57}$$

In (1.56) the term $a(x_t,t)dt$, which is of order dt, cannot be neglected despite the dominance of $b(x_t,t)\cdot dB_t$, since the average of the latter disappears due to the nonanticipating property of $b(x_t,t)$:[53] $\langle b(x_t,t)\cdot dB_t\rangle=0$.

Next we will consider any function $f(x_t)$ including the x_t as its argument and define its increment through infinitesimal time $dt(>0)$:

[53] cf. The footnote after (1.61).

$$df(x_t) \equiv f(x_{t+dt}) - f(x_t).$$

Then $df(x_t)$ can be rewritten by the following formula, called Itô's lemma:

$$df(x_t) = \left[a(x_t,t)f'(x_t) + \frac{b(x_t,t)^2}{2}f''(x_t)\right]dt + b(x_t,t)f'(x_t)\cdot dB_t, \quad (1.58)$$

where $f'(z) \equiv df(z)/dz$ and $f''(z) \equiv d^2f(z)/dz^2$.

With the recipe (1.52) in mind we can understand, though not prove, the formula (1.58) by substituting (1.56) into the (ordinary) Taylor expansion of $df(x_t) = f(x_t + dx_t) - f(x_t)$,

$$f(x_{t+dt}) = f(x_t) + f'(x_t)dx_t + \frac{f''(x_t)}{2!}(dx_t)^2 + \dots. \quad (1.59)$$

All the differences from the real analysis come out from the $\sim\sqrt{dt}$ property of dB_t, which is herited by dx_t through (1.56) and then by $df(x_t)$. We have kept the second order in dx_t because $\langle(dx_t)^2\rangle_{x_t} = [b(x_t,t)]^2dt$ (see above) indicates that $(dx_t)^2$ is also of order dt, instead of $(dt)^2$. Explicitly, the only term of $\mathcal{O}(dt)$ in $(dx_t)^2$ is $b(x_t,t)^2dt$, and we arrive at (1.58). Itô's lemma gives how the time evolution of $f(x_t)$ is generated by that of the Wiener process, dB_t. If we can invert the relation $f_t = f(x_t)$, then (1.58) gives the time evolution of f_t directly without solving (1.56).

The SDE (1.56) can be rewritten in terms of the Stratonovich-type calculus. The result writes as follows:

$$dx_t = \left[a(x_t,t) - \frac{b(x_t,t)}{2}\frac{\partial b(x_t,t)}{\partial x_t}\right]dt + b(x_t,t)\circ dB_t. \quad (1.60)$$

The derivation is given in Appendix.A.1.7. There we also show a convenient formula (A.13), that is,

$$df(x_t) = f'(x_t)\circ dx_t.$$

1.2.2.4 Langevin Equation as a Stochastic Differential Equation of the Stratonovich Type

Interpretations of $f(x_t)\xi(t)$ other than Itô or Stratonovich types are also possible, although they are not usually used. Even if we have only (1.56) and (1.60) we realize the ambiguity of writing down the differential equation for x_t in an ordinary form: $dx_t/dt = \tilde{a}(x_t,t) + b(x_t,t)(dB_t/dt)$. If $b(x_t,t)$ is constant, then $\partial b/\partial x_t = 0$ in (1.60) and the two SDEs take the same form. Still, there remains ambiguity as soon as we are to integrate in time some quantity in the form of $f(x_t)\xi(t)$. It is, therefore, indispensable to declare the type of calculus (Itô, Stratonovich, etc.) by which we write down a Langevin equation.[54]

Although the choice of calculus is a matter of convenience, the Stratonovich-type definition of Langevin equation is often better adapted to the bottom-up modeling of the equation, either by coarse graining, such as by the projection method, by an heuristic modeling based on mechanics, or by taking the continuum limit of discrete-time stochastic evolution equation.[55] For example, the SDEs which correspond to (1.31) are interpreted by Stratonovich type as [56]

$$dp = -\frac{\partial U(x,a)}{\partial x}dt - \gamma\frac{p}{m}dt + \sqrt{2\gamma k_{\mathrm{B}}T}\circ dB_t. \quad dx = \frac{p}{m}dt, \tag{1.61}$$

whether or not γ depends on x.[57]

This choice of calculus, i.e., of Stratonovich type, is widely accepted by several reasons, which are not mutually independent:

Modeling: In a physical modeling, we expect that an integral element like $f(t)\xi(t)dt$ corresponds to an element with finite time resolution. In this spirit, the interpretation of the Itô type, which treats the above quantity as strictly nonanticipating, $f(t)\,[B_{t+dt} - B_t]$, is not appropriate. For a detailed account of this issue, see [37, 38, 1].

Calculus: The fact that the formula of the real analysis is inherited by the calculus of the Stratonovich type fits well with our physical intuition.

Energetics: As we will see (Chap. 4 and later), the energetics of stochastic processes can have similar formal expressions to those in mechanics or in thermodynamics when the formalism is represented by the calculus of the Stratonovich type.

Example A rotator with a temporarily fluctuating frequency [58]

We describe a rotator whose angular velocity is modulated around ω_0 by a white Gaussian process, $\theta(t)$, satisfying $\langle\theta(t)\rangle = 0$ and $\langle\theta(t)\theta(t')\rangle = 2\alpha k_{\mathrm{B}}T\delta(t - t')$. We adopt the complex number representation in which the state of the rotator is z.[59] Physically appealing form of the evolution equation for z_t will be $\frac{dz}{dt} = i(\omega_0+\theta(t))z$,

---

[54] Or, the usage of SDE is often more convenient.

[55] See [36].

[56] Hereafter often omit the suffix "t" of x and p in SDEs.

[57] In fact, in the presence of the inertia term, the distinction among the types of calculus causes no differences: The momentum p remains finite and, from $dx = (p/m)dt$, the position variable x remains of bounded variation. It is no more the case if we neglect the inertia, as we will see in Sect. 1.3.2, or if T or γ depend on the momentum p, the cases which we excluded totally in our consideration.

[58] This example has been taken from the Lecture note by Y. Oono at UIUC.

[59] That is, $|z| = 1$ and the orientation angle of the rotator is $\arg(z)$.

where ω_0 is a constant. We like to choose the type of calculus so that the solution of this equation is $z_t = z_0 \exp(i\omega_0 t + i\sqrt{2\alpha k_B T}(B_t - B_0))$. Then the unique choice is the Stratonovich type, i.e., the interpretation of the above equation as the following SDE:

$$dz_t = iz_t \circ (\omega_0 dt + \sqrt{2\alpha k_B T} dB_t). \tag{1.62}$$

Even without solving (1.62), one can verify the relation, $d(z_t z_t^*) = z_t \circ dz_t^* + z_t^* \circ dz_t = 0$. We can rewrite (1.62) in the Itô type, and the result is

$$\begin{aligned} dz_t &= iz_t \cdot (\omega_0 dt + dB_t) - \alpha k_B T z_t dt \\ &= z_t \cdot (i\omega_0 - \alpha k_B T) dt + iz_t \cdot dB_t. \end{aligned} \tag{1.63}$$

Although (1.62) and (1.63) are mathematically equivalent, the former is more appealing than the latter *from the viewpoint of a single realization of stochastic process*. The Itô-type representation is, on the other hand, more adapted for the estimation of averaged properties: In order to understand the spurious "damping" term, $-\alpha k_B T z_t dt$ in (1.63), let us take the path average of (1.63). The result writes $d\langle z_t \rangle = \langle z_t \rangle (i\omega_0 - \alpha k_B T) dt$. The solution $\langle z_t \rangle = \langle z_0 \rangle e^{i\omega_0 t} e^{-\alpha k_B T t}$ shows the damping of the amplitude of $\langle z_t \rangle$ due to the desynchronization, i.e., the phase diffusion among the rotators of different realizations.

We note that, if we interpreted $\frac{dz}{dt} = i(\omega_0 + \theta(t))z$ as of the Itô type, i.e., as the SDE: $dz_t = i(\omega_0 dt + \sqrt{2\alpha k_B T} dB_t) \cdot z_t$, then $z_t z_t^*$ would increased exponentially, obeying $d(z_t z_t^*) = 2\alpha k_B T (z_t z_t^*) dt$.

A modified model is the amplitude modulation of the growth rate of a real quantity z_t: The model writes $dz_t = (\nu_0 dt + \sqrt{\alpha} dB_t) \cdot z_t$. It has a solution, $z_t = z_0 \exp(\nu_0 t + \sqrt{\alpha}(B_t - B_0))$, that is, $\log(z_t)$ at a given t is the Gaussian random variable with the average of $\log(z_0) + \nu t$ and the variance of αt. The probability density of z_t is then said to obey the *log-normal distribution*.

Example Correlation between the potential force and random force.

This example will show the utility of Itô-type calculus. As seen in (1.63) above, a practical advantage of Itô-type representation is that the average over the path ensemble is easily taken due to its nonanticipating property. Here is another example.

Let us consider the overdamped Langevin equation (1.32), that is,

$$0 = -\frac{\partial U(x,a)}{\partial x} - \gamma \frac{dx}{dt} + \xi(t).$$

Adopting the Stratonovich-type interpretation, we calculate

$$\int \frac{\partial U(x,a)}{\partial x}\xi(t)dt = \int \frac{\partial U(x,a)}{\partial x} \circ \sqrt{2\gamma k_{\mathrm{B}} T}dB_t$$
$$= \int \frac{\partial U(x,a)}{\partial x} \cdot \sqrt{2\gamma k_{\mathrm{B}} T}dB_t + k_{\mathrm{B}}T \int \frac{\partial^2 U(x,a)}{\partial x^2}dt, \quad (1.64)$$

where we have used (A.14). Since $\partial U(x,a)/\partial x$ on the second line is nonanticipating with respect to dB_t, the path average over the Wiener stochastic processes of this product vanishes. The path average of the above equation then yields[60]

$$\left\langle \frac{\partial U(x,a)}{\partial x}\xi(t) \right\rangle = \gamma k_{\mathrm{B}} T \left\langle \frac{\partial^2 U(x,a)}{\partial x^2} \right\rangle \qquad \text{(overdamped)}. \qquad (1.65)$$

The simplicity of the calculation owes to the Itô-type product.

When the inertia is explicitly taken into account, the second term in the second line of (1.64) is absent. Then the right-hand side of (1.65) should be replaced by 0:

$$\left\langle \frac{\partial U(x,a)}{\partial x}\xi(t) \right\rangle = 0 \qquad \text{(underdamped)}. \qquad (1.66)$$

Physical reason for the different result is that, in the case with inertia, the fluctuation of $\partial U/\partial x$, or essentially that of dx_t is smooth and, therefore, $\mathcal{O}(dt)$, not $\mathcal{O}((dt)^{\frac{1}{2}})$.

1.2.2.5 Primer of the Numerical Schemes for Solving Stochastic Differential Equations

As we have seen in the description of SDE, the short-time behavior of the white Gaussian process $\xi(t)$ requires a special attention. The numerical scheme to solve SDE requires also a particular attention about the numerical error if the scheme uses a time discretization.

Let us consider the overdamped Langevin-type equation of the form

$$\dot{x} = F(x) + \sigma\xi(t), \qquad (1.67)$$

where σ is a constant and $\xi(t)$ the normalized white Gaussian process satisfying $\langle \xi(t) \rangle = 0$ and $\langle \xi(t)\xi(t') \rangle = \delta(t-t')$. We will show below that first-order schemes of SDE can leave the errors of the 1.5th order.

The simplest first-order (explicit) Euler scheme generates $x(h)$ at time $t = h$ shortly after $t = 0$ as follows:

$$x(h) = x(0) + F(x(0))h + r(h), \qquad (1.68)$$

[60] See the question below (1.4).

where $r(h) = \int_0^h \xi(s)ds$ is zero mean Gaussian random variable with $\langle r(h)^2 \rangle = \sigma^2 h$. This discretization is correct up to $\mathcal{O}(h)$ but contains the error of $\mathcal{O}(h^{1.5})$ because the substitution of $x = x(0)$ in $F(x)$ on the right-hand side can give rise to the error of $\sim (x(h) - x(0)) = \mathcal{O}(h^{0.5})$. We will see later (Sect. 4.1.2.5 in Chap. 4) that the first-order scheme is not sufficient to reproduce the law of energy balance.

We thus motivated to go beyond the first-order Euler scheme. We can use the so-called Heun's method, a type of predictor–orrector method [39]. For the above equation, Heun algorithm is defined by first "predicting"

$$\tilde{x}(h) = x(0) + F(x(0))h + r(h), \tag{1.69}$$

and then "correcting"

$$x(h) = \frac{1}{2}\left\{\tilde{x}(h) + \left[x(0) + F(\tilde{x}(h))h + r(h)\right]\right\}. \tag{1.70}$$

This algorithm assures the second-order convergence in h for simple additive noises as above. In the case of multiplicative noise, however, this method assures only first-order convergence. We then should use improved methods (see, for example, [40–42]). The Langevin equation on manifolds (Sect. A.4.7.3) needs the multiplicative noise.

1.2.3 *Fokker–Planck Equation is the Mathematical Equivalent of the Langevin Equation when We Discuss the Ensemble Behavior*

1.2.3.1 * Fokker–Planck Equation

We have seen in Sect. 1.1.3.1 the relation between the evolution equation for $\hat{x}_t$, the position of the free Brownian particle, (1.19), and the diffusion equation (1.24) for $\mathcal{P}(X,t) \equiv \langle \delta(X - [\hat{x}_t - \hat{x}_0]) \rangle$, the probability density of $\hat{x}_t - \hat{x}_0$ at time t. As the Langevin equation is the generalization of the former equation, the partial differential equation called the Fokker–Planck equation and the Kramers equation are the generalization of the diffusion equation in the case where the Brownian particle is subject under external forces.

Now that we have the method of Itô-type product which is adapted for the ensemble description (i.e., average over path ensemble), it is better to rewrite the overdamped Langevin equation (1.32) in the standard form of SDE:

$$dx_t = a(x_t, t)dt + b(x_t, t) \cdot dB_t, \tag{1.71}$$

where

$$a(x) = -\frac{1}{\gamma}\frac{\partial U}{\partial x} + k_{\mathrm{B}}T\frac{\partial}{\partial x}\left(\frac{1}{\gamma}\right), \quad b(x) = \left(\frac{2k_{\mathrm{B}}T}{\gamma}\right)^{1/2}. \tag{1.72}$$

Using the Itô's lemma, (1.58), applied to $\delta(X - x_t)$, we obtain the following equation called the Fokker–Planck equation:[61]

$$\frac{\partial \mathcal{P}(X,t)}{\partial t} = \frac{\partial}{\partial X}\frac{1}{\gamma(X)}\left[\frac{\partial U}{\partial X} + \frac{\partial}{\partial X}k_{\mathrm{B}}T(X)\right]\mathcal{P}(X,t). \tag{1.73}$$

We refer to Appendix A.1.8 for the derivation.

When the inertia is not neglected, we start from the SDE (1.61), rather than (1.31). By similar calculations to the overdamped case, we can derive the evolution equation for the probability density, $\mathcal{P}(X, P, t) \equiv \langle \delta(X - x_t)\delta(P - p_t)\rangle$, see Appendix A.1.8 for the derivation:

$$\frac{\partial \mathcal{P}}{\partial t} = \left[-\frac{P}{m}\frac{\partial}{\partial X} + \frac{\partial}{\partial P}\left(\frac{\partial U}{\partial X} + \gamma\frac{P}{m}\right) + \frac{\partial^2}{\partial P^2}\gamma k_{\mathrm{B}}T\right]\mathcal{P} \tag{1.74}$$

This equation is called the Kramers equation [43], derived around 1940. The generalization to the higher dimensionality of the (X, P)-space is straightforward.

Throughout this book, we use the term Fokker–Planck equation to mention generically both the Fokker–Planck equation (1.73) and the Kramers equation (1.74).

1.2.3.2 General Properties of Fokker–Planck Equation

We summarize the several general properties of (1.73) [1]:

Principle of superposition: The linearity of the Fokker–Planck equation allows to construct the solution $\mathcal{P}(X, t)$ from an arbitrary initial distribution, $\mathcal{P}(X, t_0)$, as a superposition of the following form:

$$\mathcal{P}(X,t) = \int G(X, t|X_0, t_0)P(X_0, t_0)dX_0, \tag{1.75}$$

where $G(X, t|X_0, t_0)$ is the solution (called Green function) of the linear equation (1.73) satisfying the initial condition, $\mathcal{P}(X, t_0) = \delta(X - X_0)$. The same principle applies to the case where the effect of inertia is taken into account.

Equation of continuity for the probability: The Fokker–Planck equation (1.73), takes the form of equation of continuity,

---

[61] It was first written down by Einstein [9]. See, for example, the descriptions in [1].

$$\frac{\partial \mathcal{P}}{\partial t} = -\frac{\partial J_x}{\partial X}, \tag{1.76}$$

$$J_x \equiv -\frac{1}{\gamma}\left[\frac{\partial U}{\partial X} + \frac{\partial}{\partial X}k_{\rm B}T\right]\mathcal{P}(X,t), \tag{1.77}$$

where the probability flux, J_x, is the probability that flows per unit of time through the position X toward the $+X$ direction.[62] This continuity (1.76) reflects the conservation of the probability and the fact that in each path $\hat{x}_t$ is continuous in time. For the Kramers equation, (1.74), the equation of continuity and the definition of the probability flux are constructed on the 2D phase space, (X, P):

$$\frac{\partial \mathcal{P}}{\partial t} = -\frac{\partial J_x}{\partial X} - \frac{\partial J_p}{\partial P}, \tag{1.78}$$

$$J_x \equiv \frac{P}{m}\mathcal{P}(X,P,t),$$
$$J_p \equiv \left(-\frac{\partial U}{\partial X} - \gamma\frac{P}{m}\right)\mathcal{P}(X,P,t) - \frac{\partial}{\partial P}[\gamma k_{\rm B}T\mathcal{P}(X,P,t)]. \tag{1.79}$$

Equilibrium state under homogeneous temperature: If the parameter a of the potential function $U(X,a)$ is fixed and if the temperature T of the thermal environment is uniform, then the distribution $\mathcal{P}(X,t)$ approaches to the canonical equilibrium distribution, $\mathcal{P}^{\rm eq}(X,a;T)$ defined by (1.105). We can verify easily that with $\mathcal{P}(X,t) = \mathcal{P}^{\rm eq}(X,a;T)$ the probability flux J_x of (1.77) vanishes for all X. The entire vanishing of the flux is referred to as the detailed balance, which will be discussed in detail later (Sect. 3.3.1.4).[63]

In the numerical analysis of the steady state, it is sometimes important to discretize the flux J_x of the Fokker–Planck equation so that the flux-free state $J_x = 0$ is strictly compatible with the equilibrium density, $\mathcal{P}^{\rm eq}(X,a;T)$. A way to satisfy this condition is to introduce $\psi(X,t)$ by $\mathcal{P}(X,t) = e^{\psi(X,t)}$. The flux J_x then writes $J_x = \frac{1}{\gamma}e^{\psi(X,t)}\left[\frac{\partial U}{\partial X} + T\frac{\partial \psi}{\partial X}\right]$, where the discretization of ϕ matches with that of U.

Also for the Kramers equation, the probability distribution converges to the canonical equilibrium distribution completed with the Maxwell distribution of the momentum:

[62] Here the X also represents the state where the particle is at the space position, X.

[63] With the detailed balance, (i) the forward driving of $\hat{x}$ by the potential gradient is counterbalanced by the more frequent backward displacements by diffusion and (ii) the diffusive displacements of $\hat{x}$ into high-friction regions (i.e., large γ regions) are as often discouraged as do the diffusive displacements out of such regions.

$$\mathcal{P}^{\rm eq}(X,P,a;T) = \frac{e^{-(\frac{P^2}{2m}+U(X,a))/k_{\rm B}T}}{\sqrt{2\pi m k_{\rm B}T}\int e^{-U(X',a)/k_{\rm B}T}dX'}. \qquad (1.80)$$

We can verify that $\mathcal{P}^{\rm eq}(X,P,a;T)$ satisfies $J_x = J_p = 0$ uniformly.
In Sect. 1.3.2.1 we will discuss the Fokker–Planck equation for inhomogeneous temperature $T(x)$. Its steady-state probability density, $\mathcal{P}^{\rm st}(X)$, is *not* generally proportional to $e^{-U(X,a)/k_{\rm B}T(x)}$. In the general steady state the probability flux J_x defined by (1.77) is constant, may be nonzero. This condition, J_x =const., can be integrated along X-axis, and the result depends on T as well as on U, generally in a nonlocal manner.

***H**-theorem:* The approach of the probability distribution function toward the canonical equilibrium distribution under uniform temperature can be shown using the Kullback–Leibler distance (of continuum version), $D(\mathcal{P}(t)\parallel\mathcal{P}^{\rm eq})$, defined by the following[64]:

$$D(\mathcal{P}(t)\parallel\mathcal{P}^{\rm eq}) \equiv \int \mathcal{P}(X,t)\ln\left[\frac{\mathcal{P}(X,t)}{\mathcal{P}^{\rm eq}(X,a;T)}\right]dX. \qquad (1.81)$$

This entropy is nonnegative and vanishes if and only if $\mathcal{P}(t)$ coincides with $\mathcal{P}^{\rm eq}$.[65] By the uses of (1.73) one can verify that $D(\mathcal{P}(t)\parallel\mathcal{P}^{\rm eq})$ decreases strictly:

$$\frac{dD(\mathcal{P}(t)\parallel\mathcal{P}^{\rm eq})}{dt} < 0, \quad (\mathcal{P}\not\equiv\mathcal{P}^{\rm eq}). \qquad (1.82)$$

It thus implies the convergence of $\mathcal{P}(t)$ to $\mathcal{P}^{\rm eq}$ in the $t\to\infty$ limit. The functional of $\mathcal{P}(t)$ having the property of (1.82) is referred to as Lyapnov functional for $\mathcal{P}(t)$. In general, a theorem proving the approach to the equilibrium density by way of strictly nondecreasing quantity ("entropy") is called the H-theorem.

The Fokker–Planck equation and the Kramers equation are mathematically equivalent to the Langevin equations, respectively, without or with the effect of inertia in the sense that the Langevin equations can be reconstructed if we know the former ones for the probability densities. However, the Langevin equations are more fundamental than the former equations since the particular realization of stochastic process is the basis of its statistical distribution.

[64] This "distance" $D(\mathcal{P}(t)\parallel\mathcal{P}^{\rm eq})$ does not fulfill standard axioms of the distance, in particular, the reflection law. $D(\mathcal{P}(t)\parallel\mathcal{P}^{\rm eq})$ is also called the relative entropy or $\mathcal{P}$ with respect to $\mathcal{P}_{\rm eq}$. The Kullback–Leibler distance is closely related to the LDP. See [44].

[65] To verify this, we can use the inequality, $p\log(p/q)\geq -q+p$, where p and q are arbitrary nonnegative quantities. The above inequality can, in turn, be obtained by multiplying by p the simple inequality, $\log(p/q) = -\log(q/p)\geq -q/p+1$.

1.3 Physical Implications of Langevin Equations

1.3.1 Langevin Equation Transforms the Memory-Free Gaussian Noise into Variety of Fluctuations

1.3.1.1 Transformation of Stochastic Processes

The statistics of thermal fluctuations of an object in a thermal environment depends on the characteristics of the object, such as its size, shape and, more generally, on the (free-)energy landscape of its conformation and also on the way how it interacts with the environment. A kinesin molecule in a solvent fluctuates differently from a long microtubule filament because of their shapes and flexibilities. Any object, therefore, transforms the thermal fluctuations of its environment into its own fluctuation.

From this point of view, a Langevin equation realizes a transducer or converter of fluctuations. Let us take, for example, a Brownian particle under a potential: $\gamma \frac{dx}{dt} = -\frac{dU(x)}{dx} + \xi(t)$. The equation converts the white Gaussian process $\xi()$ into a stochastic process of the position of the particle $x()$. In terms of the path probability, a process $\xi()$ between $t = 0$ and $t = t$ is realized by the probability

$$P[\xi] \propto \exp\left(-\frac{1}{4\gamma k_{\mathrm{B}} T} \int_0^t \xi(s)^2 ds\right), \tag{1.83}$$

except for the normalization factor. Substituting $\xi(t)$ by $\gamma \frac{dx}{dt} + \frac{dU(x)}{dx}$ and taking care of the change of variable, the same probability is rewritten in terms of $x()$ (apart from the initial probability distribution of x):

$$P[x] \propto \exp\left(-\frac{1}{4\gamma k_{\mathrm{B}} T} \int_0^t \left\{\left[\gamma \frac{dx}{dt} + \frac{dU}{dx}\right]^2 - 2k_{\mathrm{B}} T \frac{\partial^2 U}{\partial x^2}\right\} ds\right), \tag{1.84}$$

where the term including $\partial^2 U/\partial x^2$ takes account of the transformation of variables, i.e. the Jacobian, $J = |\mathcal{D}\xi/\mathcal{D}x|$. A heuristic explanation of the Jacobian is given in Appendix A.1.9.

Equation (1.83) shows that the process $\xi(t)$ and its time-reversed process $\xi(-t)$ occurs at the equal probability. If a Langevin equation realizes an equilibrium state, then this symmetry with respect to the time reversal is preserved, whatever is the form of the potential function $U(x)$. This symmetry is called the detailed balance (DB) symmetry. Such Langevin excludes the directional motion of x on the average, since any process can be traced back with the same likeliness. The consequence of this symmetry is that we cannot extract steadily from the fluctuations of x, any systematic work. This is in accordance with the second law of thermodynamics.

If the thermal random force $\xi()$ is replaced by a process with zero average but not having the detailed balance symmetry, then the induced process $x()$ can move in one direction on the average. Hondou et al. have shown a good example where the potential energy $U(x)$ is symmetric with respect to the space inversion, $x \leftrightarrow (-x)$

[45, 46]. In these papers $\xi()$ is replaced by a chaotic noise $\eta(t)$ of zero average, $\langle \eta(t) \rangle = 0$, but there is nonzero third moment, $\langle \eta(t)^3 \rangle \neq 0$.

1.3.1.2 * Fluctuation–Dissipation Relation in Free Brownian Motion

Let us consider the free Brownian motion under a small external force $\epsilon f(t)$. The overdamped Langevin equation writes

$$\gamma \dot{x} = \xi(t) + \epsilon f(t) \tag{1.85}$$

with $\langle \xi(t)\xi(s) \rangle = 2\gamma k_B T \delta(t-s)$. For the process $x()$ a sporadic increment in $\xi(t_0)$ and a small force ϵ at the time t_0 are not distinguishable. The effect of former on $x()$ is characterized by the velocity correlation in the absence of external force,

$$C(t-s) \equiv \langle \dot{x}(t)\dot{x}(s) \rangle_{\epsilon=0}, \tag{1.86}$$

while the effect of weak external force is characterized by the linear response function $R(u)$ for the velocity ($\dot{x}$)

$$\langle \dot{x}(t) \rangle_\epsilon = \epsilon \int_{-\infty}^{+\infty} R(t-s) f(s) ds. \tag{1.87}$$

By the causality requirement, $R(u)$ satisfies $R(u) = \theta(u)\tilde{R}(u)$, where $\theta(u)$ is the step function; $\theta(u) = 1$ for $u > 0$, $\theta(u) = 0$ for $u < 0$, and $\theta(0) = 1/2$.

In the case (1.85), we have

$$R(t-s) = \theta(t-s) \cdot \frac{2}{\gamma} \delta(t-s), \qquad C(t-s) = \frac{2k_B T}{\gamma} \delta(t-s), \tag{1.88}$$

therefore

$$C(u) = k_B T [R(u) + R(-u)]. \tag{1.89}$$

The last form of the relation between the correlation function and the response function is called the fluctuation–dissipation (FD) relation (of the first kind). This relation (1.89) is closely related to the Einstein relation because the double time integral of $C(t)$ gives $D = \lim_{t\to\infty} [\langle x(t)^2 \rangle / (2t)] = k_B T/\gamma$.

By contrast, the correlation of the random force is related to the friction constant γ through

$$\langle \xi(t)\xi(t+u) \rangle = 2\gamma k_B T \delta(u).$$

The latter type of relation is called the FD relation of the second kind.

The relation (1.89) holds very generally for equilibrium fluctuations, even if we add the potential force $-dU(x)/dx$ and also the inertial term $m\ddot{x}$ to the equation of free Brownian motion. In other words, the FD relation for the free Brownian motion, which is essentially that of $\xi()$, is reserved upon the transformation of stochastic process from $\xi()$ to $x()$ through the Langevin equation (see Sects. 1.3.1.1). This is one of the main result of the *linear response theory,* including the quantum case. See, for example, [4, 2] for the general argument. In Appendix A.1.10 we describe the essence of the FD relation. We will come back to the FD relation in more general case in Part II, in the context of energetics.

1.3.1.3 * Application : Detector of Displacement

While the white Gaussian process like $\xi()$ keeps no memory, the Langevin equation transforms it into a process $x()$ with memory. For example, the trajectory of a free Brownian particle depends on its initial position. This property can be used for a single molecule to detect its own displacement.

Suppose that the molecule consists of two Brownian particles, a "motor head" (position x_1) and a "cargo" (position x_2) and that these particles are joined together by a harmonic spring or an ideal polymer chain. See Fig. 1.5. The overdamped Langevin equation for x_1 and x_2 are as follows:

$$-\gamma_1\dot{x}_1 + \xi_1(t) - K(x_1 - x_2) = 0, \qquad -\gamma_2\dot{x}_2 + \xi_2(t) - K(x_2 - x_1) = 0. \quad (1.90)$$

Here the Gaussian white noises $\xi(t)$ and $\xi_2(t)$ obey

$$\langle\xi_1(t)\xi_1(t')\rangle = 2\gamma_1 k_{\mathrm{B}}T\delta(t-t'), \qquad \langle\xi_2(t)\xi_2(t')\rangle = 2\gamma_2 k_{\mathrm{B}}T\delta(t-t'),$$

and $\langle\xi_1(t)\xi_2(t')\rangle = 0$. We will solve these equations under the initial conditions $x_1(0) = x_2(0) = 0$ at time $t = 0$, which has no bias toward $+x$ or $-x$ directions. The direct calculation shows that $x(t)$ is correlated with the *relative* displacement $w(t) \equiv x_1(t) - x_2(t)$, i.e., $\langle w(t)x_1(t)\rangle > 0$ with $\langle w(t)\rangle = \langle x_1(t)\rangle = 0$. This implies that, statistically, the motor head can detect the direction of its own displacement by monitoring the conformation of the molecule, $w(t) = x_1(t) - x_2(t)$, without directly referring to the coordinate system (Fig. 1.5). The Eq. (1.90) can be separated by introducing the coordinate of the "diffusion center," $X = (\gamma_1 x_1 + \gamma_2 x_2)/(\gamma_1 + \gamma_2)$ and $w = x_1 - x_2$:

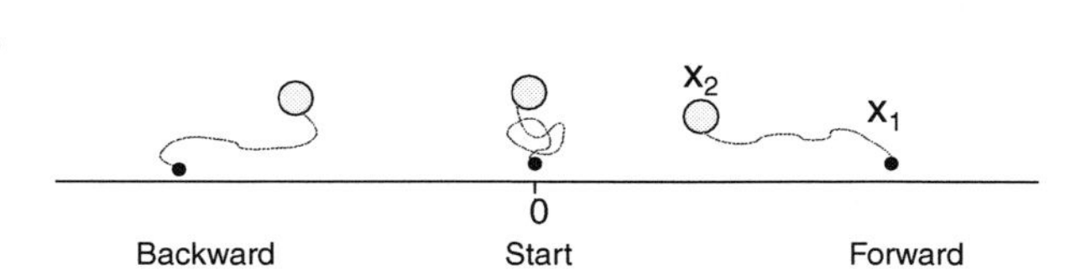

Fig. 1.5 Brownian particles "motor head" $x_1(t)$ (*small filled disc*) and "cargo" $x_2(t)$ (*big open disc*) are tied by a harmonic force and start to diffuse from $x_1(0) = x_2(0) = 0$. For $t > 0$ the relative displacement, $w(t) \equiv x_1(t) - x_2, (t)$ is statistically correlated with the absolute displacements such as $x_1(t)$

$$-(\gamma_1 + \gamma_2)\dot{X} + (\xi_1(t) + \xi_2(t)) = 0, \qquad -\zeta \dot{w} + f(t) - Kw = 0,$$

where $\zeta = (\gamma_1{}^{-1} + \gamma_2{}^{-1})^{-1}$ and $f(t) = (\xi_1(t)/\gamma_1 - \xi_2(t)/\gamma_2)/(1/\gamma_1 + 1/\gamma_2)$. The latter satisfies $\langle f(t)f(t')\rangle = 2\zeta k_{\rm B}T\delta(t-t')$ and is independent of the random force $\xi_1() + \xi_2()$. Solving these equations with the initial conditions on x_1 and x_2, we obtain

$$\langle x_1(t)w(t)\rangle = -\frac{\gamma_2}{\gamma_1}\langle x_2(t)w(t)\rangle = \frac{\gamma_2}{\gamma_1+\gamma_2}\frac{k_{\rm B}T}{K}\left(1 - e^{-\frac{2K}{\zeta}t}\right) > 0, \qquad (1.91)$$

where we have used $\langle X(t)w(t')\rangle = 0$ and $(\gamma_1 + \gamma_2)x_1 = (\gamma_1 + \gamma_2)X + \gamma_2 w$. The cargo may be another motor head if the molecule is a two-headed motor.[66] In any case, a motor head can use the information of $w(t)$ by reacting differently according to the direction of the force ($\propto -w$), e.g., by modifying its binding affinity to the filament (actin filament, microtubule, DNA, etc.).

1.3.2 Each Langevin Equation Has Its Own Smallest Space-Timescale of Applicability

We have seen that the Langevin equation is obtained as a result of elimination of the rapidly varying degrees of freedom in a more microscopic model. It is, therefore, necessary to fix a finite time resolution, Δt, of the description by the Langevin equation. Concomitantly there is a finite resolution in the variables (e.g., x or p). However, once a Langevin equation is written down, the Δt appears nowhere in the equation, and its solutions (i.e., the paths or the realizations) contain arbitrarily fine details as function of time. In general, the fine details of its solutions at timescales below Δt have nothing to do with the original microscopic model.[67]

A logical consequence of the above fact is that there can be more than one Langevin equations to describe the same phenomenon, one at a time resolution of Δt_1 and the other at $\Delta t_2 (> \Delta t_1)$, etc. The solutions of the former Langevin equation contain supplemental informations about the timescales between Δt_1 and Δt_2. It should, therefore, be possible to derive the Langevin equation with the time resolution Δt_2 from the one with better time resolution $\Delta t_1 (< \Delta t_2)$. Below we will see two examples of the different levels of Langevin equations.

---

[66] More precisely, we should study the probability density for $\hat{w}(t)$ with a given value of $x_1(t)$, where the nonzero constrained average, $\langle \hat{w}\rangle_{x_1(t)}$, is an indication of the memory. Using the probability density of $\hat{w}(t)$ (see (1.33) with $x \equiv w$ and $x_0 = 0$), we can show $\langle \hat{w}\rangle_{x_1(t)} = x_1(1+\gamma_2/\gamma_1)/(\psi(2Kt/\zeta)+\gamma_2/\gamma_1) > 0$, where $\psi(s) \equiv s/(1-e^{-s})$. For large γ_2/γ_1, the memory is maintained up to $t \sim \gamma_2/K$.

[67] One might compare this situation with the fact that the touches of a painting of a landscape on canvas do not reproduce the details of the original landscape.

1.3.2.1 Coarse Graining by Direct Short-Time Integral

We start with the Langevin equation with the inertia effect, (1.31), allowing the position dependence of the temperature, $T = T(x)$.[68] The SDE corresponding to such Langevin equation is (1.61), i.e.,[69]

$$dp = -\frac{\partial U(x,a)}{\partial x}dt - \gamma\frac{p}{m}dt + \sqrt{2\gamma k_B T(x)} \circ dB_t, \quad dx = \frac{p}{m}dt. \tag{1.92}$$

The time resolution, Δt_1, is finer than the damping time of the inertia effect, that is, $\Delta t_1 \ll m/\gamma$.

Our aim is to convert this equation to a new Langevin equation which is (only) valid for the time resolution $\Delta t \gg m/\gamma$.[70] In other words, we will eliminate the momentum, p, in (1.61) as a "fast variable." This problem has long been discussed since the landmark paper by Kramers [43], and there is generally a consensus about the result [47–49]. It is expressed as SDEs, either of Itô type or of Stratonovich type, as follows:

$$\gamma dx = \sqrt{2\gamma k_B T(x)} \cdot dB_t - \frac{\partial U(x,a)}{\partial x}dt, \tag{1.93}$$

$$\gamma dx = \sqrt{2\gamma k_B T(x)} \circ dB_t - \frac{\partial}{\partial x}\left[U(x,a) + \frac{k_B T(x)}{2}\right]dt. \tag{1.94}$$

The derivation has often been done by way of the Fokker–Planck equation (see Sect. 1.2.3.1). In Appendix A.1.11 we present a derivation that uses only the coarse graining in time along a single realization [49].

Below are the comments on the physical aspects of the above result:

1. The term $T(x)$ in (1.94) represents the effect of thermophoresis, that is, the [direct] thermal random force drives a Brownian particle toward the cooler region from the hotter region. To understand precisely what it means, suppose that there is no potential force ($U(x,a)$ =const.). From (1.93) we find

   $$\langle dx_t \rangle_{x_t} = 0, \qquad \langle (dx_t)^2 \rangle = 2(k_B T(x)/\gamma)dt,$$

   where $\langle\,\rangle_{x_t}$ is the conditional path average under a given value of $\hat{x}_t$. The former equation tells that there is equal probabilistic weights on $dx_t > 0$ and on $dx_t < 0$ up to time $t + dt$, starting from a given position, x_t, at time t. But if

[68] Up to (1.94) we assume that γ is independent of x although, naturally, γ depends on the temperature.

[69] In the Langevin form, the first equation is written as $\gamma\frac{dx}{dt} = -\frac{\partial U}{\partial x} - \gamma\frac{p}{m} + \sqrt{2\gamma k_B T(x)} \cdot \theta(t)$ with $\theta(t) = dB_t/dt$.

[70] Δt should, however, be small compared with the characteristic timescale of the change of $\partial U(x(s), a(s))/\partial x$. The opposite case is discussed in Sect. 1.3.2.2.

we compare different positions, the amplitude of random movement depends on $T(x)$ according to the latter equation.

2. If the system is bounded (i.e., on a circle), the probability density at steady-state $P_{st}(x)$ satisfies $T(x)P_{st}(x) = \text{const}$. The easiest way to see this is to reconstruct the Fokker–Planck equation from (1.93). The result writes $\frac{\partial P}{\partial t} = -\frac{\partial J_x}{\partial x}$, with $J_x = -\frac{1}{\gamma}[\frac{\partial U}{\partial x}P + \frac{\partial(k_B T(x)P)}{\partial x}]$, and the steady state of a bound system satisfies $J_x = 0$. In the presence of potential force, the value of $P_{st}(x)$ at x depends on the whole profile of $T(x)$ and of $U(x)$. Such a nonlocal aspect of nonequilibrium distribution is called Landauer's blow torch [50].
3. The Stratonovich-type SDE (1.94) shows that the apparent potential energy is the sum of $U(x, a)$ and the (equipartitioned) kinetic energy, $k_B T(x)/2$. This view will be justified later from the energetic point of view, see (4.9).
4. The trajectory x_t of (1.31) is smooth, while that of (1.93) and (1.94), i.e., the coarse-grained version of the former, has unbounded total variation. Such paradoxical roughness of the trajectory has no relevance on the solution viewed with the resolution, Δt, as discussed before (see *Remark 1* of Sect. 1.1.3).

The case where γ instead of T depends on x has been discussed in [47] (see also [48]). The incorporation of these two cases, i.e., with the inhomogeneous temperature and friction coefficient, yields the generalized SDE of the Itô and Stratonovich type:

$$\gamma(x) \cdot dx = \sqrt{2\gamma(x) k_B T(x)} \cdot dB_t - \left[\frac{\partial U(x,a)}{\partial x} + \frac{k_B T(x)}{\gamma(x)} \frac{d\gamma(x)}{dx} \right] dt. \tag{1.95}$$

$$\gamma(x) \circ dx = \sqrt{2\gamma(x) k_B T(x)} \circ dB_t - \left[\frac{\partial U(x,a)}{\partial x} + \frac{1}{2\gamma(x)} \frac{d(\gamma(x) k_B T(x))}{dx} \right] dt. \tag{1.96}$$

These two equations are equivalent. Using (1.95) we can obtain the Fokker–Planck equation corresponding to these SDEs. It writes

$$\frac{\partial P}{\partial t} = \frac{\partial}{\partial x} \frac{1}{\gamma(x)} \left[\frac{\partial U}{\partial x} P + \frac{\partial}{\partial x} (T(x) P) \right]. \tag{1.97}$$

If the temperature is uniform, $T(x) = T_0$, the stationary probability density under a fixed parameter a gives the equilibrium canonical distribution, $P_{st}(x) \propto e^{-U(x,a)/k_B T_0}$. The last result is consistent with the equilibrium statistical mechanics, which asserts that the equilibrium state is not influenced by the purely kinetic parameters like $\gamma(x)$. An illustrating example of this fact is discussed in Sect. 3.3.1.5.[71]

1.3.2.2 Spatial Coarse Graining of Langevin Equation

We start with the overdamped Langevin equation describing a Brownian particle under a periodic potential $U_m(x)$ of the period ℓ and under an almost uniform force

[71] [53] discuss its implications in the biophysical context.

potential $-a_0 x + V(x)$.

$$\gamma \dot{x} = -\frac{\partial U_m(x)}{\partial x} + a_0 - \frac{\partial V(x)}{\partial x} + [2\gamma k_B T]^{\frac{1}{2}} \xi_m(t), \tag{1.98}$$

where $\xi_m(t)$ is a Gaussian white noise with zero average and the variance, $\langle \xi_m(t) \xi_m(t') \rangle = \delta(t-t')$. The spatial variation of $V(x)$ is supposed to be much smoother than that of $U_m(x)$, while a_0 need not be smaller than the magnitude of $|\partial U_m/\partial x|$, see Fig. 1.6. The spatial resolution of (1.98), which we denote by Δx_1, is finer than the period of U_m, that is, $\Delta x_1 \ll \ell$.

Our aim is to convert this equation to a new Langevin equation which is (only) valid for the spatial resolution $\Delta x \gg \ell$. (Δx should, however, be small compared with the characteristic length scale of the change of $\partial V(x)/\partial x$.)

The first step is to analyze the case with $V(x) \equiv 0$. The spatiotemporal coarse graining of this case has been studied analytically [51, 52], and the result is written in the form of free Brownian motion under drift:

$$\dot{\tilde{x}} = v_s(a_0) + [2D_{\text{eff}}(a_0)]^{\frac{1}{2}} \xi(t), \tag{1.99}$$

where $\tilde{x}(t)$ is the position of the particle on this coarse-grained description, and $v_s(a_0) \equiv \langle \dot{\tilde{x}} \rangle_{a_0}$ is the coarse-grained steady mean velocity,[72] and $D_{\text{eff}}(a_0)$ is the coarse-grained diffusion coefficient such that

$$\lim_{t\to\infty} \frac{|\tilde{x}(t) - v_s(a_0)|^2}{2t} = D_{\text{eff}}(a_0).$$

The new Langevin equation (1.99) is valid for the spatial resolution Δx satisfying $\Delta x \gg \ell$. The quantity in the $\lim_{t\to\infty}$ approaches to $D_{\text{eff}}(a_0)$ only for $t \gg \ell^2/D_{\text{eff}}(a_0)$. A very rough estimate of $D_{\text{eff}}(a_0)$ is $\frac{k_B T}{\gamma} e^{-\Delta U_m/k_B T}$, where ΔU_m is the amplitude of the variation of $U_m(x)$, i.e., its barrier height (see Sect. 7.1.1.4).

The second step is to take into account of the spatially slowly varying potential, $V(x)$. Since $|\partial V/\partial x|$ is small, it can be treated as a local variation of the force parameter a as $a_0 \to a_0 + \Delta a$ with identifying Δa with $-\frac{\partial V}{\partial x}$. In the context of Appendix A.1.2, the effect of small change of the parameter on the mean velocity is

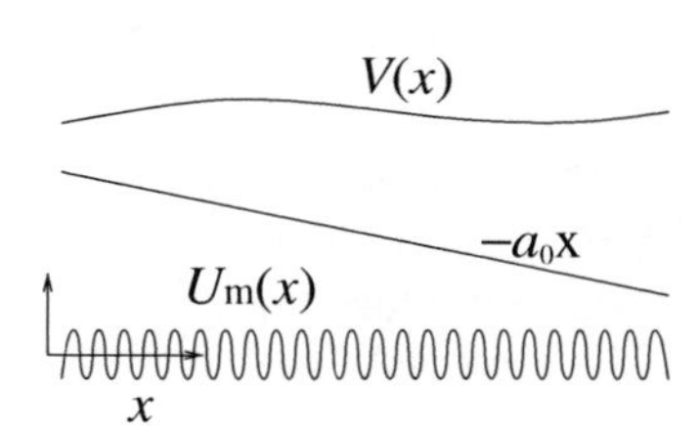

Fig. 1.6 The three potential energies in (1.98)

[72] $v_s(0) = 0$ for equilibrium.

characterized by the differential "friction constant," $\Gamma(a_0)$:

$$\frac{1}{\Gamma(a_0)} \equiv \lim_{\Delta a \to 0} \frac{\langle \dot{\hat{x}} \rangle_{a_0+\Delta a} - \langle \dot{\hat{x}} \rangle_a}{\Delta a} = \left. \frac{dv_s(a)}{da} \right|_{a=a_0}. \tag{1.100}$$

If we introduce a quantity $\Theta(a)$ [54] such that

$$\frac{1}{\Gamma(a_0)} = k_B \Theta(a_0) D_{\text{eff}}(a_0), \tag{1.101}$$

$k_B\Theta(a_0)$ plays the role of $u_a^{(1)}$ in (A.2).[73] The linear perturbation, $v_s(a) \simeq v_s(a_0) + \frac{1}{\Gamma(a_0)}\Delta a$, results in [54]

$$\dot{\hat{x}} - v_s(a_0) \simeq -\frac{1}{\Gamma(a_0)} \frac{\partial V}{\partial x} + [2D_{\text{eff}}(a_0)]^{\frac{1}{2}} \xi_m(t).$$

By (1.101), this equation takes the form of a standard overdamped Langevin equation

$$0 = -\Gamma(a_0)[\dot{\hat{x}} - v_s(a_0)] + [2\Gamma(a_0) k_B \Theta(a_0)]^{\frac{1}{2}} \xi_m(t) - \frac{\partial V}{\partial x}. \tag{1.102}$$

Especially for $a_0 = 0$, (1.102) describes a Brownian particle in the potential $V(x)$ with the spatial resolution Δx, where $v_s(0) = 0$ and $k_B\Theta(0) = k_B T$ should hold.

1.3.3 A Trajectory of Langevin Equation Over a Long Time Allows Us to Draw Some Information About the Space and the Potential

1.3.3.1 * The Residence Time Distribution with a Fixed Potential Energy Function

Suppose we follow a Brownian particle using the standard overdamped Langevin (1.32) equation, i.e., $0 = -\frac{\partial U(x,a)}{\partial x} - \gamma \frac{dx}{dt} + \xi(t)$, and assume that the parameter a, the friction coefficient γ, and the temperature T are constant.

We introduce finite time [empirical] average of the stochastic process, $\hat{x}()$, over a time interval $[0, t]$[74]:

---

[73] Generally, $v_s(a)$ is a nonlinear and complicated function of a and is not antisymmetric with respect to a.

[74] We use X for the spatial coordinate.

$$\mathcal{P}^{\mathrm{res}}_{[0,t]}(\hat{x};X) \equiv \frac{1}{t}\int_0^t \delta(X-\hat{x}(t'))\,dt'. \tag{1.103}$$

For each realization of $\hat{x}(\,)$, the functional $\mathcal{P}^{\mathrm{res}}_{[0,t]}(x;X)$ is an empirical residence time distribution function on the X-space.

If the Brownian particle can access to every point on X for which $U(X,a)$ takes a finite value, $\mathcal{P}^{\mathrm{res}}_{[0,t]}(x;X)$ will converge in the limit $t\to\infty$ to a unique density for almost all the realizations. This limit should be

$$\mathcal{P}^{\mathrm{res}}_{[0,t]}(\hat{x};X) \longrightarrow \mathcal{P}^{\mathrm{eq}}(X,a;T), \quad \text{as } t\to\infty, \tag{1.104}$$

where $\mathcal{P}^{\mathrm{eq}}(X,a;T)$ is the canonical equilibrium distribution at the temperature T which appeared in $\langle\xi(t)\xi(t')\rangle = 2\gamma k_{\mathrm{B}}T\delta(t-t')$:

$$\mathcal{P}^{\mathrm{eq}}(X,a;T) \equiv \frac{1}{Z_x}e^{-U(X,a)/k_{\mathrm{B}}T}. \tag{1.105}$$

Here $Z_x \equiv \int_{-\infty}^{\infty} e^{-U(X',a)/k_{\mathrm{B}}T}dX'$ is the configurational canonical partition function of the statistical mechanics.[75]

The convergence problem of (1.104) is an example of LDP discussed in Sect. 1.1.2.3 along the time axis. But it is understandable intuitively that the rate of convergence depends on the form of $U(X,a)/k_{\mathrm{B}}T$: For example, let us compare the case of a single-well potential $U(X,a)=U_2(X)\equiv 4k_{\mathrm{B}}TX^2$ and the double-well potential $U(X,a)=U_4(X)\equiv 100k_{\mathrm{B}}T(X^2-1)^2$. The latter potential has a barrier of $100k_{\mathrm{B}}T$ around $X=0$ and the valleys of 10 times narrower than the quadratic potential $U_2(X)$. If we take as unit of time the convergence time of $\mathcal{P}^{\mathrm{res}}_{[0,t]}(\hat{x};X)$ for the U_2 potential, then the convergence time for the U_4 potential should be much larger than unity because of the rare chance to cross the energy barrier, i.e., about $e^{100}/10 \simeq 10^{42}$.[76] If the convergence time for the U_2 potential is a nano second (10^{-9}s), the convergence time for U_4 would be $\sim 10^{24}$ years, which is practically unattainable.

This result implies that, in between nano second and 10^{24} years, there is a well-defined time range where almost all paths of $\hat{x}(\,)$ remain in one of the valleys of the U_4 potential. *Within* this time range, the empirical distribution $\mathcal{P}^{\mathrm{res}}_{[0,t]}(x;X)$ must converge to $\mathcal{P}^{\mathrm{c-eq}}(X,a;T,\hat{x}(0))$:

$$\mathcal{P}^{\mathrm{c-eq}}(X,a;T,\hat{x}(0)) \equiv \begin{cases} \frac{1}{Z_x^{\mp}}e^{-U_4(X)/k_{\mathrm{B}}T} & X/\hat{x}(0)>0 \\ 0 & X/\hat{x}(0)<0 \end{cases} \tag{1.106}$$

[75] An important corollary: The longtime average of a function of $\hat{x}$, e.g., $f(\hat{x}(t)) = \int f(X)\delta(X-\hat{x}(t))dX$, converges to its canonical average, $\int f(X)\mathcal{P}^{\mathrm{eq}}(X,a;T)dX$.

[76] The factor 10^{-1} is from the narrowness of the valley bottom, but is not essential here.

if $x(0) > 0$. Here $Z_x^+ \equiv \int_0^\infty e^{-U_4(X')/k_B T} dX'$ is the constrained equilibrium partition function in the right valley. See Fig. 1.7.

Remarks

1. The time for a Brownian particle to escape (for the first time) from a valley does not depend on the profile of the potential energy *beyond* the potential barrier. Therefore, that profile can be replaced by a sharp and infinitely deep valley. Kramers [43] used this fact to derive the efficient formula of the escape time, called the first passage time.
2. There could be a "theoretical" standpoint where one is interested only in the infinitely large time, t, to find the unique probability density, $\mathcal{P}^{eq}(X, a; T)$. In such a case, however, we implicitly assume that the potential function is realized by some material whose characteristic relaxation time is larger than this time, t. Therefore, the word "infinitely large" has only a limited meaning.
3. In Sect. 7.1.2 we will discuss the case where a system with double minimum potential represents a single bit of memory device. In that case the constrained density $\mathcal{P}^{c-eq}$ corresponds to a well-defined "bit," and the full equilibrium density $\mathcal{P}^{eq}$ represents the state where the initial memory is lost.

1.3.3.2 * Integral by the Parameter that Varies Slowly in Time

Let us continue to consider the above Langevin equation, $0 = -\frac{\partial U(x,a)}{\partial x} - \gamma \frac{dx}{dt} + \xi(t)$, but we now change the parameter a in time from a_i up to a_f. The parameter a can have more than one component. First we define a protocol $\tilde{a}(s)$ that takes a unit time, i.e., $\tilde{a}(0) = a_i$ and $\tilde{a}(1) = a_f$. We denote by τ_{rel}[77] the convergence time for the above theorem (1.104) and assume that τ_{rel} is not extremely large.

Our aim is to evaluate the following form of Stieltjes integral:

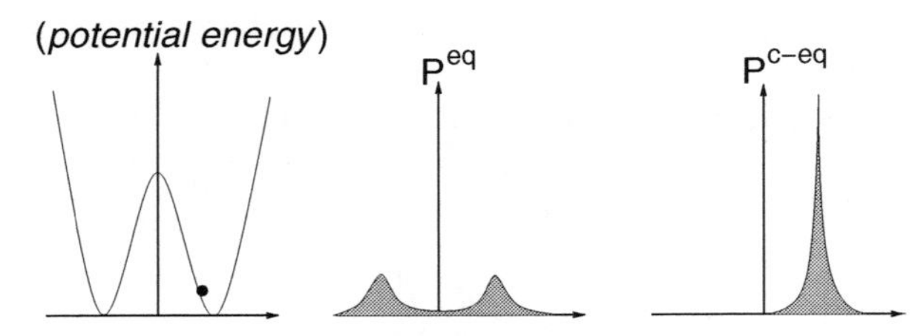

Fig. 1.7 Brownian particle in an symmetric double-well potential (*left: curve*) and the initial condition $x(0)$ (*left: dot*). At high temperature, the residence time distribution reproduces the canonical equilibrium distribution $\mathcal{P}^{eq}$ (*center*), while at low temperature it reproduces the constrained equilibrium distribution $\mathcal{P}^{c-eq}$, which depends on the initial condition (*right*)

[77] "rel" for relaxation.

$$\hat{\mathcal{I}} \equiv \int_{a=a_i}^{a=a_f} \Phi(\hat{x}(t), a(t))da(t) = \int_{a=a_i}^{a=a_f} \Phi(\hat{x}(t), a(t))\frac{da(t)}{dt}dt, \tag{1.107}$$

following the protocol $a(t) = \tilde{a}(\frac{t}{\tau_{op}})$, by taking the time $\tau_{\rm op}$.[78] Especially, we are interested in arbitrarily slow protocol, that is, the limit of $\tau_{\rm op} \to \infty$.

Leaving the details of demonstration in Appendix A.1.12, the conclusion is that, for almost all realization of $\hat{x}$,

$$\hat{\mathcal{I}} \to \int_{a_i}^{a_f} \langle \Phi(\cdot, a)\rangle_{\rm eq} da \quad \text{as } \tau_{\rm op} \to \infty, \tag{1.108}$$

where $\langle \Phi(\cdot, a)\rangle_{\rm eq}$ is the equilibrium average of $\Phi(\cdot, a)$ with a fixed value of a.

$$\langle \Phi(\cdot, a)\rangle_{\rm eq} \equiv \int \Phi(X, a)\mathcal{P}^{\rm eq}(X, a; T)dX. \tag{1.109}$$

The result can be understood intuitively: For each realization $x(\,)$, there is enough time for the particle to explore the instantaneous profile of the potential $U(\,, a)$ for *each* value of a along the protocol. By the convergence theorem of the empirical residence time distribution (1.104), the temporal integral around each a can, therefore, be replaced by the statistical average, $\langle \Phi(\cdot, a)\rangle_{\rm eq}$. This is the notion of quasistatic process, and the above formula will be fully used in Sect. 5.2.1.2.

It is noteworthy that the rate of the convergence of $\hat{\mathcal{I}}$ depends on the function $\Phi(X, a)$. If $\Phi(X, a)$ takes appreciable values for those X which are rarely realized in $\mathcal{P}^{\rm eq}(X, a; T)$, we will need extremely slow operation (i.e., extremely large $t_{\rm op}$) for the convergence.

1.3.3.3 First Passage Problem

Suppose we follow a Brownian particle that started at the position x_0 somewhere inside a domain Ω. The space x can be more than one dimension. We assume again the overdamped Langevin equation, $0 = -\frac{\partial U(x,a)}{\partial x} - \gamma \frac{dx}{dt} + \xi(t)$, and that the parameter a is fixed.

The first passage time (FPT) τ_Ω is the time when the Brownian particle arrives at the border of Ω for the first time after the start. The problem to study the FTP is called the first passage problem. See Fig. 1.8.[79] The FPT is a random variable, and x_0 is its parameter. Our aim is to find the path average of the FPT (mean first passage time; MFPT) as function of the initial position $x_0 \in \Omega$. (All the paths starting from x_0 are counted by equal weight.) Leaving the details of derivation in Appendix A.1.13, below is a recipe to calculate the MFPT:

---

[78] "op" for operation.

[79] There is a related problem called the exit problem, where we study the location on the border of Ω at FPT.

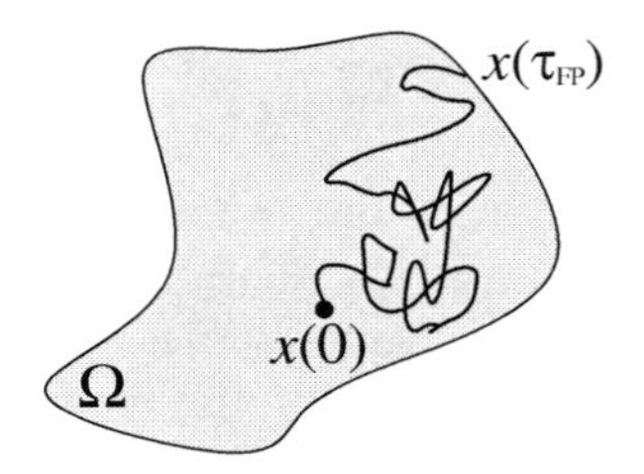

Fig. 1.8 First passage time problem. The Brownian particle started from $x(0) = x_0$ (*thick dot*) arrives for the first time on the boundary $x(\tau_\Omega) \in \partial\Omega$ at the first passage time (FPT), $t = \tau_\Omega$

(i) Solve the static differential elliptic equation $\mathcal{L}^*\phi(x) = -1$ for $x \in \Omega$ under the [Dirichlet] boundary condition, $\phi(x) = 0$ for $x \in \partial\Omega$.
(ii) $\langle\hat{\tau}_\Omega\rangle$ is given as the value of $\phi(x_0)$.

Here $\mathcal{L} \equiv \frac{\partial}{\partial x}\frac{1}{\gamma(x)}\left[\frac{\partial U}{\partial x} + \frac{\partial}{\partial x}k_B T(x)\right]$ is the operator of Fokker–lanck equation, and $\mathcal{L}^*$ is its adjoint operator, i.e., $\mathcal{L}^* = \left[-\frac{\partial U}{\partial x} + k_B T(x)\frac{\partial}{\partial x}\right]\frac{1}{\gamma(x)}\frac{\partial}{\partial x}$.[80]

For example, when we study the Brownian particle in the double-well potential energy $U_4(x)$ (see Sect. 1.3.3.1) we can use the MFPT to estimate the waiting time until the Brownian particle crosses over the potential barrier. This waiting time is the MFPT with the domain $\Omega = [0, \infty]$, for example. In such case the FPT depends little on the initial position x_0 as long as it is near the bottom of the valley.

The first passage problem is applied also to the free Brownian motion but with various shape of domain Ω. For example, the Brownian particle is a macro-ion and Ω is an absorbing surface.

1.3.4 Thermal Ratchet Models are the Examples of Nontrivial Fluctuations

1.3.4.1 Background

Models of autonomous heat engine on the mesoscopic scale have long been proposed: The Feynman ratchet (Feynman's pawl and ratchet wheel) [55] and the Büttiker–Landauer ratchet [56, 57] are among the most studied. In the former model, the system has simultaneous access to the two thermal environments of temperatures T and T', while in the latter model, the system has alternative access to one of these environments.

The aim of this subsection is to describe their models and show how it is modeled using the Langevin equations. These models have played important roles in the physics of stochastic phenomena. Parrondo [58] was the first who added a caveat on the analysis by Feynman on his ratchet model.[81] Derény, Bier, and Astumian

[80] $\mathcal{L}^*$ is also called the backward Fokker–lanck operator. In general when the operators $\mathcal{L}$ and $\mathcal{L}^*$ satisfy $\int_\Omega g(\mathcal{L}f)dx = \int_\Omega(\mathcal{L}^*g)f\,dx$ for any f and g satisfying a common homogeneous boundary condition, $\mathcal{L}$ and $\mathcal{L}^*$ are called adjoint.

[81] Soon later [60] also pointed out it independently.

[59] were among the first who argued that the overdamped limit of the Büttiker–Landauer ratchet is a singular limit. Stochastic energetics has first been applied to the Feynman's model [60]. We will come back to these models in Sect. 4.2.2.2 for the analysis from the energetics viewpoint.

1.3.4.2 Feynman Ratchet

Figure 1.9 shows the idea of Feynman pawl and ratchet wheel [55]. A ratchet wheel (center) is coaxially connected to the part that can lift a charge imposing the downward load, f. It is also coaxially connected to a vane immersed in the left thermal environment of the temperature T. We denote by x the rotation angle of the ratchet wheel.

A pawl (a thick arrow) can move back and forth along a coordinate denoted by y and under the influence of the right thermal environment of the temperature T'. The pawl is supported by a static restoring potential (symbolized by a spring in the figure), $U_2(y)$, which tends to keep the pawl in contact with the ratchet wheel.

The ratchet wheel interacts with the pawl through the potential, $U_1(x - \phi(y))$, where $\phi(x)$ represents the periodic and asymmetric profile of the ratchet tooth. The potential $U_1(z)$ represents the volume-excluding constraints between the pawl and the ratchet wheel. That is, for $x < \phi(x)$ it is strongly repulsive, while for $x \geq \phi(x)$ it exerts no forces.

The overdamped Langevin equations for the above model write [60]

$$
\begin{aligned}
0 &= -\gamma \frac{dx}{dt} - \frac{\partial U}{\partial x} + \sqrt{2\gamma k_{\mathrm{B}} T}\, \theta(t), \\
0 &= -\gamma' \frac{dy}{dt} - \frac{\partial U}{\partial y} + \sqrt{2\gamma' k_{\mathrm{B}} T'}\, \theta'(t),
\end{aligned}
\tag{1.110}
$$

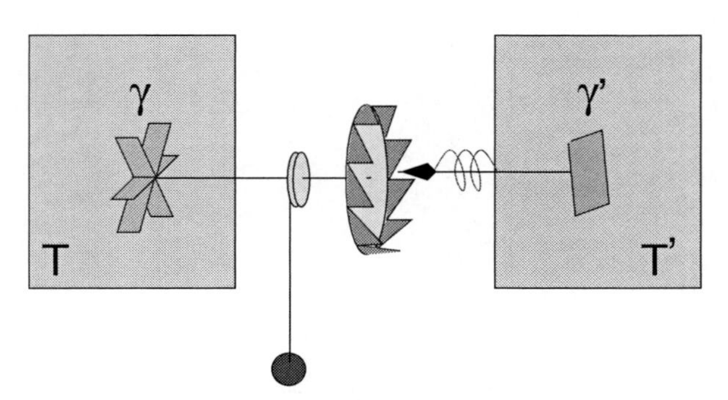

Fig. 1.9 Feynman pawl and ratchet wheel [55]. The *left heat bath* (temperature T) moves the ratchet wheel, while the *right heat bath* (temperature T') agitates the pawl. In between the elements of the potential energy $U(x, y)$ are presented, from the left to the right, as $-fx$ for the load, $U_1(x - \phi(y))$ for the pawl-ratchet wheel coupling, and $U_2(y)$ for the pawl

where $U(x, y) = U_1(x - \phi(y)) + U_2(y) - fx$, and $\theta(t)$ and $\theta'(t)$ are mutually independent white Gaussian random noises with zero mean and $\langle\theta(t)\theta(t')\rangle = \langle\theta'(t)\theta'(t')\rangle = \delta(t - t')$. Modified models have also been studied recently [61–63]. From Fig. 1.9 we expect that a light load can be lifted up if $T > T'$, because the rotation of the wheel axis in the forward (lifting-up) direction is most likely "registered" by the cool pawl, while the rotation in the opposite direction is more likely hindered by the cool pawl. For $0 < T < T'$, the wheel has a tendency to rotate in the opposite direction to the case of $T > T'$.

The "hot" pawl moves rather by thermal random force than by the potential force of $U_2(y)$. A weak back-and-forth Brownian motion of the ratchet wheel can cause the repositioning of the pawl to the neighboring spaces between the ratchet tooth. Such events occur in an asymmetric manner with respect to $\pm x$ reflection, due to the asymmetry of the profile $\phi(x)$. These results are verified numerically. It is possible to take into account of the inertia of the pawl and the ratchet wheel, and the essential behavior of the system is the same as long as $T > 0$ [62].

1.3.4.3 Büttiker and Landauer Ratchet: Inertia as a Singular Perturbation

The model proposed by Büttiker [56] and Landauer [57] contains only a single movable particle. Instead, this particle can switch the thermal environments with which it makes contact. Along a periodic potential $U(x)$, the zone of temperature T and that of T' alternates with the same period as $U(x)$. See Fig. 1.10. The Langevin equation for the position and momentum of the particle, (x, p), is

$$\frac{dp}{dt} = -\frac{\partial U(x)}{\partial x} - f - \gamma\frac{p}{m} + \xi(t), \qquad \frac{dx}{dt} = \frac{p}{m}, \tag{1.111}$$

where the thermal random force $\xi(t)$ is the white Gaussian noise with $\langle\xi(t)\xi(t')\rangle = 2\gamma k_B T(x)\delta(t-t')$.[82] f is a load on the particle toward the $(-x)$ direction. If we could justify to take the overdamping limit (see Section 1.3.2.1), the resulting Langevin equation is (1.112), that is,

$$\gamma\, dx = \sqrt{2\gamma k_B T(x)} \circ dB_t - \frac{\partial}{\partial x}\left[U(x, a) + fx + \frac{k_B T(x)}{2}\right] dt. \tag{1.112}$$

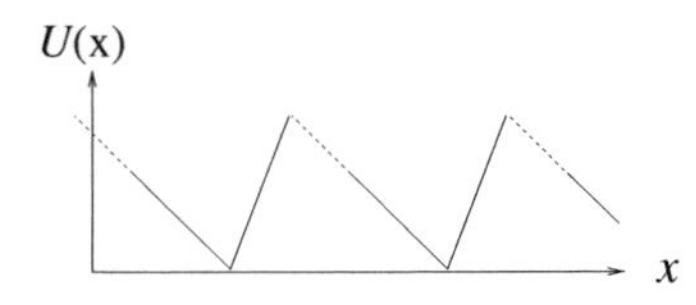

Fig. 1.10 Potential energy $U(x)$ of Büttiker and Landauer ratchet (adapted from [59]). The temperature $T(x)$ takes the value T on the *thick solid line* and T' on the *dashed line*

[82] Recall that, in the presence of the inertia, the variation of $x(t)$ during dt is $\mathcal{O}(dt)$, not $\mathcal{O}(dt^{1/2})$. The multiplicative character of the $\xi(t)$ here does not cause ambiguities [74].

In fact, however, the neglect of the inertia completely changes the energetic aspect of the model. See Sect. 4.2.2.2.

The intuitive explanation for the mechanism of this heat engine is as follows: Let us assume that $T > T'$ and that the temperature profile is as described in Fig. 1.10 and its caption. In the cool region of T' (the dashed line in the rightward downhill), the thermal activation to climb up the potential is less probable than the other side of the potential peak. The probability of barrier crossing is then less frequent from right to left than the inverse. In [59] an alternative explanation is given in terms of the effective potential, $U_{\mathrm{eff}}(x) = \frac{k_{\mathrm{B}}T(x)}{k_{\mathrm{B}}T_0}U(x)$, immersed in a fictive homogeneous thermal environment of the temperature T_0. In any case, the particle migrates toward the right on the average if $T > T'$.

1.3.4.4 "Modern" Thermal Ratchet Models and Curie Principle

A Subjective History of Thermal Ratchet Models in the Early 1990's

The latest vague of ratchet model in the early 1990s was initiated by [64] as a new principle of particle separation and transport. This arose the interest among the physicists and biophysicists, in majority from theoretical point of view. Soon after [64] this idea was reformulated as a simple physical model [65]. Different aspects of ratchet model have been developed in different field, including the nanoscale transport [66], protein motor, and biochemical reactions [67–71]. These models urged to establish the framework of stochastic energetics [60]. The machineries working under thermal fluctuations are often subtle, and the general and analytical framework helps to avoid any ambiguities in their studies. The paradox of Maxwell's demon is a symbolic example of this subtlety. See Sect. 4.2.1.2. An extensive review of ratchet models up to 2002 is in [72]. Another review from the energetics point of view is [73]. The controversy about the efficiency of ratchet models (Sect. 6.3.2) is one of the motivation to study the heat of different spatiotemporal scales, which will be described in Chap. 6.

The architecture of the initial ratchet model [64] is called the flushing ratchet, see Fig. 1.11. In this model a sawtooth-shaped potential $U(x)$ is turned on and off periodically in time. While the potential is off, a particle under no force diffuses symmetrically. When the potential is turned on, the particle is driven asymmetrically toward the potential downhill. In this way the particle is driven on the average toward the right. Even if the baseline of the potential is slightly inclined toward the right, or with $U(x) + ax$ $(a > 0)$, the particle will still be driven toward the right. The active transport will be suppressed if the thermal random force is too small, because of insufficient diffusion of particle. The thermal fluctuation plays, therefore, a constructive role in this model.[83] The overdamped Langevin equation for the particle position x writes

[83] There are ratchet models that can work without the thermal random force [80].

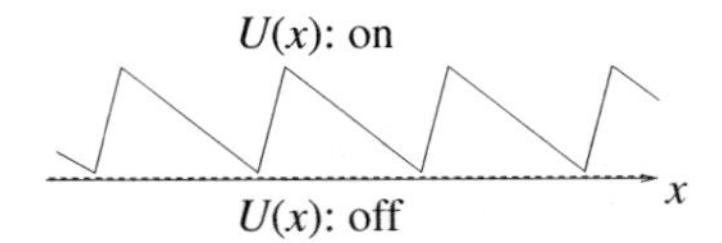

Fig. 1.11 Flushing ratchet model [64]: Two alternative potential profiles of the flushing ratchet. In the period when the sawtooth potential $U(x)$ is on, the state x is driven toward one of its valleys. While $U(x)$ is off (*horizontal dashed line*), x diffuses symmetrically. The average displacement is positive, $\langle dx/dt \rangle > 0$

$$0 = -\sigma(t)\frac{\partial U(x)}{\partial x} - \gamma\frac{dx}{dt} + \xi(t), \tag{1.113}$$

where γ and $\xi(t)$ are as usual (see (1.32)), and $\sigma(t)$ temporarily switches between 0 and 1.[84]

Symmetry Question of Ratchet Models: Curie Principle

The realization of horizontal transport by only vertical up and down of the potential $U(x)$ arose technical as well as theoretical interest. In the setup of particle separation and transport [66, 75, 76], simple on and off of electric or optic field can cause the directed transport of colloidal particles.[85] Phenomenologically, the ratchet models realize different kinds of cross-coupling between the thermodynamic forces and fluxes. In the linear nonequilibrium thermodynamics, Onsager elucidated the symmetry of cross-coupling coefficients, e.g., in the Peltier/Seebeck effect, Soret effect, magneto-optic effect [77], ion-co-transportation.

Theoretical interest is in the general conditions for the ratchet transport, not going into details of the model architectures. At the end of nineteenth century Pierre Curie [78] established the principle which can exclude some type of cross-coupling by a purely symmetry reason. The Curie principle says: "When certain causes lead to certain effects, the symmetry elements of the causes must be found in their outcomes."[86] Therefore, except for the case of spontaneous symmetry breaking, the asymmetry of the phenomena is caused by the asymmetry of the setup. This principle, which is not limited to the linear regime, was applied to the ratchet models [79]: If the dynamics of the system lacks *both* the spatial symmetry ($\pm x$) and the temporal symmetry ($\pm t$), then the principle predicts generically a nonequilibrium process accompanying directional transport. The flushing ratchet described above fulfills these two conditions. Spatially broken symmetry is not necessarily be borne by the potential energy: A directional transport is possible with a symmetric potential, $U(x) = U(-x)$, if the random force $\xi(t)$ with zero mean and $\langle \xi(t)\xi(t') \rangle = \delta(t - t')$

---

[84] $\sigma(t)$ can be a smooth function of time.

[85] The directed force is in fact applied externally through the field gradient.

[86] "*Lorsque certaines causes produisent certains effets, les éléments de symétrie des causes doivent se retrouver dans les effets produits.*" [78]

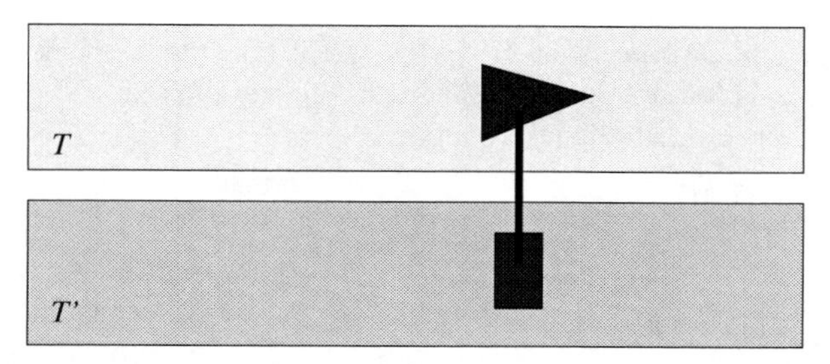

Fig. 1.12 Model of the heat engine. *Wedge-shaped* object and *rectangle-shaped* object are constrained to move together horizontally. These objects are immersed in the gas reservoirs at different temperatures. If the environment for the wedge-shaped object is hotter than that for the rectangle-shaped object, the combined objects move to the left and vice versa (Figure adapted from Fig. 1 of [25])

is a non-Gaussian process [45, 46]. A type of "super-symmetry" [81] can impose constraints on the directional transport [82, 72].

Remarks

The directed transport by spontaneous symmetry breaking is not covered by the Curie principle. Such case is demonstrated in the collective ratchet model [70]. On the fluctuating level, temporal directional transport in a particular realization is possible under the detailed balance condition, as long as the spatial $\pm x$ symmetry is broken [61–63].[87]

Another limitation of the Curie principle is that it does not tell anything about the sign and amplitude of the cross-coupling effect. Whether or not an active transport against external load occurs depends on concrete structure of the model.

Langevin equation is not the only way to model heat engines. A heat engine is modeled using the nonlinear *kinetic* couplings between the system and its environments [24, 25]. Figure 1.12 illustrates their model. Since the spatial symmetry is broken by this kinetic coupling, the model is not directly reducible to the Langevin equations. The latter have only linear kinetic coupling through the friction coefficients (γ etc.).

1.3.4.5 Formula for the Transport, $\langle \frac{dx}{dt} \rangle$

For the later use, we give the formula that expresses the average velocity $\langle \frac{dx}{dt} \rangle$ of $\hat{x}(t)$ at a given time in terms of the probability flux. It reads,

$$\left\langle \frac{dx}{dt} \right\rangle = \int \boldsymbol{J} d\Gamma, \tag{1.114}$$

[87] Spontaneous symmetry breaking along the time axis is not seen.

where the integral is done over all phase space (i.e., (x, y)-plane for Feynman ratchet, and x-axis for Büttiker and Landauer ratchet) and $\boldsymbol{J}$ is the probability flux that appears in the Fokker–Planck equations, $\frac{\partial P}{\partial t} = -\nabla \cdot \boldsymbol{J}$.

A simple derivation is to compare this equation and the identity

$$\frac{\partial \langle \delta(\boldsymbol{x} - \hat{\boldsymbol{x}}(t)) \rangle}{\partial t} = -\nabla \cdot \left\langle \frac{d\hat{\boldsymbol{x}}}{dt} \delta(\boldsymbol{x} - \hat{\boldsymbol{x}}(t)) \right\rangle,$$

knowing that $P(x, t) = \langle \delta(\boldsymbol{x} - \hat{\boldsymbol{x}}(t)) \rangle$. We then have the equation,

$$\boldsymbol{J} = \left\langle \frac{d\hat{\boldsymbol{x}}}{dt} \delta(\boldsymbol{x} - \hat{\boldsymbol{x}}(t)) \right\rangle.$$

Integration of the both sides over $\boldsymbol{x}$ yields the above formula.

1.4 Discussion

The Langevin equation is a useful method to describe the individual realization of stochastic process occurring in the thermal environment. The ensemble average viewpoint often overlooks essential features of the fluctuation phenomena, or more generally, the random phenomena. For example, the transition between the extended coil conformations and the collapsed globule conformations of a single long polymer chain has been observed by macroscopic methods and was thought to be continuous as function of temperature. But later on the transition was confirmed to be discontinuous through the observation of the single chain [83]. Another example is the phenomenon of overcharging of counterions on the surface of charged colloid. By studying the configurations of discrete charges and their Coulomb interactions, one can clearly understand how the counterion can be bound on the colloidal surface even beyond the neutralizing the colloid's charge [84]. Such phenomenon is overlooked or difficult to describe by the continuum description like the Poisson–Boltzmann equation.[88]

References

1. C.W. Gardiner, *Handbook of Stochastic Methods for Physics, Chemistry for Natural Sciences*, 3rd edn. (Springer, 2004)
2. H. Risken, T. Frank, *The Fokker-Planck Equation: Methods of Solutions and Applications*, 2nd edn. (Springer, 1996)
3. R. Zwanzig, J. Stat. Phys. **9**, 215 (1973)
4. R. Kubo, M. Toda, N. Hashitsume, *Statistical Physics II: Nonequilibrium Statistical Mechanics*, 2nd edn. (Springer-Verlag, Berlin, 1991)

[88] In the latter approach the probability density of a single counterion is assumed to be proportional to the charge density of counterion.

5. B. Alberts et al., *Essential Cell Biology*, 3rd edn. (Garland Pub. Inc., New York & London, 2009)
6. K. Sato, Y. Ito, T. Yomo, K. Kaneko, PNAS **100**, 14086 (2003)
7. N.G. van Kampen, *Stochastic Processes in Physics and Chemistry*, Revised ed. (Elsevier Science, 2001)
8. D.S. Lemons, A. Gythiel, Am. J. Phys. **65**, 1079 (1997)
9. A. Einstein, Ann. Phys. (Leipzig) **17**, 549 (1905)
10. A. Rahman, Phys. Rev. **136**, A405 (1964)
11. B.J. Alder, T.E. Wainwright, Phys. Rev. Lett. **18**, 988 (1967)
12. R.D. Vale, D.R. Soll, I.R. Gibbons, Cell **59**, 915 (1989)
13. K. Tawada, K. Sekimoto, J. Theor. Biol. **150**, 193 (1991)
14. A. Einstein, *Investigation on the theory of the Brownian Movement*, (original edition, 1926) ed. (Dover Pub. Inc., New York, 1956), chap.V-1, pp. 68–75
15. L.D. Landau, E.M. Lifshitz, *Fluid Mechanics (Course of Theoretical Physics, Volume 6)*, 2nd edn. (Reed Educational and Professional Publishing Ltd, Oxford, 2000)
16. E. Entchev, A. Schwabedissen, M. González-Gaitán, Cell **103**, 981 (2000)
17. F. Perrin, Le Journal de Physique et Le Radium **5**, 497 (1934)
18. F. Perrin, Le Journal de Physique et Le Radium **7**, 1 (1936)
19. C. Ribrault, A. Triller, K. Sekimoto, Phys. Rev. E **75**, 021112 (2007)
20. S.C. Harvey, J. Garcia de la Torre, Macromolecules **13**, 960 (1980)
21. Y. Han et al., Science **314**, 626 (2006)
22. A.J. Levine, T.B. Liverpool, F.C. MacKintosh, Phys. Rev. E **69**, 021503 (2004)
23. M.X. Fernandes, J. Garcia de la Torre, Biophys. J. **83**, 3039 (2002)
24. C. Van den Broeck, R. Kawai, P. Meurs, Phys. Rev. Lett. **93**, 090601 (2004)
25. C. Van den Broeck, R. Kawai, Phys. Rev. Lett. **96**, 210601 (2006)
26. G. Ford, M. Kac, P. Mazur, J. Math. Phys. **6**, 504 (1965)
27. D.A. Beard, T. Schlick, Biophys. J. **85**, 2973 (2003)
28. G. Goupier, M. Saint Jean, C. Guthmann, Phys. Rev. E **73**, 031112 (2006)
29. G.E. Uhlenbeck, L.S. Ornstein, Phys. Rev. **23**, 823 (1930)
30. N.G. Van Kampen, *Stochastic Processes in Physics and Chemistry* (North-Holland, Amsterdam, 2007)
31. H. Mori, Prog. Theor. Phys. **33**, 423 (1965)
32. K. Kawasaki, J. Phys. A **6**, 1289 (1973)
33. R. Zwanzig, J. Chem. Phys. **33**, 1338 (1960)
34. K. Itô, *Stochastic Processes (Lecture Notes from Aarhus University 1969)* (Springer Verlag, New York, 2004)
35. G. Gallavotti, *Statistical Mechanics, A Short Treatise*, 1st edn. (Springer-Verlag, New York, 1999)
36. N.G. van Kampen, J. Stat. Phys. **24**, 175 (1981)
37. E. Wong, M. Zakai, Ann. Math. Statist. **36**, 1560 (1965)
38. R. Kupferman, G. Pavliotis, A. Stuart, Phys. Rev. E **70**, 036120 (2004)
39. R. Mannella, Int. J. Mod. Phys. C**13**, 1177 (2002)
40. P.E. Kloeden, E. Platen, H. Schurz, *Numerical Solution of SDE Through Computer Experiments* (Springer, New York, 1994)
41. P.E. Kloeden, E. Platen, *Numerical Solution of Stochastic Differential Equations* (Springer, New York, 1992)
42. J. Qiang, S. Habib, Phys. Rev. E **62**, 7430 (2000)
43. H.A. Kramers, Physica **7**, 284 (1940)
44. Y. Oono, Progr. Theor. Phys. Suppl. **99**, 165 (1989)
45. T. Hondou, J. Phys. Soc. Jpn. **63**, 2014 (1994)
46. T. Hondou, Y. Sawada, Phys. Rev. Lett. **75**, 3269 (1995) (Erratum; T. Hondou, Y. Sawada, Phys. Rev. Lett. **76** 1005 (1996))
47. J.M. Sancho, M.S. Miguel, D. Dürr, J. Stat. Phys. **28**, 291 (1982)

48. M. Doi, S.F. Edwards, *The Theory of Polymer Dynamics*, 1st edn. (Oxford Science Pub, Oxford, 1986), Sect. 3.3
49. K. Sekimoto, J. Phys. Soc. Jpn. **68**, 1448 (1999)
50. R. Landauer, Phys. Rev. A **12**, 636 (1975)
51. G. Costantini, F. Marchesoni, Europhys. Lett. **48**, 491 (1999)
52. P. Reimann et al., Phys. Rev. Lett. **87**, 010602 (2001)
53. K. Sekimoto, A. Triller, Phys. Rev. E **79**, 031905 (2009)
54. K. Hayashi, S. Sasa, Phys. Rev. E **69**, 066119 (2004)
55. R.P. Feynman, R.B. Leighton, M. Sands, *The Feynman Lectures on Physics – vol.1* (Addison Wesley, Reading, Massachusetts, 1963), Sects.46.1–46.9
56. M. Büttiker, Zeitschrift der Physik B **68**, 161 (1987)
57. R. Landauer, J. Stat. Phys. **53**, 233 (1988)
58. J. Parrondo, P. Español, Am. J. Phys. **64**, 1125 (1996)
59. I. Derényi, M. Bier, R.D. Astumian, Phys. Rev. Lett. **83**, 903 (1999)
60. K. Sekimoto, J. Phys. Soc. Jpn. **66**, 1234 (1997)
61. N. Nakagawa, T. Komatsu, Progr. Theor. Phys. Suppl. **161**, 290 (2006)
62. N. Nakagawa, T. Komatsu, Physica A **361**, 216 (2006)
63. N. Nakagawa, T. Komatsu, J. Phys. Soc. Jpn. **74**, 1653 (2005)
64. A. Ajdari, J. Prost, C. R. Acad. Sci. Paris. Sér. II **315**, 1635 (1992)
65. M.O. Magnasco, Phys. Rev. Lett. **71**, 1477 (1993)
66. J. Prost, J. Chauwin, L. Peliti, A. Ajdari, Phys. Rev. Lett. **72**, 2652 (94)
67. S. Leibler, D.A. Huse, J. Chem. Biol. **121**, 1357 (1993)
68. M.O. Magnasco, Phys. Rev. Lett. **72**, 2656 (1994)
69. R.D. Astumian, M. Bier, Phys. Rev. Lett. **72**, 1766 (1994)
70. F. Jülicher, J. Prost, Phys. Rev. Lett. **75**, 2618 (1995)
71. H. Zhou, Y. Chen, Phys. Rev. Lett. **77**, 194 (1996)
72. P. Reimann, Phys. Rep. **361**, 57 (2002)
73. J.M.R. Parrondo, B.J. De Cisneros, Appl. Phys. **A75**, 179 (2002)
74. M. Matsuo, S. Sasa, Physica A **276**, 188 (2000)
75. J. Rousselet, L. Salomé, A. Ajdari, J. Prost, Nature **370**, 446 (1994)
76. L.P. Faucheux, L.S. Bourdieu, P.D. Kaplan, A.J. Libchaber, Phys. Rev. Lett. **74**, 1504 (1995)
77. L.D. Landau, E.M. Lifshitz, L.P. Pitaevskii, *Electrodynamics of Continuous Media*, 2nd edn. (Elsevier Butterworth-Heinemann, Oxford, 2004)
78. P. Curie, J. Phys. (Paris) 3ème série **3**, 393 (1894)
79. F. Jülicher, A. Ajdari, J. Prost, Rev. Mod. Phys. **69**, 1269 (1997)
80. M. Porto, M. Urbakh, J. Klafter, Phys. Rev. Lett. **85**, 491 (2000)
81. M. Bernstein, L.S. Brown, Phys. Rev. Lett. **52**, 1933 (1984)
82. R. Kanada, K. Sasaki, J. Phys. Soc. Jpn. **68**, 3579 (1999)
83. I. Nishio, S.-T. Sun, G. Swislow, T. Tanaka, Nature **281**, 208 (1979)
84. R. Messina, C. Holm, K. Kremer, Phys. Rev. E **64**, 021405 (2001)

Chapter 2
Structure of Macroscopic Thermodynamics

Macroscopic thermodynamics deals with "energy" and materials stocked in and exchanged by the system. The system can be surrounded by environment(s) and external system(s). When the exchange of energy or materials takes place, several important constraints, i.e., the laws of thermodynamics are imposed. Thermodynamics allows for different point of views: view from the system, from the external system, or from the environment. Thermodynamics is also relevant for the study of (free) energy conversion, the thermal analysis of chemical reactions and phase transitions.

Stochastic energetics combines the description of stochastic processes and thermodynamic processes. In order to prepare for the main chapters (from Chap. 4) this chapter reminds the readers of the basic concepts and relations of thermodynamics and introduces the terminology and notations of thermodynamics. This chapter does not cover all the aspects of thermodynamics, but it focuses on those aspects of thermodynamics which will be relevant in later chapters. Those who know macroscopic equilibrium thermodynamics well may skip this chapter and come back when necessary.

2.1 Basic Concepts of Thermodynamics

*2.1.1 * Terminology of Thermodynamics Includes System, Environments, and External System*

The basis of thermodynamics is empirical. The definitions of elementary concepts rely on empirical or undefined notions. One can axiomatize it [1, 2] but cannot derive it. The framework of thermodynamics is supposed to be asymptotically exact for the large homogeneous system based on experiments.

System: Of the whole world, a part which is properly cut out is called the system. In this chapter we suppose that a system contains a macroscopic number (say, $\sim 10^{23}$) of constituting elements of the same kind like particles, spins, etc. If a system is divided into more than one part by some criterion, each

Sekimoto, K.: *Structure of Macroscopic Thermodynamics*. Lect. Notes Phys. **799**, 67–92 (2010)
DOI 10.1007/978-3-642-05411-2_2

part is called a subsystem. A system having no contact with any other system is called isolated system.

Internal energy of a system: If one can access the microscopic Hamiltonian of the system, or if one can observe or calculate the mechanical energy, either classical or quantum, then we naturally identify this energy as the internal energy of the macroscopic thermodynamics.

Extensive variables: When a system and its identical copy are combined to make a new system, certain variables take the values twice as large as the original ones before combining with its copy. Such variables are called extensive variables. For a system consisting of single-component particles, the internal energy, E, the volume, V, the number of particles, N (as well as the entropy, S, in the equilibrium states, see below), are extensive variables.

The other types of variables which do not change their values are called intensive variables of the system, but the latter refers to the equilibrium states (see below).

Environment: Those background systems to which the system of our interest is connected are called the environments, reservoirs, or baths. According to the entity that is exchanged with the system, the environment is called thermal environment or heat reservoir/bath (for energy exchange), particle environment or particle/chemical reservoir (for mass exchange) or pressure environment (for volume exchange).[1]

As an environment forms the background of a system and not the main object of study, they are considered to satisfy several simplifying assumptions.

(1) We ignore the interaction energy between the system and its environments. (In case that it is not realistic assumption, we try to define the interface layer of interaction as a new system.)
(2) Environments are big in the sense that the conservation laws for the total energy, mass, volume, or momentum have no effect on the state of the environment.
(3) Environments return instantaneously to their equilibrium states (see below), and keep no memories of the system's action in the past. In other words, any action on the environments is supposed to be quasistatic from macroscopic point of view.[2]

Because of these simplifications, environments are characterized only by their intensive parameters, that is, temperature, chemical potential (or the density of particles), and pressure.

Heat exchange: It is the form of exchange of energy not through work, or involving mass exchange. This process occurs either between a system and its environment, or among different subsystems.

[1] In this book we will use the words "heat bath" and "thermal environment" interchangeably.

[2] In other words, we do not discuss the irreversible entropy production proper to the environments.

Mass exchange: Mass is exchanged through the transport of particles between a system and its environment, or among different subsystems. Mass exchange can also take place through chemical reactions.[3]

External system: An agent which is capable of macroscopically controlling the system. The system can exchange energy and/or volume (and sometimes mass) with the external system. The role of external system and environment is somehow interchangeable in the sense that the definition of a system can include the environment through the Legendre transformation (see below). For example, the external system that controls the system's volume can be replaced by a pressure environment. In the latter case, the system's internal energy counts also as the potential energy of the pressing agent (a weight). The energy is then called enthalpy.

Work: The form of exchange of energy which is done by the control of external systems through the change of system's volume, external force, electric field, etc.

Macroscopic character of external system: The external system is allowed neither to make use of the microscopic information of a system nor to react adaptively to the microscopic change of the state of the system. This condition, on the one hand, excludes the intervention of the Maxwell's demon, a hypothetical agent which cools down the system by letting only rapidly moving particles to escape through a hole.[4] On the other hand, the above condition does not exclude certain devices having molecular selectivity of chemical species such as semipermeable membranes. The reason is that selectivity in the latter case is static with no adaptive functions to the particles which arrive at the membrane and also that the work is done only through the macroscopic displacement of the membrane as a whole.[5]

Sign conventions: Throughout this book, unless explicitly defined otherwise, we assign a positive sign to whatever quantities which the *system receives*.[6] For example, if an external system brings energy to a system through work, we say that a positive work, $W > 0$, is done. Also, if an environment brings energy to a system through heat, we say that positive heat ($Q > 0$) is transferred.

Some aspects of macroscopic thermodynamics are peculiar to the fact that the system is macroscopic.

1. When the system consists of several macroscopic subsystems, a subsystem can do work on the other subsystems as an external system.
2. The presence of extensive variables and the negligence of interaction energy between the system and environments suppose that the *surface* contributions

[3] More about the macroscopic chemical thermodynamics will be described in the next chapter.

[4] For a review, see [3]. cf. Sect. 4.2.1.2

[5] The author acknowledges Izumi Ojima for his comment on this point.

[6] Note that certain textbooks of thermodynamics adopt the opposite sign convention.

to those variables are negligible compared with the *bulk* contributions. If the surface-to-volume ratio is larger than some inverse characteristic length (e.g., the interaction range of particles), this hypothesis is no more valid.

3. The extensive character implies the absence of fluctuations. It is related to the law of large numbers. It is often called the self-averaging property of macroscopic system. Entropy and temperature which are ensemble-based concept can be assigned to a single macroscopic system. Systems at critical point of phase transition, or with long-range interactions, cannot be described by usual thermodynamics.

2.1.2 * Some Laws of Macroscopic Thermodynamics Distinguish Thermodynamics from Mechanics

About the process of macroscopic thermodynamic systems, several universal laws apply independent of the details of interactions among the constituent materials. The laws deal with the relationship among the change of energy, the heat transferred, and the work done during a process of a system. We mention the four principal laws below. They do not constitute a complete set of axioms of the macroscopic thermodynamics, see [1, 2].

Zeroth law: If a system is left isolated for sufficiently long time from any environment and from any external system, it will reach a state with no further macroscopic changes.[7] Such a state is said to be in an equilibrium state. For a single-component gas isolated in a single compartment, for example, the set of energy, volume, number of gas particles, (E, V, N), is sufficient to characterize its equilibrium state, and the entropy S is then a function of these variables. These variables are called thermodynamic variables, and the functions which relate the thermodynamic variables in an equilibrium system are called thermodynamic functions. If any process occurs slowly enough so that the system remains almost in equilibrium at each instant, the process is said to be quasistatic (or quasiequilibrium) process.

First law: We cannot realize a perpetual machine of the first kind, that is, there is no autonomous system, isolated from the environment, that produces work (i.e., $W < 0$) through a cyclic process. The inverse process is also not possible. More concretely, a balance of energy should always be established among the changes in the internal energy of the system, ΔE, the work, W, and a heat, Q, during any process:

$$\Delta E = W + Q. \tag{2.1}$$

[7] Strictly speaking, there may remain global motions due to conserved total momentum or total angular momentum of the whole system.

Second law: One cannot realize a perpetual machine of the second kind, that is, being in contact with a single thermal environment at constant temperature, there is no autonomous cyclic engine which converts heat into work. The inverse process from work to heat is possible. More precisely, there is a thermodynamic function of equilibrium systems called entropy (S) which never decreases during processes of an isolated system. Entropy is an extensive variable.

Third law: This law, also called the Nernst–Planck's theorem, concerns the limit to absolute zero temperature. At absolute zero, the quantum interference is believed to impose a unique ground state at least in most cases, and the extensivity of the macroscopic thermodynamics should be carefully tested [4]. As quantum fluctuations are beyond the scope of this book, we will not go into the detail of this law.

"Fourth law" [8]: Any thermodynamic variable characterizing an equilibrium state is either extensive one or intensive one. For the system whose extensive variables are (E, V, N) and S, the intensive variables are temperature T, pressure p, and chemical potential μ. The intensive variables are the homogeneous function of the extensive variables of zeroth order. The intensive variables characterize the equivalence relationship "$\sim$" of equilibrium. If an equilibrium system A and another one B satisfy $A \sim B$, it means that the contact of these two systems cause no changes in their extensive variables. The system B can be an environment with which the system A is in contact.

The transitive property of this equivalence relationship, $(A \sim C) \wedge (B \sim C) \Rightarrow A \sim B$, allows to compare, for example, the temperature T of two systems A and B using the "thermometer" C. The reflective property, A$\sim$A, implies that the intensive variables remain unchanged upon combining the two identical systems A. The symmetric property, A$\sim$B$\Leftrightarrow$ B$\sim$A, implies that the assignment of the measuring system and the measured system is relative in the context of macroscopic equilibrium. Further properties of the extensive and intensive variables will be described below.

If the system is not macroscopic, the intensive variables can still be used to characterize the environment. But they are not the state variables of the small system.

2.1.3 Thermodynamic Relations Come from Several Different Aspects of Thermodynamic Systems

In this section we will describe several basic consequences of the laws of thermodynamics.

One important aspect of the thermodynamics, especially from the zeroth and fourth laws, is the presence of thermodynamic function by which various

[8] There is no general consensus on which law should be put as the fourth law. Here we take up a version.

thermodynamics variables are derived. Its consequence is the fundamental relation and Maxwell relation.

Another important aspect, especially from the first and second laws, is the homogeneity of those thermodynamics function related to its extensive character. Its consequence is the Euler relations and Gibbs–Duhem relation.

2.1.3.1 Fundamental Relation

For a simple gas system, the equilibrium states are completely describable either by $\{S, V, N\}$ or by $E(S, V, N)$. In respective case, E or S are the functions of the first three extensive variables, i.e., $E{=}E(S, V, N)$ or $S{=}S(E, V, N)$. These are represented as a curved surface in 4D space coordinated by (S, E, V, N). Each point on this surface represents an equilibrium state. On a point on this surface, say (S, E, V, N), the equation of the tangent plane is written as

$$dE = T\,dS - p\,dV + \mu\,dN, \quad \text{or} \quad dS = \frac{1}{T}dE + \frac{p}{T}dV - \frac{\mu}{T}dN, \tag{2.2}$$

where "differentials," (dS, dE, dV, dN), are the local variables along (S, E, V, N), with the origin relocated to the tangent point, and the coefficients $(1/T, p/T, -\mu/T)$ are independent of these differentials. Thus we can have essentially all relations between the extensive variables and the intensive ones, such as $\frac{1}{T} = \frac{\partial S}{\partial E}_{V,N}$ for these equations. We call the relations (2.2) the fundamental relations. Those thermodynamic functions that generate the fundamental relations, such as $S(E, V, N)$ or $E(S, V, N)$, are called the complete thermodynamic functions. The pairs of thermodynamic derivatives associated with the partial derivative like $T(E, V, N)^{-1} = \frac{\partial S}{\partial E}_{V,N}$ are called conjugate variables. (T, S), $(-p, V)$, and (μ, N) are pairs of conjugate variables.[9]

Complete thermodynamic functions are not unique, and therefore, the fundamental relations are not either. However, all the fundamental relations of a system represent the same interdependencies. To obtain a new complete thermodynamic function, we must apply the Legendre transformation to the original complete thermodynamic function. For example, we can obtain the complete thermodynamic function, "Helmholtz free energy," $F(T, V, N)$ of a simple gas system by the Legendre transformation of $E(S, V, N)$: We solve $T = \partial E(S, V, N)/\partial S$ to find (formally) $S(T, V, N)$. (This function, $S(T, V, N)$, is not a complete thermodynamic function.) Then we substitute this into $E(S, V, N) - T(S, V, N)S$ to obtain $F(T, V, N) = E(S(T, V, N), V, N) - T(S(T, V, N), V, N)\,S(T, V, N)$. The variable S is now found through $(\partial F/\partial T)_{V,N} = -S$, instead of $(\partial E/\partial S)_{V,N} = T$. The exchange of the conjugate pair, (S, T), as independent–dependent variables

[9] Incomplete thermodynamic functions can also be made. For example, we can solve $T(E, V, N)^{-1} = \frac{\partial S}{\partial E}_{V,N}$ for V and substitute the result in $S(E, V, N)$. Then we have a function $S = S(E, V(E, T, N), N)$. Such function, although correct, cannot generate all the other thermodynamic variables; there remains an undetermined additive constant.

requires the subtraction of the term TS from the original complete thermodynamic function.[10]

Physically, the subtraction of a product of conjugate thermodynamic variables (e.g., TS) in the transform (e.g., $E \mapsto F = E - TS$) implies the attachment or detachment of an environment. For example, suppose that an isolated simple gas system has the complete thermodynamic function, $S(E, V, N)$. We now attach a thermal environment to this system. Since the combined system is a new isolated system, we may use the entropy of the whole system $S(E, V, N) + S_{\text{res}}(E^{\text{tot}} - E)$ as a new thermodynamic function. Since the system can exchange energy with this environment satisfying the law of energy conservation, we control no more the energy E, but control the energy of the whole system, E^{tot}. Since the reservoir retains a constant temperature T_{res} by definition, we choose as a new independent thermodynamic variable T_{res} rather than E^{tot}, where T_{res} is given by $\partial S_{\text{res}}/\partial E^{\text{tot}} = 1/T_{\text{res}}$.[11] Then we have

$$
\begin{aligned}
S(E, V, N) + S_{\text{res}}(E^{\text{tot}} - E) &\simeq S(E, V, N) + S_{\text{res}}(E^{\text{tot}}) - \frac{E}{T_{\text{res}}} \\
&= -\frac{1}{T_{\text{res}}}(E - T_{\text{res}} S) + \text{const.}
\end{aligned} \tag{2.3}
$$

The repartitioning of the energy into E and $E^{\text{tot}} - E$ is self-adjusted so that the temperatures are equilibrated, $T = T_{\text{res}}$. The new complete thermodynamic function $F{=}E - TS$ then describes the combined system (the system and the thermal environment) with $\{T, V, N\}$ as independent variables.

As a consequence of the second law of thermodynamics, the work W done to the system through arbitrary isothermal system is bounded below by the change in the Helmholtz free energy, $\varDelta F = \varDelta E - T\varDelta S$[12]:

$$W \geq \varDelta F.$$

An ensemble of a very large number of identical copies of a small system obeys macroscopic thermodynamics. As there are no interactions among those copies, the

---

[10] Mathematically the Legendre formulation is defined as follows: Take a fundamental relation, $df = \sum_{j=1}^{n} R_j dx_j$, where $f(x_1, \ldots, x_n)$ is the complete thermodynamic function and $R_j \equiv \partial f/\partial x_j$. In order to derive the new fundamental relation with $\{R_1, \ldots, R_k\}$ and $\{x_{k+1}, \ldots, x_n\}$ as the independent variables, where $0 < k < n$, we solve $R_j = \partial f/\partial x_j$ for $j = 1, \cdots, k$ to represent $\{x_1, \ldots, x_k\}$ as functions of $\{R_1, \ldots, R_k, x_{k+1}, \ldots, x_n\}$. Then the identity $d\Big[f - \sum_{j=1}^{k} R_j x_j\Big] = \sum_{j=1}^{k}(-x_j)dR_j + \sum_{j=k+1}^{n} R_j dx_j$ is the new fundamental relation which is generated by a new complete thermodynamic function $\tilde{f} \equiv f - \sum_{j=1}^{k} R_j x_j$ as function of new independent variables, $\{R_1, \ldots, R_k, x_{k+1}, \ldots, x_n\}$.

[11] Since we simplify the characterization of the environments, we do not explicitly take into account the volume and the particle number of the heat bath.

[12] The result is obtained from $\varDelta S_{\text{res}} + \varDelta S \geq 0$, $\varDelta S_{\text{res}} = -Q$, and $\varDelta E = W + Q$.

extensive character of the variables like E, V, and N is evident. By the law of large numbers (Sect. 1.1.2.3) we can define the energy, etc. *per* system. However, the results which we obtain from this macroscopic framework concern only the statistical average over the individual systems or the property characterizing the ensemble.

2.1.3.2 Maxwell Relation

If the equation of one plane (2.2) is the fundamental relation associated with a monovalent function, $S = S(E, V, N)$, the coefficients of the differentials, i.e., $(1/T, p/T, -\mu/T)$, should satisfy certain conditions.[13] For example, $(\partial\mu/\partial V)_{S,N} = -(\partial p/\partial N)_{S,V}$ is derived from (2.2) since

$$\left(\frac{\partial\mu}{\partial V}\right)_{S,N} = \frac{\partial^2 E}{\partial V \partial N} = \left(\frac{\partial(-p)}{\partial N}\right)_{S,V}.$$

These types of equations are called the Maxwell relation. Inversely, if the coefficients of the Eq. (2.2) satisfy the Maxwell relation, these local equations can be integrated to define a surface. See Appendix A.2.1 for a brief proof.

2.1.3.3 Euler Relations of Thermodynamic Variables

The complete thermodynamic potentials, $E = E(S, V, N)$ or $S = S(E, V, N)$, are the relations among the extensive thermodynamic variables. If a system in equilibrium and its copy is combined to make a new system, all of these variables should become twice as large as the original. In general, we expect

$$E(\lambda S, \lambda V, \lambda N) = \lambda\, E(S, V, N), \qquad S(\lambda E, \lambda V, \lambda N) = \lambda\, S(E, V, N), \tag{2.4}$$

for $\lambda > 0$. This is a mathematical representation of the extensivity of the macroscopic thermodynamics. If we draw a graph, for example, of E as function of (S, V, N), the extensive character of the function $E(S, V, N)$ imposes a particular geometrical property on this graph: the graph consists only of the straight lines passing through the origin $(0, 0, 0, 0)$. Figure 2.1 illustrates such geometry for the case with three extensive variables. (One can imagine an umbrella cloth when it is folded loosely.)

An important consequence of this constraint is the following relationship among the thermodynamic variables:

$$TS - pV + \mu N = E. \tag{2.5}$$

---

[13] In other words, if the continuous functions $(T(E, V, N), p(E, V, N), \mu(E, V, N))$ satisfy no mutual relations, a monovalent graph S vs. (E, V, N) cannot be constructed through the connection of the tangent planes (2.2).

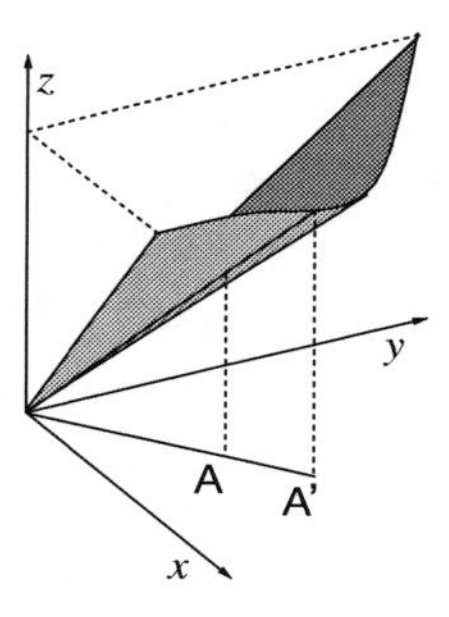

Fig. 2.1 The graph of extensive thermodynamic function $z = f(x, y)$ as function of the two extensive thermodynamic variables, x and y

It is easily derived from either one of the relations in (2.4) by differentiating with respect to λ and then putting $\lambda = 1$. This type of identity is called the Euler's theorem.[14,15]

Euler's theorem applies also to the expressions of the intensive variables, e.g., $\mu = \partial E/\partial N = \mu(S, V, N)$. Since the value of μ should remain the same when the system and its copy is combined, we have

$$\mu(\lambda S, \lambda V, \lambda N) = \mu(S, V, N).$$

By the same token to the derivation of (2.5), we have

$$S\frac{\partial \mu}{\partial S} + V\frac{\partial \mu}{\partial V} + N\frac{\partial \mu}{\partial N} = 0 \tag{2.6}$$

and similar identities.

2.1.3.4 Gibbs–Duhem Relation

Extensivity of the complete thermodynamic function imposes further relationship among the intensive thermodynamic variables. Let us take the differential of (2.5), that is, $d(TS - pV + \mu N) = dE$. This expression imposes a linear relation among the first-order differentials of each variables when an equilibrium state, $(E(S, V, N), S, V, N)$, is displaced infinitesimally to another equilibrium state:

$$SdT + TdS - (pdV + Vdp) + \mu dN + Nd\mu = dE.$$

---

[14] Mathematically, when the function $f(x_1, \ldots, x_n)$ has a homogeneity of kth order, i.e., $f(\lambda x_1, \ldots, \lambda x_n) = \lambda^k f(x_1, \ldots, x_n)$, the derivation of this equation by λ at $\lambda = 1$ yields an identity $\sum_{i=1}^{n} x_i\, R_i = kf$ valid for any $(x_1, \ldots, x_n)$, where $R_i = \partial f/\partial x_i$.

[15] Each new Legendre transformation on the complete thermodynamic function reduces by one of the independent extensive variables. We cannot have a complete thermodynamic function in which all the extensive variables are eliminated, since then the complete thermodynamic function would be $E - TS + pV - \mu N \equiv 0$ because of (2.5), or because the intensive variables along cannot fix the size of the system.

By subtracting the fundamental relation, (2.2), we obtain

$$SdT - Vdp + Nd\mu = 0. \tag{2.7}$$

This relation is called the Gibbs–Duhem relation.

When the macroscopic thermodynamic system consists of many copies of a small system, the Euler relations and Gibbs–Duhem relation may bring little constraint or information about the individual system.

2.1.4 * Heat of Macroscopic Open System Consists of Two Terms Both Including Entropic Parts

An open system exchanges not only heat but also particles with its environment. See Fig. 2.2.[16] Thermodynamics of open system is an important issue in biology since protein molecular motors are usually open systems which consume source molecules like ATP in the environment. The problem of organic or inorganic adsorption and diffusional transport across membranes are also the subject of open system. In this section and in Sects. 2.3.2 and 2.3.3, we deal explicitly with an open system.

For a closed system, we can identify the system with the particles, etc., that are therein. An open system, however, is rather a container of particles, and a particle belongs to the system while it is in the container.[17]

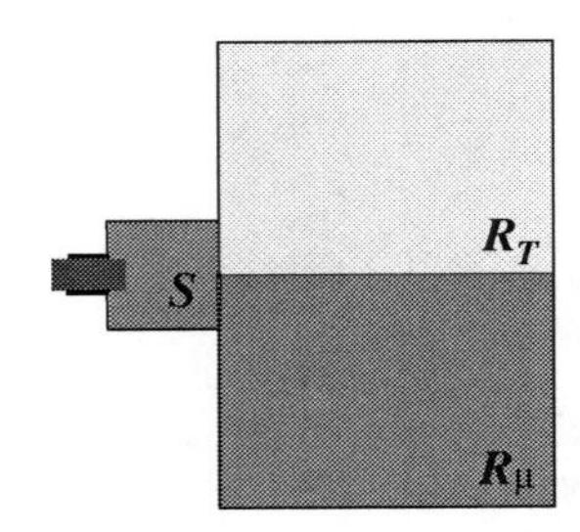

Fig. 2.2 An open system containing gas particles, S, in contact with a thermal environment, R_T, and a particle environment, R_μ. The *thick lines* and *filled rectangle* are isolating walls and piston, respectively, and the thin walls are thermally conducting walls. A *thick dashed line* is the wall permeable for the gas particle

2.1.4.1 Energetics of the Open System

The energy E of an open system can change not only due to heat and work but also due to the exchange of particles between the system and the particle environment.

[16] In quantum physics, often the same word means the systems in contact with (only) a thermal environment.

[17] In the problem of adsorption, the container is the ensemble of adsorption sites. Those particles on the adsorption sites belong to the open system.

Let us consider a gas system in which work is done through the change of volume, $d'W = -pdV$. The first law of thermodynamics reads

$$dE = -pdV + d'Q^{\text{tot}}, \tag{2.8}$$

where $d'Q^{\text{tot}}$ denotes *all* the energy entering into the system except for the work $-pdV$.[18] We suppose that the system is in contact with a thermal environment of temperature T and also with a particle environment of chemical potential μ. Therefore, if a small number of particles, dN, have entered into the open system in keeping the equilibrium, and the volume is increased by dV, then the change in the internal energy is given by the fundamental relation, $dE{=}TdS - pdV + \mu dN$. The substitution of this expression for dE in (2.8) yields

$$d'Q^{\text{tot}} = TdS + \mu dN. \tag{2.9}$$

In general, both dN and dS are nonzero when a particle enters or leaves an open system. In other words, the chemical potential μ is not the (only) energy carried by a particle upon its migration across the system's border:[19] When the only change is the immigration of a single particle, $(dN)_1 = 1$, we will denote by $(dS)_1$ the associated change of entropy. The "heat" associated with this immigration is then the sum, $(d'Q^{\text{tot}})_1 = T(dS)_1 + \mu$. When the particle is an ideal gas particle, such energy changes should not involve the concentration of the particle in the particle environment simply by the definition of ideal particle. On the other hand, the particle concentration of the particle environment c enters in μ in entropic form, $k_B T \log c$. In order that the $(d'Q^{\text{tot}})_1$ is independent of the particle concentration of the environment, the last entropic term should be cancelled by a similar term in $T(dS)_1$.

2.1.4.2 Complete Thermodynamic Function for an Open System

When a system contains N_α particles of the species α, the pertinent complete thermodynamic function J is defined as follows:

$$J(T, a, \{\mu_\alpha\}) \equiv E - TS - \sum_\alpha \mu_\alpha N_\alpha, \tag{2.10}$$

[18] Hereafter we will use d' to denote the infinitesimal transfer of work, $d'W$, and of (generalized) heat, $d'Q$ to distinguish from d which is related to the thermodynamic variables.

[19] Statistical mechanics shows that the chemical potential is related to the fugacity, i.e., the probability to escape, of *anonymous* particles. This is why μ depends on the concentration of the particles even if the particles do not interact among each other.

where μ_α is the chemical potential of the particle environment of the species α, and a denotes the extensive parameter that is controlled by an external system (e.g., $a = V$ if the volume is controlled). J gives rise to the following fundamental relation:

$$dJ = -SdT + \frac{\partial J}{\partial a}da - \sum_\alpha N_\alpha d\mu_\alpha. \tag{2.11}$$

The work done to the open system through the change of the parameter a is calculated from $\int (\partial J/\partial a)da$. When it is done quasistatically, the work, which we denote by $W|_{T,\{\mu_\alpha\}}$, is equal to the increment of J:

$$W|_{T,\{\mu_\alpha\}} = \Delta J|_{T,\{\mu_\alpha\}} \qquad \text{(quasistatic)}. \tag{2.12}$$

The chemical reaction is similar to the open system in the context that discrete atomic masses are transferred between one state to the other. For an open system the transfer is between a system and its particle environment, while for a chemical reaction it is between a molecular species and the another. These two processes can coexist. Typical examples are surface catalytic reactions and motor proteins[20] or the transport of molecules across a membrane channel. In the latter case, we can regard the transport as unimolecular reaction from one "species" to another "species," such as "cytoplasm"$\rightleftharpoons$ "nucleus." If both sides of a membrane are macroscopic, the channels on the membrane can be also described as an open system that exchanges particles with two particle environments.[21]

2.2 Free Energy as an Effective Potential Energy for the External System

In Sect. 2.1.3.1 we have shown that the complete thermodynamic function $F = E - TS$, i.e., the Helmholtz free energy, describes the combined system, the system and the thermal environment, with $\{T, V, N\}$ as independent variables. Here we describe another aspect of this thermodynamic function from the operational point of view. The Helmholtz free energy acts as an effective potential energy under the quasistatic operations by the external system. Figure 2.3 (top) shows a system of the gas confined in a cylinder with a piston, which an external system can move. When the external system pushes the piston, the first law of thermodynamics (2.1)

[20] Different species of molecules can share the same space as distinct particle environments if the spontaneous reaction between these species is negligible.

[21] In the master equation formalism and related discrete Langevin equation discussed in the next chapter, the change in closed systems and the migration of particles in open systems are both described as a state transition of the whole system.

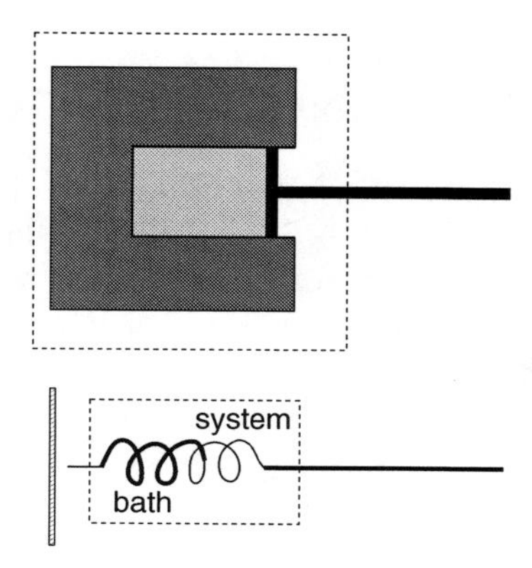

Fig. 2.3 Thermodynamic system viewed from the external system. Under quasistatic operations of the piston (*top figure*), the "black box" (*dashed rectangle*) which consists of a gas in the cylinder and the heat environment behaves like a spring (*bottom*). The temperature, instead of the force of spring, is equilibrated between the system and the thermal environment (heat bath)

tells us that the associated work W done onto the system is partitioned into the increment of the energy of the system, ΔE, and the heat released to the thermal environment, $(-Q)$.[22] For the external system, the system *and* the thermal environment, therefore, look like a single black box. The external system measures only the relation between the work W and the displacement of a piston.

To characterize this black box, let us denote by a the position of the piston and by $W(a_1; a_2)$ the work done through the displacement of the piston from a_1 to a_2. If this work can be described by a potential, $\phi(a)$, such that $W(a_1; a_2) = \phi(a_2) - \phi(a_1)$, then one might regard this black box as an elastic spring whose potential energy is $\phi(a)$ up to an additive constant (see Fig. 2.3 (bottom)). Macroscopic thermodynamics tells that (1) such $\phi(a)$ exists if the parameter a is changed quasistatically and (2) this potential energy, $\phi(a)$, is the Helmholtz free energy, $F = E - TS$, up to an arbitrary additive constant.[23] Thus the word "free energy" for F is justifiable because it behaves like a potential energy (for "energy"), and it is available as work (for "free"). A similar interpretation can be done for the free energies $G = E - TS + pV (= \mu N)$ or $J = E - TS - \mu N (= -pV)$ when the system undergoes a process keeping in contact with a pressure environment (for G) or with a particle environment (for J), respectively.

Let us summarize how the quasistatic operation of the piston in the Fig. 2.3 can be described differently according to the standpoint of the observer: the Ext who sees the system only through the parameter a, the Sys who has access to both E and

[22] Note the sign convention about the work and heat, see Sect. 2.1.1.

[23] $dE = TdS - pdV + \mu dN$ under the constraints, $dT = dN = 0$, yields $d(E - TS) = dF = -pdV = d'W$.

S, and the Tot who regards the totality of the system and the thermal environment as an isolated system.

Ext: The work W is apparently stored in the system as the increment of the potential energy, ΔF $(= \Delta E - T\Delta S)$.[24] If the data of ΔF are accumulated at different temperatures, the change of the (internal) energy ΔE can be obtained by the following formula called the van't Hoff equation (see, e.g., [5]):

$$\Delta E = \Delta F - T\,\Delta\left(\frac{\partial F}{\partial T}\right)_a, \tag{2.13}$$

where the suffix "$_a$" indicates that the initial and final values of a should be fixed upon taking the derivative.[25] Equation (2.13) comes from the relation: $E = F + TS = F - T\partial F/\partial T$.

Sys: The work W is partly stored in the system as ΔE, while the rest is transferred to the thermal environment as heat, $-T\Delta S$.[26] The actual partitioning of the work into ΔE and $(-T\Delta S)$ depends on the details of the system. For example, ΔF should be ΔE for a system made of metallic spring but it should be $(-T\Delta S)$ for a system of ideal gas or ideal polymer network.

Tot: The work W is stored in the isolated system as the increment of (true) energy, $\Delta E + T\Delta S_{\mathrm{env}}$, where the entropy of the thermal environment, S_{env}, changes so that $\Delta S + \Delta S_{\mathrm{env}} = 0$.

For nonquasistatic processes the second law of thermodynamics tells us that the entropy of the isolated system is nondecreasing: $\Delta S + \Delta S_{\mathrm{env}} \geq 0$. This inequality relates the work done to the system, $W = \Delta E - (-T\Delta S_{\mathrm{env}})$, [27] and the change of the Helmholtz free energy, $\Delta F = \Delta E - T\Delta S$, by the inequality,

$$W \geq \Delta F. \tag{2.14}$$

In other words, the work obtainable from the system $(-W)$ is limited by the decrease of the free energy of the system $(-\Delta F)$:

$$(-W) \leq (-\Delta F). \tag{2.15}$$

---

[24] The quantities denoted like "ΔM" implies the increment of M through the change of the parameter a.

[25] Remember that the partial derivative can give different results depending on what variables are fixed: For example, $p = -(\partial U/\partial V)_S$ is rewritten as $p = -(\partial U/\partial V)_T + T(\partial p/\partial T)_V$. Generally, given the fundamental relation, $f(x_1, \ldots, x_n)$ and $R_j \equiv \partial f/\partial x_j$, we have a formula, $(\partial f/\partial x_i)_{R_j} = (\partial f/\partial x_i)_{x_j} - R_j(\partial R_i/\partial R_j)_{x_i}$.

[26] Remember the sign convention: the heat, $Q = T\Delta S$, is positive if the energy is transferred from the thermal environment to the system.

[27] Note that the thermal environment is supposed to be in equilibrium and, therefore, $Q = -T\Delta S_{\mathrm{env}}$ holds always.

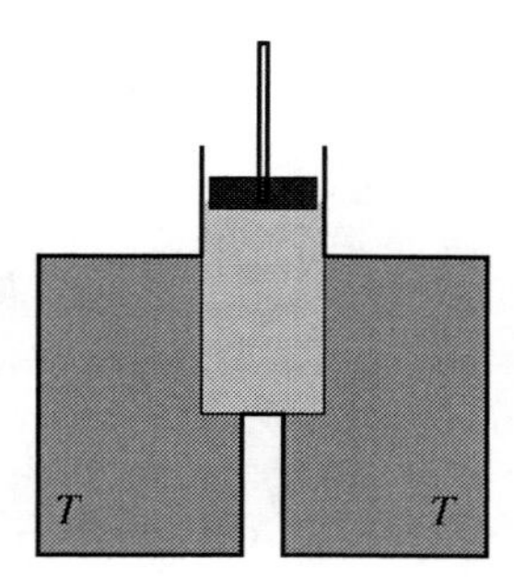

Fig. 2.4 A gas in cylinder (*central rectangle*) in contact with two heat baths of the same temperature, T

In the description of the energetics of fluctuating systems, the consciousness of the standpoint is as important as the macroscopic thermodynamics.

We should note that macroscopic thermodynamics cannot analyze the *kinetic* aspects of the quasistatic processes. First, thermodynamics gives no criterion for the quasistatic operation by an external system. Thermodynamics is not the framework to determine the characteristic timescale beyond which the excess work, i.e., the difference between the actual work and the quasistatic work, is sufficiently small. To answer such a question, we do not need a knowledge of the microscopic mechanics, but need an energetics of fluctuations in Chap. 4. Second, thermodynamics cannot access a slow kinetic process. Let us suppose a gas is confined in a cylinder and a piston, which is in contact with two thermal environments at the same temperature, T (Fig. 2.4). If we press down the piston quasistatically, the partition of the heat released to these environments could be experimentally observable.[28] The linear nonequilibrium thermodynamics [6, 7] incorporates the irreversible flux between the two equilibrium[29] systems when their intensive variables have a small gradient.

2.3 * Free-Energy Conversion

Free-energy conversion is one of the principal subjects of thermodynamics, since its advent by Carnot, until the recent interest in the context of molecular motor [7]. We, therefore, introduce several important notions and concepts related to the free-energy conversion.

2.3.1 Unused Work Is Not Released Heat

We here discuss reaction heat and the work produced by a heat engine.

A thermal environment is supposed to be an inexhaustible buffer of energy. Because of this character, the isothermal processes exhibit sometimes

[28] If the (almost) quasistatic process takes an extremely large time t, diffusive exchange of heat between the two thermal environments via the system will be of order $\sim t^{\frac{1}{2}}$. We neglect this as compared to the systematic part of $\sim t$.

[29] It is called “local equilibrium”.

counterintuitive behaviors about their energetics. The rubber band mentioned above is an example: little heat is released from an ideal rubber band when it freely shrinks from a stretched state, while a quasistatic shrinkage can absorb heat. Free expansion of an ideal gas behaves similarly: although a quasistatic expansion from the volume V_{ini} to V_{fin} can do the reversible work of $k_{\mathrm{B}}T \log(V_{\mathrm{fin}}/V_{\mathrm{ini}})$, the work-free expansion (Joule–Thomson process) absorbs no heat.

In the biochemical context, the hydrolysis reaction of an ATP (adenosine triphosphate) with a water molecule into an ADP (adenosine diphosphate) and an inorganic phosphate (Pi), $\mathrm{ATP} + \mathrm{H_2O} \to \mathrm{ADP} + \mathrm{P_i}$, can do about $20k_{\mathrm{B}}T$ of work under usual physiological conditions. It does not mean that we will observe this amount of dissipative heat when no work is extracted through this reaction. (See below.) We might also recall the existence of endothermic chemical reaction – while the reaction proceeds with consuming the relevant free energy (see below), the system absorbs heat, instead of absorbing heat. The point is that, for open systems, the heat released to the thermal environment, $(-Q)$, is not the total increase of entropy of the environment ΔS_{env}, that is $Q \neq -T\Delta S_{\mathrm{env}}$: The total change ΔS_{env} generally includes a purely combinatorial part (mixing entropy change).

The chemomechanical coupling is a mechanism which converts a chemical free energy into work under constant temperature and pressure. We will see the interplay of heat and work in this mechanism. Figure 2.5 shows a schematic setup of chemomechanical coupling. In this setup, the material before the reaction (called the substrate) and the one after the reaction (called the product) are separated as subsystems in order to avoid a work-free reaction. As we are interested in the process under constant pressure p, the energy of the pressure environment is included in the energy of the system. That is, we use enthalpy, $H \equiv E + pV$, instead of E, where V is the volume. By adopting enthalpy as the energy of the system, the work due to volume change is not counted in W. Finally there is an engine that works between the substrate and the product subsystems. The two subsystems and the engine are immersed in a thermal environment (Fig. 2.5). The combined system, therefore, consists of four subsystems, i.e., the two particle subsystems, the pressure reservoir, and the chemical engine.

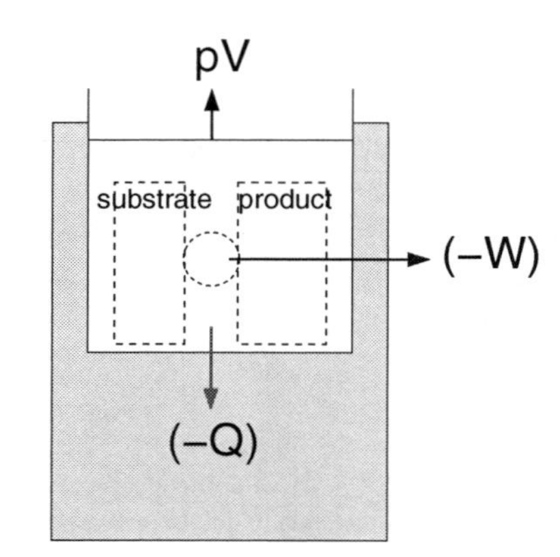

Fig. 2.5 Schematic setup of chemomechanical coupling. The system consists of the substrate subsystem, the product subsystem, the ambient pressure (*piston*), and the chemical engine (*circle*), all immersed in a thermal environment

In the case of the quasistatic process, the analysis is done by the aid of a thought experiment called a van't Hoff reaction box (see e.g., [5]). Here we include the general case of a nonquasistatic process and consider what occurs upon one cycle of operation of the engine. Let us denote by $(-Q)$ the heat released from the combined system to the thermal environment and by $(-W)$ the work done by the combined system to the external system. Let us also denote by ΔS, ΔV, and $\Delta H(= \Delta E + p\Delta V)$, the changes in the entropy, volume, and the enthalpy of the combined system.[30]

Application of the first law of thermodynamics: In the present context this law gives $\Delta E = W + (-p\Delta V) + Q$, or by noting that p is constant,

$$(-\Delta H) = (-W) + (-Q). \tag{2.16}$$

The minus signs are incorporated in the above equation so that we can use the extracted work, $(-W)$, and the released heat, $(-Q)$. There are two extreme cases:

Case where no work is extracted: We substitute $W = 0$ in (2.16) to have $(-Q) = (-\Delta H)$, which we shall call the heat of Joule–Thomson, $(-Q)_{\mathrm{JT}}$. That is,

$$(-\Delta H) = (-Q)_{\mathrm{JT}}, \qquad W = 0. \tag{2.17}$$

Therefore, this reaction is either exothermic or endothermic, depending on whether $(-\Delta H) > 0$ or $(-\Delta H) < 0$.

Case where the maximal work is extracted: The released heat, $(-Q) = T\Delta S_{\mathrm{env}}$, is related to ΔS through $\Delta S + \Delta S_{\mathrm{env}} = 0$. We shall call this the reversible extracted heat, $(-Q)_{\mathrm{rev}} = -T\Delta S$. In this case, the extracted work $(-W)$ is the reversible work, $(-W)_{\mathrm{rev}} = -\Delta H + T\Delta S$. Using the relation, $\Delta H = \Delta G + T\Delta S$, we have included the substrate and product subsystems as a part of the combined system. This justifies the Gibbs free energy as the fundamental function of the present problem:

$$\begin{aligned} (-\Delta H) - (-\Delta G) &= (-Q)_{\mathrm{rev}} : \text{reversible extracted heat} \\ (-\Delta G) &= (-W)_{\mathrm{rev}} : \text{reversible extracted work.} \end{aligned} \tag{2.18}$$

In case where the particle environments are treated as environment, the chemical potentials should be the independent variables. See Sect. 2.1.4.2.

General case: In the general case, the total entropy is nondecreasing, $\Delta S + \Delta S_{\mathrm{env}} \geq 0$, by the second law of thermodynamics. Therefore,

$$\begin{aligned} (-W) &\leq (-W)_{\mathrm{rev}} \\ (-Q) &\geq (-Q)_{\mathrm{rev}}. \end{aligned} \tag{2.19}$$

[30] The changes associated to the engine are 0 by definition of the cycle.

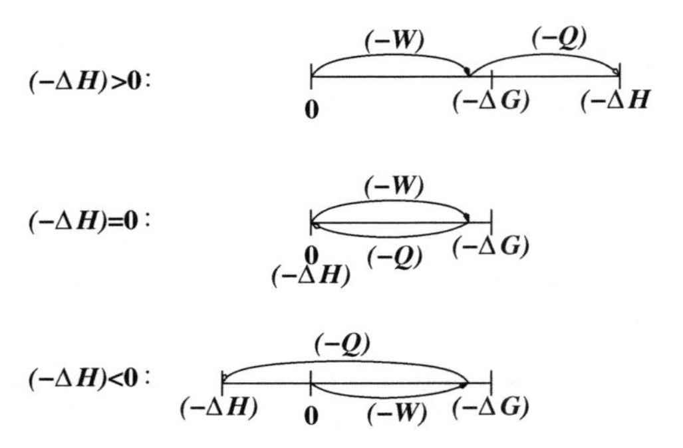

Fig. 2.6 The partition of the extracted work, $(-W)$, (*filled arrows*) and the released heat, $(-Q)$, (*open arrows*) under constant temperature and pressure. They obey the first law; $(-\Delta H) = (-W) + (-Q)$. The reaction is either exothermic ($(-\Delta H) > 0$, *top*), athermal ($(-\Delta H) = 0$, *middle*), or endothermic ($(-\Delta H) < 0$, *bottom*). The figures show only the case of "engine", i.e., $(-\Delta G) > 0$

Figure 2.6 summarizes the above relations for the cases with $(-\Delta H) > 0, (-\Delta H) = 0$, and $(-\Delta H) < 0$, respectively.[31]

Remark: We should avoid confusions like the following: "If the extracted work $(-W)$ is less than its maximum available value, $(-\Delta G)$, then the difference $(-\Delta G)-(-W)$ should be measured as the released heat." The correct statement is "..., then we have the irreversible heat, $(-\Delta G)-(-W) \geq 0$, in addition to the reversible one, $(-\Delta H) - (\Delta G)$."

Chemical pump: In the reverse case, that is, when the work is done to the system in order to drive the reaction or transport against the natural tendency, the Gibbs free energy increases, $\Delta G > 0$.

The equations and inequalities above are always valid with appropriate reinterpretation. For example, $(-W) \leq (-W)_{\text{rev}}$ of (2.19) should be read as $W \geq W_{\text{rev}}$, that is, the work done to achieve a cycle of engine is no less than the reversible one, W_{rev}, which in turn is ΔG according to (2.18).

Finally we note that the model of autonomous chemomechanical coupling is beyond the scope of macroscopic thermodynamics because it concerns the dynamics of the system.

2.3.2 Chemical Coupling Is a Transversal Downhill of a Gibbs Free-Energy Surface

We formulate the notion of chemical coupling (or conjugation) in the context of macroscopic thermodynamics. We will take up a very simple example: we suppose

---

[31] If the volume, instead of the pressure, of the system is fixed, all the enthalpy H in the above formulas should simply be replaced by the energy E.

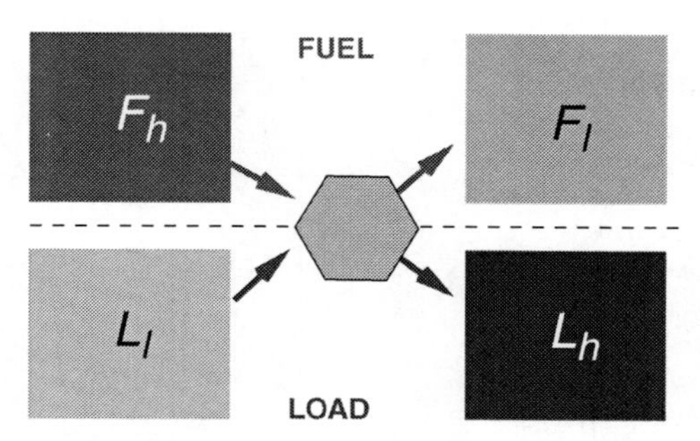

Fig. 2.7 Two reservoirs of the fuel (F) particles and two others of the load (L) particles with the high (h) and low (ℓ) chemical potentials. The passive diffusion of the fuel particles along their chemical potential gradient (*two arrows* in the *top layer*) is coupled by the chemical engine (*hexagon* at the *center*) to the active transport of the load particles against their chemical potential gradient (*two arrows* in the *bottom layer*)

two species of particles, the fuel (F) and the load (L) particles. For each species of particles, we prepare two particle reservoirs with high and low densities, which we distinguish by the suffixes, $_h$ and $_\ell$, respectively (see Fig. 2.7). For example, F_h denotes the high-density reservoir of fuel particles. We will denote the chemical potential of each reservoir by μ_{Fh}, etc. By definition, $\mu_{Fh} > \mu_{F\ell}$ and $\mu_{Lh} > \mu_{L\ell}$. A chemical engine (denoted by a hexagon at the center of Fig. 2.7) enables the active transport of the load particles from L_ℓ to L_h at the expense of the passive transport of the fuel particles from F_h to F_ℓ.

Gibbs free-energy surface: As in the previous section, we regard the four particle reservoirs and the chemical engine as a combined system working at a constant temperature and pressure. The relevant thermodynamic potential is then the Gibbs free energy,

$$G_{\text{tot}} = (\mu_{Fh} - \mu_{F\ell})N_{Fh} + (\mu_{Lh} - \mu_{L\ell})N_{Lh} + \text{const.}, \tag{2.20}$$

where N_{Fh} [N_{Lh}] are the number of the fuel [load] particles in their respective high-density reservoirs, and we have used the fact that the total number of the fuel [load] particles are constant.[32] Equation (2.20) defines an inclined plane in the 3D space of $(N_{Lh}, N_{Fh}, G_{\text{tot}})$. See Fig. 2.8. We shall call this plane the Gibbs free-energy surface. The thermodynamically allowed processes are those which decrease this free energy, G_{tot}. Suppose that the reactions $L_\ell \to L_h$ and $F_h \to F_\ell$ are coupled at the ratio of $n_L{:}n_F$ particle transport on the average. This imposes a condition on (N_{Lh}, N_{Fh}) that

[32] After each cycle of the chemical engine, the number of the particles in the machinery returns to the same value. We then ignore the cyclically varying part of G_{tot}.

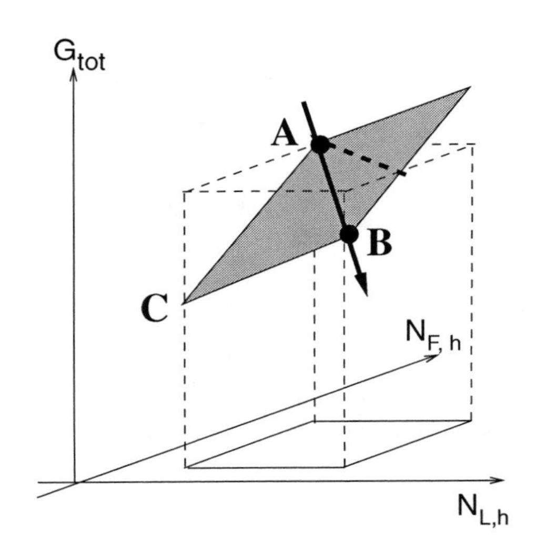

Fig. 2.8 Chemical coupling and Gibbs free energy. *The shaded plane* shows the total Gibbs free-energy surface (2.20) in the $(N_{\mathsf{L}h}, N_{\mathsf{F}h}, G_{\text{tot}})$. Chemical coupling is realized along the *thick arrowed line*, which satisfies (2.21) as well as (2.20). The process $\boldsymbol{A} \to \boldsymbol{B}$ consists of the downhill diffusion of the fuel, $\boldsymbol{A} \to \boldsymbol{C}$, and the uphill diffusion of the load, $\boldsymbol{C} \to \boldsymbol{B}$. The total Gibbs free energy G_{tot} decreases as a result. The $G_{\text{tot}} = \text{const.}$ line (*thick dashed line* on the Gibbs free-energy surface) is a visual guide

$$\frac{N_{\mathsf{L}h} - N^0_{\mathsf{L}h}}{N_{\mathsf{F}h} - N^0_{\mathsf{F}h}} = \frac{n_{\text{L}}}{n_{\text{F}}}, \tag{2.21}$$

where $(N^0_{\mathsf{L}h}, N^0_{\mathsf{F}h})$ are constants. While the chemical engine consumes the n_{F} fuel particles, decreasing the free energy by $\Delta G_{\text{F}} = (\mu_{\mathsf{F}h} - \mu_{\mathsf{F}\ell})(-n_{\text{F}}) < 0$, it pumps up the n_{L} load particles, increasing the free energy by $\Delta G_{\text{L}} = (\mu_{\mathsf{L}h} - \mu_{\mathsf{L}\ell})n_{\text{L}} > 0$. Figure 2.8 shows how the active transport, $\Delta G_{\text{L}} > 0$, is realized while satisfying the second law of thermodynamics, $\Delta G_{\text{tot}} \equiv \Delta G_{\text{F}} + \Delta G_{\text{L}} < 0$.

2.3.3 The Efficiencies of Heat Engine and Chemical Engine are Limited by the Second Law of Thermodynamics

In most cases, the phrase "efficiency of a heat engine" would refer to the Carnot cycle. When Carnot discussed his famous cycle, people had not yet recognized that heat can be transferred without transporting any material, nor had they discovered the second law of thermodynamics. It was, therefore, natural that the efficiency was defined so that it is unity when all the heat from the high-temperature environment is turned into a useful work. After thermodynamics was established, the efficiency of a heat engine was redefined in reference to the Carnot's theoretical maximum, allowed by the second law of thermodynamics.

For a single thermodynamic process, it is easy to introduce the efficiency of energy conversion in reference to the second law of thermodynamics. For example, for an isothermal process of changing the parameter a from a_1 to a_2, we can define the efficiency Θ as

$$\Theta = \frac{(-W)}{(-\Delta F)} \quad \text{(Isothermal process)}, \tag{2.22}$$

where $(-W)$ is the actually extracted work and $(-\Delta F) = F(a_1) - F(a_2)$ is the decrement in the Helmholtz free energy, that is, the maximally available work of this process (Sect. 2.1.2). By definition, Θ satisfies $(0 \leq)\Theta \leq 1$.

When a heat engine converts energy indefinitely, it must undergo a cycle of thermodynamic processes. The ideal cycle requires generally four quasistatic processes, as is the case with the ideal Carnot cycle: two processes in contact with the two thermal environments, respectively, one to absorb heat and the other to discard heat, and the other two processes for adjusting the system between these environments. The Carnot cycle is in this sense a minimal cyclic process to avoid irreversible losses. Below we derive the maximal work available from one cycle of Carnot heat engine. Later we will discuss the Carnot cycle on the fluctuating scale (Sect. 8.1.1).

When the cycle is ideal, each of the above mentioned thermodynamic processes must be reversible. Moreover, the attachment and detachment between the consecutive processes should cause no irreversibility. This requires that, at the end of each adiabatic process, the temperature of the system should be equal to the temperature of the thermal environment with which the system will be attached.[33] Let us denote the temperatures of the hot [cool] thermal environment by T_h $[T_\ell]$, respectively. For the isothermal process in contact with the hot environment the extracted work, $(-W_h)$, is the change of Helmholtz free energy of the system, $\Delta F_h = \Delta E_h - T_h \Delta S_h$, and similarly for the isothermal process with cool environment, i.e.,

$$(-W_h) = (-\Delta F_h) = (-\Delta E_h) + T_h \Delta S_h,$$
$$(-W_\ell) = (-\Delta F_\ell) = (-\Delta E_\ell) + T_\ell \Delta S_\ell.$$

During adiabatic processes the extracted work is equal to the change of the energy of the system:

$$(-W_{\mathrm{ad}:h\ell}) = (-\Delta E_{\mathrm{ad}:h\ell}), \qquad (-W_{\mathrm{ad}:\ell h}) = (-\Delta E_{\mathrm{ad}:\ell h}),$$

where the suffixes $_{\mathrm{ad}:h\ell}$ and $_{\mathrm{ad}:\ell h}$ indicate, respectively, the adiabatic cooling and heating processes. Because of the cyclicity of the processes, we impose

$$(-\Delta E_h) + (-\Delta E_{\mathrm{ad}:h\ell}) + (-\Delta E_\ell) + (-\Delta E_{\mathrm{ad}:\ell h}) = 0.$$

---

[33] We can assume the continuity of the temperature of the system from an isothermal process to the following adiabatic process.

Reversibility implies

$$(-\Delta S_h) + (-\Delta S_\ell) = 0.$$

The maximal available work per a cycle of the ideal heat engine, $(-W^{\text{tot}})_{\text{max}}$, is then

$$\begin{aligned}(-W^{\text{tot}})_{\text{max}} &\equiv (-W_h) + (-W_\ell) + (-W_{\text{ad}:h\ell}) + (-W_{\text{ad}:\ell h}),\\ &= (T_h - T_\ell)\Delta S_h. \end{aligned} \tag{2.23}$$

The efficiency of an actual heat engine, which we denote by Θ, should be the ratio of the actual extracted work, $(-W^{\text{tot}})$, to the above maximum reversible value, $(-W^{\text{tot}})_{\text{max}}$:

$$\Theta \equiv \frac{(-W^{\text{tot}})}{(-W^{\text{tot}})_{\text{max}}} = \frac{(-W^{\text{tot}})}{(T_h - T_\ell)\Delta S_h} \le 1. \tag{2.24}$$

The traditional definition of Carnot efficiency, η, as distinguished from our efficiency Θ, is $\eta = (-W^{\text{tot}})/(T_h \Delta S_h)$. For the reversible process, when $\Theta = 1$, the Carnot efficiency takes the well-known maximum, $\eta_{\text{rev}} = (T_h - T_\ell)/T_h$.

A similar argument can be applied to chemical engine consisting of an open system and two particle reservoirs, see Fig. 2.5.[34] In this case the system is in contact with a single thermal environment at temperature T throughout the cycle. Let us denote the chemical potentials of dense [dilute] particle reservoir by μ_h [μ_ℓ], respectively. An ideal cycle consists of two quasistatic processes under constant chemical potential and the intervening two quasistatic (closed) isothermal processes. The extracted work in the former processes reads $(-W_h) = (-\Delta E_h) + T\Delta S_h + \mu_h \Delta N_h$, etc.[35], where ΔN_h, etc., denote the changes in the number of particles in the system in contact with a particle reservoir. The extracted work during the isothermal processes is $(-W_{\text{it:h}}) = (-\Delta E_{\text{it:h}}) + T\Delta S_{\text{it:h}}$, etc., where the suffix $_{\text{it}}$ stands for closed isothermal processes. The condition of the cyclicity is imposed on the energy of the system as above and the number of particles in the system; $\Delta N_h + \Delta N_\ell = 0$. The condition of reversible process imposes $\Delta S_h + \Delta S_{\text{it:h}} + \Delta S_\ell + \Delta S_{\text{it:}\ell} = 0$. The maximal extracted work, $(-W^{\text{tot}})_{\text{max}}$ then is

$$(-W^{\text{tot}})_{\text{max}} = (\mu_h - \mu_\ell)\Delta N_h. \tag{2.25}$$

This result can be rewritten in terms of the total Gibbs free energy G^{tot} defined by $G^{\text{tot}} = \mu_h N_h^{\text{res}} + \mu_\ell N_\ell^{\text{res}} + G^{\text{engine}}$, where N_h^{res}, etc., denote the number of particles

[34] In the present context the substrate and product, Fig. 2.5, should be reread as high-density reservoir and low-density reservoir, respectively.

[35] "etc." implies the change of suffix, $_h \mapsto {}_\ell$.

in a particle reservoir and G^{engine} is that of the chemical engine. Since a balance of the number of particles transferred between a reservoir and the chemical engine imposes $\Delta N_h^{\text{res}} + \Delta N_h = 0$ and $\Delta N_\ell^{\text{res}} + \Delta N_\ell = 0$, and since the system undergoes a cycle, $\Delta G^{\text{engine}} = 0$, we arrive at $(\mu_h - \mu_\ell)\Delta N_h = -\Delta G^{\text{tot}}$. Therefore, the relevant definition of the efficiency of the chemical engine is

$$\Theta = \frac{(-W^{\text{tot}})}{(-\Delta G^{\text{tot}})}. \tag{2.26}$$

Remark: Any arbitrary positive quantity whose upper bound is unity does not necessarily deserve to be called efficiency: if we have an inequality in the form of $A - B - C \leq 0$ about the cycle of an energy converter, we might ask ourselves which of $A/(B + C)$, $(A - B)/C$, or $(A - C)/B$ is the most appropriate definition of the efficiency of energy conversion.

General remark: It is only the difference that the thermodynamic functions appeared in the thermodynamic efficiency Θ or in the Carnot efficiency η. It was so also in the heat and work of reversible reactions (2.18). It is because the thermodynamic functions have the arbitrariness of an additive constant and the experimentally observable results do not depend on that constant. So as to be compatible with the extensive character of these functions, this arbitrariness is reduced to the arbitrary additive constants in the molar [specific] energy and entropy. All the thermodynamic laws and relations are invariant under the change of these constants. See Appendix A.2.2.

The analysis of a Carnot cycle by conventional macroscopic thermodynamics is tautological, since macroscopic thermodynamics is constructed to explain the Carnot cycle. The above review will be still useful as a reference frame when a similar cycle is studied on the scale of thermal fluctuations.

The cost of the operations, i.e., of connecting or disconnecting with the environments, macroscopic thermodynamics is supposed to be negligible. For mesoscopic systems, irreversibilities associated to the operation on the system require attention. This issue will be discussed later (Chap. 7).

2.3.4 Discontinuous Phase Transition Accompanies the Compensation Between Enthalpy and Entropy

Phase transition can occur if a system takes two thermodynamically distinguishable equilibrium states (phases) at a temperature T and a pressure p of the environment. For a single-component system the transition occurs along a coexistence curve in the (T, p)-plane.[36] Typical examples are the boiling of water or precipitation of vapor.

[36] The generalization to many component systems is known as the Gibbs phase rule.

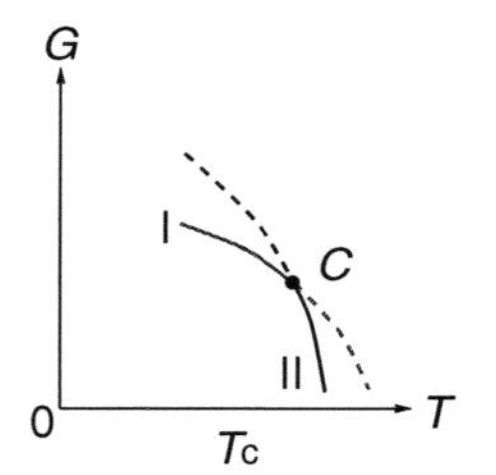

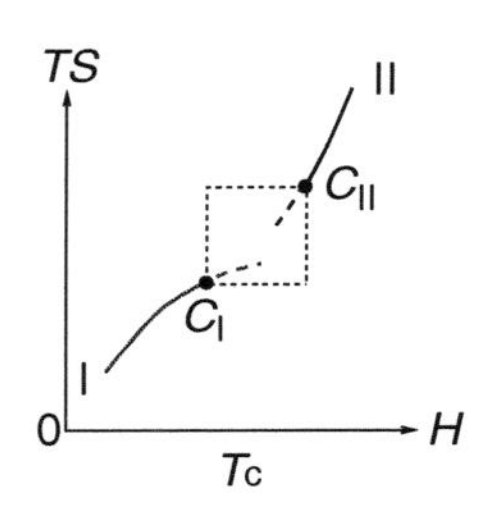

Fig. 2.9 (*Left*) Schematic representation of Gibbs free energy of two phases vs. T. From the gradient $\partial G/\partial T$ the entropy $S = -T\partial G/\partial T$ is estimated, while from the intersection at $T = 0$ of the tangent lines, the enthalpy, $H = G - T\partial G/\partial T$ is obtained. The smaller of $G_{\rm I}$ and $G_{\rm II}$ is realized; i.e., $G_{\rm I}$ for $T < T_c$ and $G_{\rm II}$ for $T > T_c$. (Right) The jump of H and of TS across the phase transition $C_{\rm I} \leftrightarrow C_{\rm II}$ on the (H, TS)-plane. The gradient of the jump $C_{\rm I}$–$C_{\rm II}$ (a diagonal of the *dashed square*) is 45° because of the H–S compensation (2.28)

In thermodynamics this discontinuous transition, called the first-order phase transition, is characterized by the equality of the Gibbs free energy per mass of the two states, $G_{\rm I}$ and $G_{\rm II}$:

$$G_{\rm I} = G_{\rm II} \qquad \text{(at transition)}, \tag{2.27}$$

where we distinguish the two phases by I and II, see Fig. 2.9 (left), where the pressure p is fixed. Exactly at the transition point, either one of the states is realized or the two phases coexist macroscopically in the system.

Gibbs free energy consists of enthalpy (i.e., the internal energy of the system *plus* the pressure environment) H and the entropic term, $-TS$, that is $G = H - TS$. In the case of the liquid–vapor transition of water, the liquid phase (I) gains the enthalpic part of $G_{\rm I}$ by its cohesive energy, while the vapor phase (II) gains the entropic term ($-TS$) by having a large specific volume per molecule. As a result, the enthalpy of these two phases satisfies $H_{\rm II} - H_{\rm I} > 0$, while their entropy obeys $T(S_{\rm II} - S_{\rm I}) > 0$. At the phase transition (2.27) tells that the differences of these two terms must balance (" Enthalpy–entropy (H–S) compensation"):

$$H_{\rm II} - H_{\rm I} = T(S_{\rm II} - S_{\rm I}) \qquad \text{(at transition)}. \tag{2.28}$$

Figure 2.9 (right) shows schematically the condition (2.28).

When the system is infinitely large, the zeroth law of thermodynamics is compatible with the switching of two phases at the transition, because the "sufficiently longtime" at the transition point cannot be realized. By contrast, for a finite system, the equilibrium state is unique and there are no singularities in phase transition. Some proteins can take clearly distinguishable shapes ("conformations") over the timescale of msec. Upon the conformation transition, a phenomenon similar to the above mentioned H–S compensation has been observed [8]. The differences of enthalpy and, therefore, of the entropy between the two conformations are large when the conformation change accompanies the binding/release of the surrounding water molecules.[37] Biological molecules are thought to use such transition-like phenomena. Some soft materials often undergo the transition under physiological conditions.

The H–S compensation has sometimes been an object of controversy because it can also arise as an experimental artifact [10]. In fact, if the term $-T\partial G/\partial T$ is predominant in the formula $H = G - T\partial G/\partial T$ over the magnitude of G, we have $TS = -T\partial G/\partial T \simeq H$. Therefore, we would have $\Delta H \simeq T\Delta S$ for any change of parameters, including T, within a relatively narrow range. The estimation of this artifact may be a good exercise for error analysis problems.[38]

2.4 Notes on the Extension of the Thermodynamics

It is being recognized that thermodynamics is an asymptotic mathematical structure where certain variables obey a set of universal relations. This structure appears in the phenomena consisting of a large number of similar objects. The conventional macroscopic thermodynamics, either classical or quantum mechanical, is not the only example. Thermodynamic relations that we will explore in this book from Chap. 4 are the another example. We mention below several extensions of the thermodynamics.

The geometrical aspect of thermodynamics has been studied under the name of *metric geometry of equilibrium thermodynamics* [11–13]. A framework to describe a large number of "thermodynamic elements" has been sought for by analogy to the theory of electric circuit, called *network thermodynamics* [14]. In linear nonequilibrium thermodynamics (see Sect. 2.2), subsystems are supposed to maintain local equilibrium within themselves and the transport and dissipation among the subsystems obey the Onsager (force–flux) relation [15, 16, 6, 7]. The Navier–Stokes equation [17] incorporates this framework.

More fundamental project has recently been proposed toward the *axiomatization of thermodynamics* [1, 2]. The latter [2] constructs a set of axioms for the conventional thermodynamics. The former [1] studies the axioms which are valid for any steady states including conventional thermodynamics. The subject is currently a hot topic [18–23].

In quantum information theory, people look for the *thermodynamics of quantum entanglement* [24]. The thermodynamic structure of chaotic systems has also been formulated [25, 26].

There are also attempts to remove some of the fundamental requirements of thermodynamic phenomena: in the *thermodynamics of small systems* [27], one does

[37] Upon a principal conformation change of a myosin molecule (called the isomerization), about 120 water molecules leave from the surface, according to the analysis based on dielectric measurements [9]. There, the enthalpy changes of about $H = +51$(kJ/mol) is mostly compensated, leaving little Gibbs energy difference between the conformations. (cf. $k_{\mathrm{B}}T = 2.4$(kJ/mol).)

[38] [29] is a good introduction of error analysis.

without the extensivity of the system behavior. In the *theory of glass [transition]*, the lack of quasistatic processes gives rise to many interesting phenomena, such as memory effect, aging, and plasticity. The related issue will be addressed in Sects. 7.1.2 and A.7.2. The so-called effective temperature [28] of glassy states indicates the (decaying) memory of the initial preparation.

References

1. Y. Oono, M. Paniconi, Prog. Theor. Phys. Suppl. **130**, 29 (1998)
2. E. Lieb, J. Yngvanson, Phys. Rep. **310**, 1 (1999)
3. H.S. Leff, A.F. Rex, *Maxwell's Demon: Information, Entropy, Computing* (A Hilger (Europe) and Princeton U.P. (USA), 1990)
4. M. Toda, R. Kubo, N. Saitô, *Statistical Physics I: Equilibrium Statistical Mechanics*, 2nd edn. (Springer-Verlag, Berlin, 1992)
5. E. Fermi, *Thermodynamics*, 1st edn. (Dover Publications, New York, 1956)
6. S.R. De Groot, P. Mazur, *Non-Equilibrium Thermodynamics* (Dover Publications, New York, 1984)
7. T.L. Hill, *Free Energy Transduction and Biochemical Cycle Kinetics*, dummy edn. (Springer-Verlag, New York, 1989)
8. E. Grunwald, L.L. Comeford, in *Protein-Solvent Interactions*, ed. by R.G. Gregory (Marcel Dekker, Inc., New York, 1995), Chap.10, p.421
9. M. Suzuki et al., Biophys. J. **72**, 18 (1997)
10. A. Cornish-Bowden, J. Biosci. **27**, 121 (2002)
11. F. Weinhold, J. Chem. Phys. **63**, 2479 (1975)
12. F. Weinhold, J. Chem. Phys. **63**, 2488 (1975)
13. F. Weinhold, J. Chem. Phys. **63**, 2495 (1975)
14. G. Oster, A. Perelson, A. Katchalsky, Nature **234**, 393 (1971)
15. L. Onsager, Phys. Rev. **37**, 405 (1931)
16. L. Onsager, Phys. Rev. **38**, 2265 (1931)
17. L.D. Landau, E.M. Lifshitz, *Fluid Mechanics (Course of Theoretical Physics, Volume 6)*, 2nd edn. (Reed Educational and Professional Publishing Ltd, Oxford, 2000)
18. M. Paniconi, Y. Oono, Phys. Rev. E **55**, 176 (1997)
19. K. Sekimoto, Prog. Theor. Phys. Suppl. **130**, 17 (1998)
20. T. Hatano, S. Sasa, Phys. Rev. Lett. **86**, 3463 (2001)
21. S. Sasa, H. Tasaki, J. Stat. Phys. **125**, 125 (2006)
22. T. Komatsu, N. Nakagawa, Phys. Rev. Lett. **100**, 030601 (2008)
23. T. Komatsu, N. Nakagawa, S. Sasa, H. Tasaki, Phys. Rev. Lett. **100**, 230602 (2008)
24. M. Horodecki, J. Oppenheim, R. Horodecki, Phys. Rev. Lett. **89**, 240403 (2002)
25. Y.G. Sinai, Sov. Math. Dokl. **4**, 1818 (1963)
26. D. Ruelle, *Thermodynamic Formalism* (Addison-Wesley, Reading, 1978)
27. T.L. Hill, *Thermodynamics of Small Systems, Parts I and II* (Dover Publications Inc., New York (Original: W.A. Benjamin, Inc. New York, 1963, 1964), 1994)
28. L.F. Cugliandolo, J. Kurchan, L. Peliti, Phys. Rev. E **55**, 3898 (1997)
29. J.R. Taylor, *An Introduction to Error Analysis*, (Sect. 8.5) 2nd edn. (University Science Books, California, 1997)

Chapter 3
Fluctuations in Chemical Reactions

We will survey several concepts and notions related to chemical reactions at various scales and from various viewpoints. In the later chapters we will use them in the energetics of the fluctuating world.

The notion of a molecule belongs to the far from equilibrium states. Chemical reactions are then the transitions among different steady states of molecules, which often accompany the conversion of degrees of freedom between the translational degrees of freedom and the intramolecular ones.

In the survey of macroscopic reaction theory we limit ourselves to the case where the reactions are described in terms of the concentrations of the chemical components and the rate constants. Chemical equilibrium is related to the equilibrium thermodynamics. Inversely, macroscopic open systems can be described as reaction systems. Even within a macroscopic description of chemical reactions, the characteristic scale is a useful notion. We mention as examples the buffer solutions and the Michaelis–Menten kinetics.

At more microscopic scales chemical reactions are described in terms of the number of molecules (i.e., integers) of each chemical species. Stochastic processes of discrete systems are the general framework of chemical reactions on this scale. The master equation is the most often used description of discrete stochastic processes. The *detailed balance* condition is the equilibrium condition expressed by the probabilities of discrete states and the transition rates among these states. The discrete version of Langevin equation is an alternative and equivalent method to the master equation. The continuous Langevin equation (Chap. 1) can be represented as a limit of discrete processes. We take the same examples as those we have taken in the macroscopic description: $\mathrm{A} + \mathrm{B} \rightleftharpoons \mathrm{AB}$, the open system, and the Michaelis–Menten kinetics. It will be useful to find the similarities and differences between the two different scales of description.

3.1 * Background of Chemical Reactions

There are different mechanisms for chemical reactions: the transition through quantum tunneling, the quantal energy injection like photo activations, the molecular collisions induced by thermal motion or by forced molecular injection, the effect of

Sekimoto, K.: *Fluctuations in Chemical Reactions*. Lect. Notes Phys. **799**, 93–131 (2010)
DOI 10.1007/978-3-642-05411-2_3

intermediating substances like catalysts, thermal fluctuations causing large conformational changes in a protein, to mention a few. Each of these mechanisms could be best described on certain appropriate scales of space, time, or energy. Such appropriate scales may range over many orders of magnitude from one case to another. For example, quantum chemistry deals with the order of femtoseconds (10^{-12}s), while the isomerization (conformational change) of protein occurs on the order of milliseconds (10^{-3}s). In this book we limit ourselves to either macroscopic or mesoscopic descriptions where quantum mechanisms do not appear on the surface. We will begin by examining the notion of a molecule.

3.1.1 "Molecule" in Chemical Reaction Is a Nonequilibrium State with Internal Degrees of Freedom

3.1.1.1 "Molecule" as Nonequilibrium State

When one says there is a water molecule, it means there is a stable spatial aggregation of two hydrogen atoms and one oxygen atom. The notion of the macromolecule such as a polymer chain or a protein molecule implies that the aggregation of constituent atoms is maintained stably. Actually, however, no molecular state is stable forever. What we call molecule is usually a nonequilibrium state which is at best metastable or transiently stable. In other words, the (nonequilibrium) notion of the molecule is meaningful only when we are interested in timescale where the molecule maintains its identity.

The importance of the timescale is not particular to the notion of molecules: We can discuss the equilibrium thermodynamics of the glassy material if it has an extremely long though finite relaxation time. Even when we discuss the equilibrium state of a monoatomic ideal gas in a container, we assume implicitly that the container stays stable after infinitely long time when the equilibrium state of the gas is established. In other words, we limit our discussion to the timescale when the container remains in its nonequilibrium state. (This argument can be generalized: The notion of atom also requires the transient stability of the atom.)

In Boltzmann statistical mechanics, the statistical entropy of a thermodynamic state is the logarithm of the whole phase space volume which the system can visit. When the thermodynamic state is defined, the extent of the word "whole" (phase space) should be properly limited.[1] For example, we should specify which kinds of chemical reaction are admitted and which are not within a given timescale.

While such limitations of timescale do not cause problems usually, the exception is when the observation timescale and the relaxation timescale of the system reverse their relative magnitudes. For example, upon raising/lowering the temperature across the glass transition point of a material, or upon addition/depletion of the catalyst of a reaction, the ratio r of these two timescales can switch from

---

[1] See, Sect. 1.3.3.1.

$r \ll 1$ to $r \gg 1$ and vice versa. Such case will be discussed in detail later (Chap. 7). In this section we assume that the containers of the chemical reaction and the molecules are stable *except* through the chemical reactions which we describe explicitly.

3.1.1.2 State of Molecules

There is a characteristic timescale of environment at which a reactant molecule interacts critically with other reactant molecules or with the environment, through molecular collisions. Another timescale of the molecule is related to its internal dynamics such as vibrational and rotational motions. If the latter dynamics takes place sufficiently rapidly as compared to the former timescale, then we can justify the statistical description of each molecule, and it can be unambiguously represented by its chemical formula or molecular formula (like H_2O).

By contrast, if the time evolution of the internal state of a molecule is slower than the timescale of reactive molecular interactions, we need more parameters to characterize the internal states of the molecule. For example, the flexible polymeric chains or proteins, the deformable tethered membrane, or the soft network of gel, etc., require the conformational parameters other than their chemical formula to correctly describe the chemical reactions.

3.1.1.3 Molecular Reaction Viewed from the Degrees of Freedom

"Molecular reaction" replaces a group of transiently stable aggregated states of atoms by a new group of aggregated states. By taking the reaction, $A + B \to AB$ as an example, we can characterize the change of molecular states in two different ways.

Transfer of degrees of freedom: The translational degrees of freedom decrease from six to three. The three translational degrees of freedom are compactified. The molecule AB then possesses three internal degrees of freedom, one for vibration and two for rotation.

Correlation among translational degrees of freedom: Among the six independent translational degrees of freedom of monoatomic molecules, A and B, three of them become spatially correlated in the molecule AB due to the binding potential between the atoms. If the reaction takes place in a gas chamber or in water, this forced correlation reduces the gas pressure or the osmotic pressure, respectively.

The consequences of the forced correlation or the compaction of translational degrees of freedom are not limited to the entropic effect but there is also an energetic effect: monovalent cations could make dimers if the electrostatic intradimer repulsion is stabilized by a molecular binding force. Since the divalent cations feel less repulsion among each other than do the monovalent cations of the same charge density, the divalent cation dimers form a more dense counterionic cloud around an anion than monovalent cations do. This is a qualitative way to understand why the overcharging around a charged colloid is more effective by multivalent counterions than the monovalent ones [1].

The stability of molecular states is due to the high free-energy barrier (relative to $k_B T$) associated with the destabilizing reactions. Since the reaction rate is mainly governed by the exponential Boltzmann factor, the reactivity is efficiently changed if the free-energy barrier is modified. This exponential dependence enables the near discrete switching on/off of reactions in inorganic and organic matter. The energetic aspect of controlling the barrier height will be discussed later (Chap. 7).

On the molecular level, the description of chemical reactions as well as open system has to incorporate the transfer of degrees of freedom. Concomitantly the energetics of chemical reaction and open system on the fluctuation level must take into account the change in degrees of freedom (see Sect. 4.2.3).

3.2 Macroscopic Description

We survey below briefly the framework of the macroscopic reaction theory of dilute solutions, using again the example of the reaction, $A + B \to AB$. The solution is assumed to be spatially homogeneous.

3.2.1 * Law of Mass Action Relates the Rate Constants of Reaction to the Canonical Equilibrium Parameters

3.2.1.1 Rate Equation and Rate Constant

For dilute solutions of molecules having low molecular weight, the speed of reaction depends only on the concentrations of the chemical components and the kinetic parameters called the (reaction) *rate constants*. We will represent by [A], [B], and [AB] the concentration of the A, B, and AB molecules, respectively. The rate of production of the AB molecules by the (irreversible) reaction, $A + B \to AB$, is

$$\frac{d[\mathrm{AB}]}{dt} = k[\mathrm{A}][\mathrm{B}], \tag{3.1}$$

where the parameter k is the rate constant of this reaction. The macroscopic equation for the rate of production, like (3.1), is called a rate equation. $k[\mathrm{A}][\mathrm{B}]$ indicates the frequency of the "collisions," or encounter, between an A molecule and a B molecule. The dimension and magnitude of k depend on the choice of the unit of concentration, [A], etc. In physical chemistry one usually uses the molar concentration, or molarity, [no. of mols of solute]/[no. of liters of solution].[2] In this book, however, we will adopt the following unit, unless stated otherwise explicitly:

[2] The use of mol in the definition is reasonable in the sense that macroscopic observers cannot count the number of solute molecules.

$$[\mathrm{X}] \equiv \frac{(\text{no. of X molecules})}{(\text{volume in cm}^3\text{ of solution})}. \tag{3.2}$$

We choose this unit with a view to maintain continuity with the more microscopic level discussions in later sections.

If we also take into account the reverse reaction, i.e., $\mathrm{A}+\mathrm{B} \rightleftharpoons \mathrm{AB}$, we introduce another rate constant, k', which has different dimensionality from that of k:

$$\frac{d[\mathrm{AB}]}{dt} = k[\mathrm{A}][\mathrm{B}] - k'[\mathrm{AB}]. \tag{3.3}$$

The second term on the right-hand side, $-k'[\mathrm{AB}]$, implies that an isolated AB molecule has an average inverse lifetime, k'. The law (3.3) is valid in the limit of (i) dilute solution and (ii) near equilibrium. We will discuss a more general case of nondilute solution in Sect. 3.3.3.

3.2.1.2 Chemical Equilibrium

A reaction system is said to be closed (with respect to the chemical reactions) if (i) there is no exchange of molecules A, B, or AB between the system and the outside and also (ii) there is no mechanism to change the rate constants.[3] If the above system, $\mathrm{A}+\mathrm{B} \rightleftharpoons \mathrm{AB}$, is closed, the change of [A] and of [B] are given by[4]

$$d[\mathrm{A}] = d[\mathrm{B}] = -d[\mathrm{AB}]. \tag{3.4}$$

Application of the zeroth law of thermodynamics to the whole system, the closed reaction system *plus* the thermal environment, implies that this isolated system will reach the thermal equilibrium after infinitely long time. The steady state of (3.3), i.e.,

$$\frac{d[\mathrm{AB}]}{dt} = \frac{d[\mathrm{A}]}{dt} = \frac{d[\mathrm{B}]}{dt} = 0 \tag{3.5}$$

then gives the chemical equilibrium and satisfies

$$\frac{[\mathrm{AB}]}{[\mathrm{A}][\mathrm{B}]} = \frac{k}{k'} \qquad \text{(equilibrium).} \tag{3.6}$$

This type of equilibrium condition, where the powers of concentration appear in the denominator and numerator, is called the *law of mass action*. The ratio of the kinetic parameters on the right-hand side is called the *equilibrium constant*, which depends only on the thermodynamic parameters, as we will see below.

---

[3] For example, the sedimentation of [AB] would decrease k'.

[4] We omit the dt in the denominator.

If the reaction is not closed, for example, under the steady injection of A and B and the removal of AB, the stationary condition, (3.5), does not give equilibrium.

3.2.1.3 Chemical Equilibrium and Macroscopic Thermodynamics

If the law of mass action characterizes a thermal equilibrium, the relation (3.6) should be equivalent to the equilibrium condition of macroscopic thermodynamics. In the latter framework, we require the minimum of the Gibbs free energy $G(T, p, N_{\mathrm{A}}, N_{\mathrm{B}}, N_{\mathrm{AB}})$ with respect to N_{AB}.[5] This minimization yields

$$\mu_{\mathrm{A}} + \mu_{\mathrm{B}} = \mu_{\mathrm{AB}}. \tag{3.7}$$

On the other hand, the chemical potential of the solute molecules in a dilute solution is

$$\mu_{\mathrm{A}} = \mu_{\mathrm{A}}^0 + k_{\mathrm{B}}T \ln[\mathrm{A}], \qquad \text{etc.} \tag{3.8}$$

with μ_{A}^0 being a constant. Therefore, the equilibrium condition (3.7) is $[\mathrm{AB}]/([\mathrm{A}][\mathrm{B}]) = \exp[-(\mu_{\mathrm{AB}}^0 - \mu_{\mathrm{A}}^0 - \mu_{\mathrm{B}}^0)/k_{\mathrm{B}}T]$. Identifying this expression with (3.6), we reach the following relation between the rate constants and the thermodynamic parameters:

$$\frac{k}{k'} = \exp\left[\frac{\mu_{\mathrm{A}}^0 + \mu_{\mathrm{B}}^0 - \mu_{\mathrm{AB}}^0}{k_{\mathrm{B}}T}\right]. \tag{3.9}$$

3.2.1.4 Equilibrium with a Particle Reservoir – Open System

When the reaction system, $\mathrm{A} + \mathrm{B} \rightleftharpoons \mathrm{AB}$, allows for the exchange of the AB molecules with its particle reservoir, the rate equations are

$$\begin{aligned} \frac{d[\mathrm{AB}]}{dt} &= k[\mathrm{A}][\mathrm{B}] - k'[\mathrm{AB}] - k_{\mathrm{out}}[\mathrm{AB}] + k_{\mathrm{in}}, \\ \frac{d[\mathrm{A}]}{dt} &= -k[\mathrm{A}][\mathrm{B}] + k'[\mathrm{AB}] = \frac{d[\mathrm{B}]}{dt}. \end{aligned} \tag{3.10}$$

Here the exchange with the particle reservoir of [AB] is characterized by the two rate constants, $-k_{\mathrm{out}}$ and k_{in}. The stationary condition, (3.5), yields the following two independent relations as the law of mass action:[6]

---

[5] Note that $N_{\mathrm{A}} = [\mathrm{A}]V$ etc., where V is the volume of the system. Under the isothermal and isobaric condition, the second law requires $\Delta(E - TS + PV) = \Delta G \le W = 0$ for any spontaneous changes. dN_{A} and dN_{B} are dependent on dN_{AB} according to (3.4).

[6] If the particle reservoir exchanges (only or also) A molecules, the equation for $d[B]/dt$ still assures the law of mass action. If the system exchanges all the species of molecules, the law of mass action will no more hold.

$$\frac{[A][B]}{[AB]} = \frac{k'}{k}, \quad [AB] = \frac{k_{\text{in}}}{k_{\text{out}}} \quad \text{(equilibrium)}. \tag{3.11}$$

At the equilibrium, the total Gibbs free-energy, $G(T, p, N_{\rm A}, N_{\rm B}, N_{\rm AB}) + G_{\rm res}(T, p, N_{\rm AB,res})$, should be minimized under the constraints of particle conservation, $dN_{\rm A} + dN_{\rm AB} + dN_{\rm AB,res} = 0$ and $dN_{\rm B} + dN_{\rm AB} + dN_{\rm AB,res} = 0$. This yields[7]

$$\mu_{\rm A} + \mu_{\rm B} = \mu_{\rm AB} = \mu_{\rm AB,res}, \tag{3.12}$$

where $\mu_{\rm AB,res}$ is the chemical potential of the particle environment of AB molecules. With (3.11) we have the following relations:

$$\frac{k}{k'} = \exp\left[\frac{\mu^0_{\rm A} + \mu^0_{\rm B} - \mu^0_{\rm AB}}{k_{\rm B}T}\right], \quad \frac{k_{\text{in}}}{k_{\text{out}}} = \exp\left[\frac{\mu_{\rm AB,res} - \mu^0_{\rm AB}}{k_{\rm B}T}\right]. \tag{3.13}$$

Remarks

The chemical reaction theory emphasizes the transformation of the mass, but the energetic aspects (change of free energy, endothermic, or exothermic, etc.) are treated separately using thermodynamics. If we use stochastic energetics, both mass transformation and energetics are discussed on the basis of a single event of the reaction.

The chemical potential is a quantity on the level of the description where the particles are anonymous. The chemical potential (e.g., (3.8)) does not represent the free energy carried by an individual molecule. $\mu_{\rm A}$ is the energetic interpretation of the relative probability of the arrival of anonymous molecule A, $\propto [A]e^{\mu^0_{\rm A}/k_{\rm B}T}$, using the form of the Boltzmann factor, $e^{\mu_{\rm A}/k_{\rm B}T}$.

3.2.2 * *Large Separation of the Rate Constants Causes Different Regimes of Reaction and Rate-Limiting Processes*

It often occurs that two or more reactions related to the same molecular species have widely different equilibrium constants. That the equilibrium constants depend exponentially on the chemical parameters, like $\mu^0_{\rm A}$ in (3.9) and (3.13), explain this.

Because of this aspect, we expect that a chemical reaction system can exhibit qualitatively different behaviors in different regimes. The following two simple examples demonstrate how the phenomena of different scales are treated.

[7] We may use the method of *Lagrange multiplier*: The requirement of $d[G(x, y, z) + G_{\rm res}(\tilde{z})] = 0$ under the constraints, $d(x + z + \tilde{z}) = 0$ and $d(y + z + \tilde{z}) = 0$, is equivalent to the constraint-free requirement, $d[G(x, y, z) + G_{\rm res}(\tilde{z}) - \lambda_x(x + z + \tilde{z}) - \lambda_y(y + z + \tilde{z})] = 0$, with λ_x and λ_y being unknown, called the *Lagrange multipliers*. The parameter set, $(x, y, z, \tilde{z})$, stands for $(N_{\rm A}, N_{\rm B}, N_{\rm AB}, N_{\rm AB,res})$. Then $\mu_{\rm A} = \lambda_{\rm x}$, $\mu_{\rm B} = \lambda_{\rm y}$, $\mu_{\rm AB} = \lambda_{\rm x} + \lambda_{\rm y}$, and $\mu_{\rm res} = \lambda_{\rm x} + \lambda_{\rm y}$.

3.2.2.1 Titration and Buffer Solution

The object is to study the titration and the buffer solution of acid–base system starting from a unified framework. The solution (1ℓ in total) is prepared from a_0 mol of acid, HA, b_0 mol of a base, BOH, and pure water. The base is assumed to be strong base, so that it 100% dissociates ($BOH \to B^+ + OH^-$) into B^+ (b_0 mol) and OH^- (b_0 mol). We study how the amount of the hydronium ion $h = [H_3O^+]$ or equivalently $pH \equiv -\log_{10} h$ is related to the initial amount of the base b_0. This relation results from the two mechanisms: the water dissociation equilibrium, $[H_3O^+][OH^-] = K_w$,[8] or

$$h\,(b_0 + h - y) = K_w, \tag{3.14}$$

and the acid dissociation equilibrium, $[H_3O^+][A^-]/[HA] = K_a$, or

$$\frac{hy}{a_0 - y} = K_a, \tag{3.15}$$

where $y = [A^-] = a_0 - [HA]$, and K_w and K_a are the equilibrium constants of the water and the acid, respectively, at a given temperature. Elimination of y from (3.14) and (3.15) yields

$$b_0 - a_0 = \frac{K_w}{h} - C(h, a_0, K_a)\,h, \tag{3.16}$$

where $C(h, a_0, K_a) = 1 + \frac{a_0}{K_a + h}$. Equation (3.16) is called the Charlot equation. If the acid HA is also a strong acid ($K_a \gg a_0$) which dissociates 100% into ions through $HA + H_2O \to H_3O^+ + A^-$, then $C_\infty \equiv \lim_{K_a \to \infty} C(h, a_0, K_a) = 1$. In this case the solution behaves, $a_0 - b_0 \simeq h$ for $h \gg \sqrt{K_w}$ and $b_0 - a_0 \simeq K_w/h$ for $h \ll \sqrt{K_w}$, with very narrow range of crossover, $\Delta|b_0 - a_0| \sim \sqrt{K_w}$, around the equivalence point, $b_0 = a_0$ with $h = \sqrt{K_w}$.

If the acid HA is a weak acid (not very large K_a), but if its dissociation, $HA + H_2O \to H_3O^+ - A^-$, is much stronger than the dissociation of water ($2H_2O \to H_3O^+ + OH^-$), there can arise the situation where the nondimensional parameters, $X \equiv K_w/a_0^2$ and $Y \equiv K_a/a_0$ satisfy $X \ll Y \ll 1$. Figure. 3.1 shows an example of pH vs. $b_0 - a_0$ for $a_0 = 1$, $K_w = 10^{-14}$ and $K_a = 10^{-5}$. In this case, one regime $h/K_a \ll 1$ is again the crossover between $a_0 - b_0 \simeq h/Y$ for $h \gg \sqrt{K_w K_a/a_0}$ and $b_0 - a_0 \simeq \frac{K_w}{h}$ for $(0 <) h \ll \sqrt{K_w K_a/a_0}$, across the equivalence point $h = \sqrt{K_w K_a/a_0}$ (i.e., $h/a_0 = \sqrt{XY}$). At the equivalence point, the pH is shifted toward basic by $-\frac{1}{2}\log_{10} Y$. The consistency of the assumption $h/K_a \ll 1$ is verified since $\frac{h}{K_a} = \left\{\frac{X}{Y}\right\}^{1/2} \ll 1$ at the equivalence point.

[8] In molar unit, what would be the denominator, $[H_2O] = 55.5$, is roughly of order 1 and conventionally suppressed.

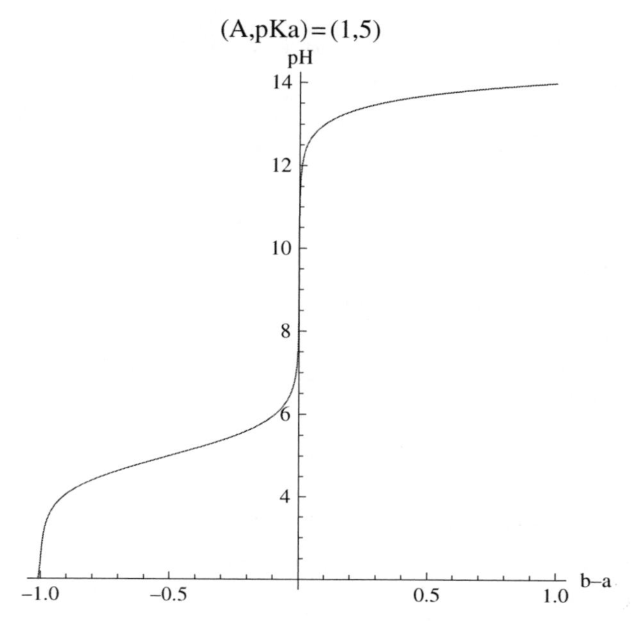

Fig. 3.1 pH vs. $b_0 - a_0$ for $a_0 = 1$, $K_w = 10^{-14}$ and $pK_a = 5$, i.e., $K_a = 10^{-5}$

The new regime, called the *buffer regime*, is $h/K_a \equiv 10^{-\mathrm{pH}+\mathrm{p}K_a} \sim 1$ in (3.16). If we assume $h \ll a_0$ and $\frac{K_w}{h} \ll a_0$ in (3.16), h/K_a is written as $\frac{h}{K_a} \simeq \frac{b_0-a_0}{b_0}$. It implies that, as far as $\frac{b_0-a_0}{b_0} \sim 1$, the pH of the solution is kept at around pH $\simeq$ p$K_a \equiv -\log_{10} K_a$. The above assumptions are consistent since $\frac{h}{a_0} \sim \frac{K_a}{a_0} = Y \ll 1$ and $\frac{K_w}{ha_0} \sim \frac{K_w}{K_a a_0} = \frac{X}{Y} \ll 1$. The buffering regime is, therefore, realized due to the double inequalities, $X \ll Y \ll 1$.

3.2.2.2 Michaelis–Menten Kinetics

Michaelis–Menten kinetics is one of the fundamental reaction schemes in biochemistry, because it describes a catalytic (enzyme) reaction and it also applies to many practical situations. We describe this kinetic scheme below and discuss the generality of this kinetics from the viewpoint of characteristic scales or the rate-limiting processes.

Reaction with a Catalyst

A catalyst is a chemical substance which is not consumed in the reaction but increases the rate constant of the reaction in both the forward and the backward directions by the same factor. The catalyst, therefore, does not change its equilibrium constant defined in Sect. 3.2.1.4. In biology, catalytic proteins are called the enzymes.

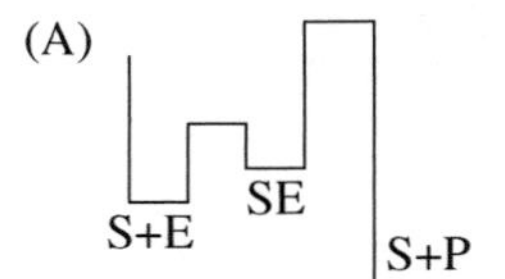

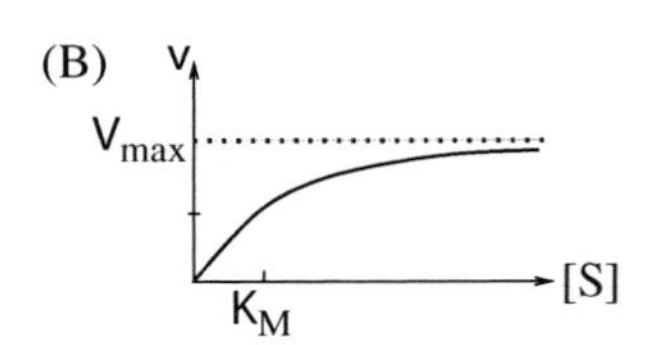

Fig. 3.2 (**A**) Schema of Michaelis–Menten reaction. (**B**) Rate of production v as function of substrate concentration, [S]. See (3.21)

A simple 1:1 reaction between a Substrate (S) molecule and a Product (P) molecule catalyzed by an Enzyme (E) is $\mathrm{E} + \mathrm{S} \rightleftharpoons \mathrm{E} + \mathrm{P}$. When we are interested in the dependence on the enzyme concentration, [E], we do not simplify this scheme as $\mathrm{S} \rightleftharpoons \mathrm{P}$. One of the representative schemas of enzymic reaction is the Michaelis–Menten kinetics which has been introduced around 1913. The schema of Michaelis–Menten kinetics is

$$\mathrm{E} + \mathrm{S} \rightleftharpoons \mathrm{ES} \rightarrow \mathrm{E} + \mathrm{P}. \tag{3.17}$$

In this schema we assume the following circumstance, see Fig. 3.2 (A):

1. The direct reaction, $\mathrm{S} \rightleftharpoons \mathrm{P}$, is slow enough to be ignored.
2. There is a transition state called Enzyme–Substrate complex (ES).
3. The total concentration of the enzyme, $[\mathrm{E}]^{\mathrm{tot}} = [\mathrm{E}] + [\mathrm{ES}]$, is finite.
4. At most one substrate particle, S, can interact at any time with an enzyme particle (protein).
5. [P] is much smaller than the equilibrium value, so that the backward reaction, $\mathrm{ES} \leftarrow \mathrm{E} + \mathrm{P}$, can be neglected.

These hypotheses might look very particular among many other possibilities. Nevertheless we will argue later that the above schema (3.17) is rather general from the viewpoint of the timescales of reaction.

Michaelis–Menten Equation

Let us denote by k_{cat} the rate constant of the forward reaction, $\mathrm{ES} \rightarrow \mathrm{E} + \mathrm{P}$. See Fig. 3.3 (*left*). We seek the rate of production,

$$\frac{d[\mathrm{P}]}{dt} = k_{\mathrm{cat}}[\mathrm{ES}], \tag{3.18}$$

when the preceding step of complex formation has reached the steady state: [9]

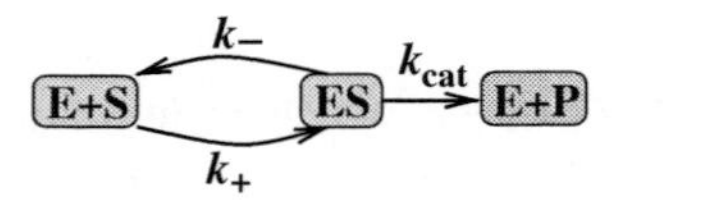

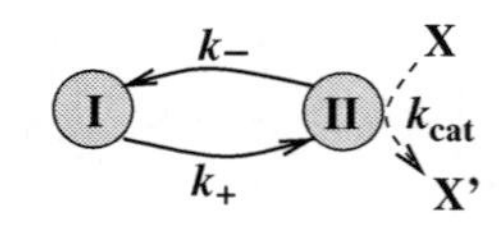

Fig. 3.3 Michaelis–Menten reaction (*left*) and its generalized form (*right*). The change $X \rightarrow X'$ requires the transition $\mathrm{I} \rightarrow \mathrm{II}$

[9] This condition is often attributed to Briggs and Haldene.

$$\frac{d[\mathrm{ES}]}{dt} = k_+[\mathrm{E}][\mathrm{S}] - (k_- + k_{\mathrm{cat}})[\mathrm{ES}] = 0. \tag{3.19}$$

This equation leads to the relation[10]

$$\frac{[\mathrm{E}][\mathrm{S}]}{[\mathrm{ES}]} = \frac{k_- + k_{\mathrm{cat}}}{k_+}. \tag{3.20}$$

From this equation, we eliminate [E] using $[\mathrm{E}] = [\mathrm{E}]^{\mathrm{tot}} - [\mathrm{ES}]$ and then we substitute the resulting [ES] as function of [S] into (3.18). We then obtain the desired result, which is called the Michaelis–Menten equation (See Fig. 3.2 (B)),

$$v \equiv k_{\mathrm{cat}}[\mathrm{ES}] = \frac{V_{\mathrm{max}}[\mathrm{S}]}{K_{\mathrm{M}} + [\mathrm{S}]}, \tag{3.21}$$

where we have introduced the maximal production rate, V_{max}, for $[\mathrm{S}] = +\infty$, and the so-called Michaelis–Menten constant, K_{M}:

$$V_{\mathrm{max}} \equiv k_{\mathrm{cat}}[\mathrm{E}]^{\mathrm{tot}}, \quad K_{\mathrm{M}} \equiv \frac{k_- + k_{\mathrm{cat}}}{k_+}. \tag{3.22}$$

When $[\mathrm{S}] = K_{\mathrm{M}}$, we have $v = \frac{1}{2}V_{\mathrm{max}}$. To estimate these parameters from experimental data of v and [S], the sigmoidal curve, (3.21), is often replotted as the linear plot between v^{-1} and $[\mathrm{S}]^{-1}$ or between $v^{-1}[\mathrm{S}]$ and [S].[11]

As a model of chemical reaction, the Michaelis–Menten scheme is related to the transient state [Eyring's] theory (see, for example, [2] Chap. 28) and also to Kramer's theory [3]. The latter refined the notion of the transition state as a group of states between which the molecule diffuses. The identification of the transient state(s) in complex reactions is among contemporary topics [4].

Generality of the Michaelis–Menten Kinetics

When many enzymatic reactions are mutually linked and form a network (e.g., the metabolic network), there can be several key reactions – the rate-limiting reactions – which determines the global rate of the reactions. In order for a reaction to be the key reaction, there must not be other important bypassing or substituting reactions. Moreover, it is necessary that (i) the activation energy barrier of the reaction is high (even in the presence of enzyme) and/or (ii) the number of enzymes is limited and many substrate molecules are waiting for the unbound enzyme. The substrate S for this key reaction is a product of the preceding (upstream) reaction(s),

---

[10] Although (3.20) takes of the form of the law of mass action, the system is not in equilibrium. The exception is when $k_{\mathrm{cat}} \ll k_-$ holds. The approximation of neglecting k_{cat} in (3.19) is called the rapid-equilibrium assumption.

[11] The latter representation is called the Hanes–Woolf plot.

and the product P for this key reaction is a substrate of the following (downstream) reaction(s). Thus, if the product of the upstream reaction is supplied faster than it is converted into P, then the substrate S stays among E + S and ES states, and the quasi-equilibrium (rapid equilibrium), E+S $\rightleftharpoons$ ES, is realized. Also, if the substrate of the downstream reaction is consumed faster than it is converted from S, then the product P has little probability to be bound to E to form ES, and the unidirectionally reaction, ES $\to$ E+P, is realized. These two features are what constitute the schema of Michaelis–Menten kinetics. Therefore, the Michaelis–Menten schema describes a general feature of the key reactions. The formula (3.21) shows that the key reaction is *controlled* by the total number of the enzyme, $[E]_{tot}$.[12]

If a reaction has the substrate concentration, [S], much smaller than the K_M, then this reaction cannot be controlled by $[E]_{tot}$, and the reaction is not the key reaction. If a lot of enzyme $[E]_{tot}$ is injected in a key reaction, then there would be a shortage of the substrate and the process is no more controlled by this reaction, i.e., the reaction ceases to be the key reaction. In brief, the Michaelis–Menten kinetics works when and where the reaction in question is among the key reactions in the network. In biochemical reaction network the change of the *activity* of enzymes[13] may change dynamically the locations of the key reactions.

The notion of the key reaction described above is not limited to the bulk enzymatic reaction in solution. The production rate of the form of (3.21) is found in other conditions: for the reaction with a surface catalyst the rate of production obeys the form of (3.21). If the concentration [S] is higher than K_M, then most catalyst molecules are occupied, and it limits the production rate. For crystal growth from vapor or from solution competition between bulk diffusion and surface kinetics leads to the growth velocity in the form of (3.21).

There is another interpretation of the Michaelis–Menten equation, which also explains why Michaelis–Menten-type behavior is found in a variety of situations. See Fig. 3.3 (right). If we look back to the derivation of (3.21), we find that the mathematical origin of its saturating feature is found simply in the bistable transition between the "states," E+S and ES. See Fig. 3.2. Wherever this (quasi) equilibrium to-and-fro exists, any observable linearly related to the occupied fraction of the ES state should show Michaelis–Menten type saturation.[14] For example, if S and E are, respectively, the ATP-hydrolyzing motor protein and its filament, then the ATP consumption rate should obey the curve of (3.21) since ATP hydrolysis is catalyzed by the motor–filament interaction. In this case the products, ADP and inorganic phosphate (Pi), are not fragments of a motor protein.

[12] Describing the dynamics of a reaction network by focusing on the rate-limiting reactions is similar in idea to what is done in the statistical dynamics, that is, reducing dynamical variables by focusing on the slowly varying ones.

[13] i.e., The ability to function as enzyme.

[14] The fraction of the occupied enzyme, $[ES]/[E]_{tot}$, is $[S]/(K_M + [S])$, which saturates for $[S] \gg K_M$.

3.3 Stochastic Description

3.3.1 Stochastic Transitions Among Discrete States Are Described by Master Equation or Discrete Langevin Equation

Rate equations are not the fundamental equations but can be derived from more microscopic levels. In order to address the chemical reaction on more microscopic scales, we survey the framework of stochastic processes over discrete states. The master equation describes the evolution of instantaneous probabilities, as the Fokker–Planck equation does for continuous states. As the latter was derived from the Langevin equation, the master equation can be derived from the discrete Langevin equation. Gillespie algorithm is an efficient method to generate discrete stochastic processes.

We specify the class of master equations which allow *detailed balance (DB)* among the discrete states. For this class of systems the global steady state is the canonical equilibrium. The transition rates and the equilibrium probability can then be represented in terms of a potential (free-energy) landscape.

Stochastic description of chemical reactions usually uses the master equation, where the discrete states are distinguished by the number of molecules of each chemical species. The (continuous) Langevin equation can be regarded as a limit of discrete stochastic process. The condition of the detailed balance is related to the Einstein's relation.

3.3.1.1 * Basic Concepts

Discrete States

Unlike the quantum level description, where the microscopic states of a finite system are essentially discretized, we refer in this book the approximately discretized groups of continuous states as the discrete states: Suppose that the state of a system undergoes, for most of the time, small fluctuations around one of the discrete representative states and undergoes, only occasionally, rapid jumps from around a representative state to the domain of other representative state. In such cases, we can simplify the description of the evolution of the system by the use of coarse graining both in state and time. For example, if a conformational change of a protein is described as $\mathrm{S} \rightleftharpoons \mathrm{P}$, it implies that many substates of the protein are represented by either S or P and that the substates joining these two groups of substates are ignored.[15]

Hereafter we represent by S_j, etc., the discrete states of a system (obtained by the above mentioned coarse graining), where the suffix j takes, for example, an integer number or a set of integer numbers.

[15] One can imagine S=coiled state and P=globular state.

State Transition and Markov Approximation

Temporal change among discrete states is called state transition. We will regard transitions as instantaneous events. That is, we ignore the time lapse of each transition. Such description presupposes that (1) in most cases the system stays in the same state for much longer time than the time of transition and (2) the time resolution of the description is coarser than the time of transition.

We further suppose that the consecutive state transitions are Markovian, that is, the statistical characteristic of the transition from the present state S_j is independent of the previous transition to this state, see Sect. 1.2.1.4. Such description presupposes that in most cases the system stays in the same state for the time long enough so that the intrastate fluctuations erase the memory of the system's previous state.

Transition Rate

We denote by $I_i = \{j\}_i$ the set of indices of the states $\{S_j\}$ that the state S_i of the system can make direct transitions.[16] The so-called transition rate, $w_{i \to j}$, from the discrete state S_i to the discrete state S_j ($j \in I_i$) is defined as follows:

- *Suppose that a system is in the state S_i at a time t.*
- *The (conditional) probability that the system makes the transition to a different state S_j during an infinitesimal time lapse, $dt(> 0)$, is $w_{i \to j}dt$.*

As a result, the probability that the system remains in the same state S_i at $t + dt$ is $(1 - \sum_{j \in I_i} w_{i \to j} dt)$.

We recall the first passage time, introduced in Sect. 1.3.3.3: The transition from the state i to the other state can correspond to the exit from a basin Ω of a potential energy $U(x)$. This was the idea of Kramers to calculate the transition rate [3].[17] The average first passage time (FPT) is related to $(\sum_{j \in I_i} w_{i \to j})^{-1}$. To assess the individual transition rate, $w_{i \to j}$, we will need to solve the first passage problem under constraints. In the context of FPT, the Markov approximation amounts to the neglect of the initial position dependence of the first passage time.

3.3.1.2 * Statistical Approach – Master Equation

Probability Flux

The transition rate $w_{i \to j}$ characterizes the redistribution of the (conditional) probability from the state S_i to S_j per unit of time. We can generalize this notion to the case where we find a system in the state S_i with the probability of P_i at present time. Then the redistribution of the probability through all the possible transitions during infinitesimal time, $dt(> 0)$, constitutes a network of fluxes of probabilities. Between

---

[16] By definition, $i \notin I_i$.

[17] A review [5] surveys many papers after [3] up to 1990.

an arbitrary pair of states, S_i and S_j, there is the flow of probability $P_j w_{i\to j}$ from S_i to S_j and $P_i w_{j\to i}$ from S_j to S_i. We call the net flow of the probability per unit of time the (net) probability flux and denote it by $J_{i\to j}(= -J_{j\to i})$:

$$J_{i\to j} = P_i w_{i\to j} - P_j w_{j\to i} = -J_{j\to i}. \tag{3.23}$$

For later convenience, we complement the definitions; $J_{i\to j} \equiv 0$ $(\forall j \notin I_i)$, especially, $J_{i\to i} \equiv 0$.

For a finite time interval, Δt, the redistribution of the probabilities from S_i to S_j corresponds to many different types of state transitions, such as $S_i \to S_k \to S_j$ or $S_i \to S_j \to S_i \to S_j$, in addition to the direct one. In the limit of $\Delta t \to 0$, however, all the indirect transitions have probabilities of higher order of Δt, and the redistribution of the probability is described only by the direct transitions, $J_{i\to j}$.

Master Equation

Let us denote by $P_i(t)$ the probability to find the system at time t in the state S_i. Given the concept of the probability flux (3.23), the redistributed probabilities $\{P_i(t+dt)\}$ should satisfy

$$P_i(t+dt) = P_i(t) - \sum_j J_{i\to j}\, dt.$$

Then we have the following evolution equation for $\{P_i(t)\}$ called the *master equation*:

$$\frac{dP_i(t)}{dt} = -\sum_j J_{i\to j}, \tag{3.24}$$

where the sum runs for all states, S_j.[18] Because of the identity, $J_{j\to i} = -J_{i\to j}$, the total probability, $\sum_i P_i(t)$, is conserved: $\frac{d}{dt}\sum_i P_i(t) = 0$. Different approximation methods to deal with the master equation are found in, for example, the textbooks [6, 7] or monographs [8, 9].

Steady State

The steady state (or the stationary state) of the master equation is defined such that $dP_j(t)/dt = 0$ for every state, S_j. The steady state does *not* imply the flux-free state: $J_{i\to j} = 0$. A simplest model may be the three-state system with the transition rates being $w_{1\to 2} = w_{2\to 3} = w_{3\to 1} \equiv w > 0$ and $w_{2\to 1} = w_{3\to 2} = w_{1\to 3} \equiv w' > 0$

[18] The expression (3.24) has the same structure as the equation of continuity or the mass conservation, where the sum on the right-hand side is the divergence of the flux.

($w \neq w'$). The steady state of this model is $P_1 = P_2 = P_3 = 1/3$ but the probability flux is nonzero; $J_{1\to 2} = J_{2\to 3} = J_{3\to 1} = (w - w')/3$.

Convergence to a Steady State

It is often observed the situation where the evolution of the probabilities $\{P_j(t)\}$ is convergent to a nonequilibrium steady state, like the above simple example. The convergence to an equilibrium state has long been understood using variational inequality about the entropy, $-\sum_j P_j \ln P_j$, under appropriate constraints (on energy, volume, etc.) [10]. Recently, variational inequality has been developed also for the nonequilibrium steady states. It is the minimization of the so-called the Kullback–Leibler distance or the relative entropy (see, for example, [11]). For two sets of normalized probabilities, $\boldsymbol{P} \equiv \{P_j\}$ and $\boldsymbol{Q} \equiv \{Q_j\}$ ($j = 1, \dots, n$), we denote by $D(\boldsymbol{P} \parallel \boldsymbol{Q})$ the Kullback–Leibler distance of $\boldsymbol{P}$ relative to $\boldsymbol{Q}$ and define as follows:

$$D(\boldsymbol{P} \parallel \boldsymbol{Q}) \equiv \sum_{i=1}^{n} P_i \ln \frac{P_i}{Q_i} (\geq 0). \tag{3.25}$$

The continuous version of this quantity has been introduced in Sect. 1.2.3.2, where the nonnegativity of this quantity has been shown.[19] When the time-discretized probability $\boldsymbol{P}$ converges to the steady-state probability $\boldsymbol{Q}$, the $D(\boldsymbol{P} \parallel \boldsymbol{Q})$ monotonically decreases to 0. Below is a brief derivation. The technical details are given in Appendix A.3.1.

1. If the time and the states are discretized, the evolution of the probability $\boldsymbol{P}$ by the master equation can be formally written as a *Markov chain*, i.e., the discrete-time discrete state Markov process,[20]

$$\boldsymbol{P} \equiv \{P_i\} \mapsto \mathsf{K}\boldsymbol{P} \equiv \{\sum_{j=1}^{n} K_{ij} P_j\} \quad (i = 1, \dots, n), \tag{3.26}$$

   where the $n \times n$ matrix $\mathsf{K} \equiv \{K_{ij}\}$ ($i, j = 1, \dots, n$) has nonnegative components, $K_{ij} \geq 0$, and satisfies the sum rule, $\sum_{i=1}^{n} K_{ij} = 1$ for $j = 1, \dots, n$. In order to recover the continuous time version, we identify $\boldsymbol{P}(t + dt) = \mathsf{K}\boldsymbol{P}(t)$, where $\boldsymbol{P}(t + dt) = \{P_i(t + dt)\}$ ($i = 1, \dots, n$) and $\mathsf{K} = 1 + \mathcal{O}(dt)$.
2. With this Markov chain, the following inequality holds:

$$D(\boldsymbol{P} \parallel \boldsymbol{Q}) \geq D(\mathsf{K}\boldsymbol{P} \parallel \mathsf{K}\boldsymbol{Q}). \tag{3.27}$$

[19] Despite the word "distance," this quantity is not symmetric; $D(\boldsymbol{P} \parallel \boldsymbol{Q}) \not\equiv D(\boldsymbol{Q} \parallel \boldsymbol{P})$.

[20] For the Markov process, see Sect. 1.2.1.4.

The proof is given in Appendix A.3.1.[21]

3. Suppose that $\boldsymbol{Q}$ is a steady state of this Markov chain, $\mathsf{K}\boldsymbol{Q} = \boldsymbol{Q}$. Then (3.27) implies

$$D(\boldsymbol{P} \parallel \boldsymbol{Q}) \geq D(\mathsf{K}\boldsymbol{P} \parallel \boldsymbol{Q}) \qquad (\mathsf{K}\boldsymbol{Q} = \boldsymbol{Q}). \tag{3.28}$$

4. If the Markov chain, $\{\boldsymbol{P}, \mathsf{K}\boldsymbol{P}, \mathsf{K}^2\boldsymbol{P}, \ldots\}$, converges to this steady state $\boldsymbol{Q}$, then $D(\mathsf{K}^m\boldsymbol{P} \parallel \boldsymbol{Q})$ must decrease monotonically to 0.

The inequality (3.27) can be interpreted in the context of Stein's lemma [11], which states the following: Given n events obeying the i.i.d. probability $\boldsymbol{P}$, the chance of mistaking $\boldsymbol{P}$ for another probability $\boldsymbol{Q}$ is given by the following formula of LDP (see Sect. 1.1.2.3): $\sim \exp[-nD(\boldsymbol{P}\|\boldsymbol{Q})]$. According to this lemma, the inequality (3.27) implies that the pair $\mathsf{K}\boldsymbol{P}$ and $\mathsf{K}\boldsymbol{Q}$ is less distinct as compared to the pair $\boldsymbol{P}$ and $\boldsymbol{Q}$, and, therefore, there is more chance of mistaking.

Notice that, if a system does not tend to a steady state, the above result does not apply directly. For example, the time evolution of (x, p) by a Hamiltonian $H(x, p)$ is a Markov process. But its evolution starting from a definite initial condition, $P(x, p, 0) = \delta(X - x(0))\delta(P - p(0))$, does not attain any steady state.

In conclusion, the variational principle of the steady state is not a unique property of the equilibrium distribution. Which aspects of the equilibrium thermodynamics are conserved in the nonequilibrium steady states is the question under active field research. See, for example, [12–15].

3.3.1.3 * Single-Process Approach I – Discrete Langevin Equation

In order to motivate the introduction of discrete Langevin equation, we recall that the Fokker–Planck equation gives the evolution of ensemble probability at time t while the Langevin equation generates a single realization of stochastic process. The former can be derived by the latter, and the latter can be deduced from the former. Similar duality of description is widely found in physics

Heisenberg picture ⟷ Schrödinger picture of quantum mechanics
Lagrange picture ⟷ Euler picture of hydrodynamics
Langevin equation ⟷ Fokker–Planck equation
Discrete Langevin equation ⟷ Master equation

In the above the left-hand side follows some observables, while the right-hand side observes at a fixed point. As the white Gaussian noise was the elementary source of Langevin equation, the stochastic process called the Poisson process/noise plays an elementary role in the discrete Langevin equation. Poisson noise is also the base of the Gillespie algorithm (see below). We, therefore, start by the definition of Poisson noise.

---

[21] That $\mathsf{K}\boldsymbol{P}$ approaches $\mathsf{K}\boldsymbol{Q}$ does not imply the uniformization of $\boldsymbol{P}$ and $\boldsymbol{Q}$ by the Markov evolution, K.

Poisson Noise

A particular realization of Poisson noise $\hat{\zeta}()$ is as follows:

$$\zeta(t) = \sum_{\alpha} \delta(t - t_\alpha), \tag{3.29}$$

where $\delta(z)$ is the Dirac delta function and $\{t_\alpha\}$ with $t_\alpha < t_{\alpha+1}$ represents the time of spike events, which take place randomly. The spike events are a Markov processes and, therefore, characterized only by the mean rate of spiking per unit of time, w. Within a very small time interval, $(t, t + \Delta t]$, the normalized probability for having n spikes is given by the Poisson distribution:

$$P\left[\int_t^{t+\Delta t} \zeta(s) ds = n\right] = \frac{e^{-w\Delta t}}{n!}(w\Delta t)^n. \tag{3.30}$$

For $\hat{n} \equiv \int_t^{t+\Delta t} \hat{\zeta}(s)ds$ we can verify $\langle \hat{n} \rangle = w\Delta t$. We generalize this definition to all the time slices, and we allow for the dependence of the mean spiking rate $w(a)$ on the external parameter, a. When a is varied as a function of time, $\langle \hat{\zeta}(t) \rangle = w(a(t))$.[22]

If $\hat{\zeta}_1(t)$ and $\hat{\zeta}_2(t)$ are two independent Poisson noises, we can identify $\hat{\zeta}_1(t)\hat{\zeta}_2(t)=0$ within any integration over time t. The reason is that, for a given time interval, $(t, t + \Delta t]$, the probability that *both* processes give rise to at least one spike is $\mathcal{O}(\Delta t^2)$. In the limit of $\Delta t \to 0$ such events are negligible (i.e., measure 0). Hereafter we often omit "^" for stochastic processes for simplicity of notation.

To apply the Poisson process for transition between discrete states, we assign i.i.d. Poisson process to each distinct transition:

$$\hat{n}_{i,j}(t, t + \Delta t) = \int_t^{t+\Delta t} \zeta_{i,j}(s) ds, \tag{3.31}$$

and $\zeta_{i,j}(s)$ are independent Poisson noises with $\langle \zeta_{i,j}(t) \rangle = w_{i\to j}(a(t))$.[23]

We denote, following [16], the states that the system can take by the base vectors $|i\rangle$, etc., i is a discrete index. These states can depend on the system's parameter a. We introduce also the dual base vectors, $\langle i|$, etc., so that $\langle i|j\rangle = \delta_{ij}$ for the same parameter a.

By $|i(t)\rangle$ we denote the state of the system at time t. $i(t)$ is among the index mentioned above. We define that, unless the system's state undergoes transition, $\frac{\partial}{\partial t}|i(t)\rangle = 0$ even if the system's parameter a depends on time.

---

[22] We can verify the following formula for the characteristic function: $\left\langle e^{i\phi\hat{n}} \right\rangle = e^{(e^{i\phi}-1)w\Delta t}$. The characteristic functional of the Poisson noise $\hat{\zeta}()$ is $\left\langle e^{i\int_0^t \phi(t')\zeta(t')dt'} \right\rangle = \exp\left[\int_0^t (e^{i\phi(t')} - 1)w(a(t'))dt'\right]$, where $\phi(t)$ is an arbitrary smooth real function.

[23] The characteristic function $\left\langle e^{i\int_0^t \phi_{i,j}(t')\zeta_{i,j}(t')dt'} \right\rangle$ is equal to $\exp\left[\int_0^t (e^{i\phi_{i,j}(t')} - 1)w_{i\to j}(a(t'))dt'\right]$.

Using the above notation, the discrete Langevin equation is as follows:

$$\frac{d}{dt}|i(t)\rangle\rangle = \sum_j (|j\rangle - |i(t)\rangle\rangle) \cdot \zeta_{i(t),j}(t), \tag{3.32}$$

where $\frac{d}{dt}|i(t)\rangle\rangle \equiv (|i(t+dt)\rangle\rangle - |i(t)\rangle\rangle)/dt$, and the multiplicative Poisson noise, $\zeta_{i(t),j}(t)$, obeys (3.31). The symbol "$\cdot$"means the Itô-type multiplication: The vector $|i(t)\rangle\rangle$ on the right-hand side is nonanticipating with respect to the variation of $\zeta_{i(t),j}(t)$.

In order to see how (3.32) works, let us assume that the first future spike in the Poisson noise among $\{\zeta_{i(t),j}(t)\}$ is at $t = t^*$ with $j = j^*$. Then the integration of (3.32) from the present time t up to $t^* + 0$ yields

$$|i(t^*)\rangle\rangle - |i(t)\rangle\rangle = (|j^*\rangle - |i(t)\rangle\rangle) \times 1,$$

where the last factor 1 comes from the time integration of $\delta(t - t^*)$. We then update the system's state to $|i(t^*)\rangle\rangle = |j^*\rangle$.

Another confirmation is that (3.32) reproduces the master equations (3.24) and (3.23). We denote the probability for the state $|i\rangle$ at the time t by $P_j(t) \equiv \langle\delta_{j,i(t)}\rangle$ with $\delta_{j,i(t)} = \langle j|i(t)\rangle\rangle$. Then we have

$$\begin{aligned}\frac{dP_j}{dt} &= \sum_\ell \langle(\delta_{j,\ell} - \delta_{j,i(t)}) \cdot \zeta_{i(t),\ell}(t)\rangle \\ &= \sum_\ell \sum_k \langle(\delta_{j,\ell} - \delta_{j,i(t)})\delta_{k,i(t)} \cdot \zeta_{k,\ell}(t)\rangle \\ &= \sum_k [P_k w_{k\to j}(a(t)) - P_j w_{j\to k}(a(t))].\end{aligned} \tag{3.33}$$

Equation (3.32) is analogous to the Langevin equation and the characteristic functional of its noise $\xi(t)$. Gillespie has proposed an approximate approach with finite Δt and then coarse grained it to have a Langevin equation [17].

3.3.1.4 * Single-Process Approach II – Gillespie Algorithm

The discrete Langevin equation (3.32) is not common in the literature. But we actually solve this when we simulate the master equation (3.24). From practical viewpoint, the temporal discretization of (3.32) is not a very efficient method, especially when the transitions occur rarely. A better idea is to focus on the waiting time, Δt^* $(0 \le \Delta t < \infty)$, with which a system ceases to stay the present state S_i to make the *first* transition to the new state S_{j^*} ("event driven method"). With this idea Gillespie [18–20] formulated an efficient simulation algorithm to generate the particular realizations of (3.32). It consists of the following two steps:

(1) We take a sample of the set of independent random variables, $\{\hat{y}_j\}_{j\in I_i}$, which are uniformly distributed on the domain, [0, 1]. From a sample $\{y_j\}_{j\in I_i}$, we define $\{\tau_j\}_{j\in I_i}$ such that $e^{-w_{i\to j}\tau_j} \equiv y_j$.
(2) We define Δt^* as the minimum among $\{\tau_j\}_{j\in I_i}$ and identify j^* so that τ_{j^*} gives this minimum of $\{\tau_j\}_{j\in I_i}$.[24]

The derivation is a little technical but at the same time pedagogical. We, therefore, summarize its outline below. The key quantities are

$p_{ii}(\Delta t)$: the probability that the system stays continuously in the state between the time t and $t + \Delta t$.
$p_{ij}(\Delta t)$ $(j \in I_i)$: the probability that the system has ceased to stay in the state S_i to make the *first* transition to the state S_j between the time t and $t + \Delta t$.

By definition, the initial conditions for p_{ii} and p_{ij} are

$$p_{ii}(0) = 1, \qquad p_{ij}(0) = 0 \quad (j \in I_i), \tag{3.34}$$

and they should obey the following evolution equations:

$$\frac{dp_{ii}(\Delta t)}{d\Delta t} = -p_{ii}(\Delta t)\sum_{j\in I_i} w_{i\to j}, \qquad \frac{dp_{ij}(\Delta t)}{d\Delta t} = p_{ii}(\Delta t)w_{i\to j}, \quad j \in I_i. \tag{3.35}$$

The solution for $p_{ii}(t)$ and $p_{ij}(\Delta t)$ $(j \neq i)$ with any $\Delta t(\geq 0)$ are

$$\begin{aligned} p_{ii}(\Delta t) &= \exp\{-\sum_{j''\in I_i} w_{i\to j''}\Delta t\}, \\ p_{ij}(\Delta t) &= \frac{w_{i\to j}}{\sum_{j'\in I_i} w_{i\to j'}}[1 - \exp\{-\sum_{j''\in I_i} w_{i\to j''}\Delta t\}]. \end{aligned} \tag{3.36}$$

These probabilities satisfy the sum rule, $p_{ii}(\Delta t) + \sum_{j\in I_i} p_{ij}(\Delta t) = 1$.

We can verify that those $(\Delta t^*, j^*)$ defined by (1)–(2) reproduces (3.36). Using the rules (1)–(2), the probability corresponding to $p_{ij}(\Delta t)$ writes

$$\mathrm{Prob}\big[T_j(\hat{y}_j) \leq \Delta t, \text{ and } T_{j'}(\hat{y}_{j'}) \geq T_j(\hat{y}_j)(\forall j' \in I_i)\big] \equiv \tilde{p}_{ij}(\Delta t), \tag{3.37}$$

where we defined the functions $T_j(y)$ by $e^{-w_{i\to j}T_j(y)} = y$. Using the uniform distribution of $\{\hat{y}_j\}_{j\in I_i}$, we have

[24] That is, $\Delta t^* = \min_{j\in I_i} \tau_j = \tau_{j^*}$.

$$\tilde{p}_{ij}(\Delta t) = \int_0^1 \theta(\Delta t - \tau(y_j)) \left\{ \prod_{j' \in I_i (j' \neq j)} \left[\int_0^1 \theta(\tau(y_{j'}) - \tau(y_j)) dy_{j'} \right] \right\} dy_j,$$

where $\theta(z) = 0$ for $z < 0$ and $\theta(z) = 1$ for $z \geq 0$. The integrations are simple to do,[25] and we find for $\tilde{p}_{ij}(\Delta t)$ the identical expression $p_{ij}(\Delta t)$ of (3.36).

3.3.1.5 * Detailed Balance and Equilibrium

Detailed balance adds stringent conditions on the steady state of master equation. The unique steady state with detailed balance can be regarded as a thermal equilibrium state. The equilibrium probability is reconstructed using the detailed balance conditions, and the set of rate constants that enable the equilibrium state have simple interpretations in terms of the free-energy landscape.

State of Detailed Balance

We consider the steady state of a master equation that satisfies more stringent conditions:

$$J_{i \to j} = 0, \quad \text{for } \forall i, \forall j \text{ (detailed balance)} \tag{3.38}$$

or, equivalently,

$$P_i w_{i \to j} = P_j w_{j \to i}, \quad \text{for } \forall i, \forall j \text{ (detailed balance).} \tag{3.39}$$

We call such steady state the state of detailed balance. Whether a system has the state of detailed balance depends on the transition rates because there are more equations $\{J_{i \to j} = 0\}$ than the number of components of $\{P_j\}$.

* Equilibrium State of Master Equation

We call the steady state of master equation the equilibrium state if this state satisfies the detailed balance. As mentioned in Sect. 3.3.1.4, not all steady states satisfy detailed balance. Even if the probabilities $\{P_i(t)\}$ evolving according to a master equation converge to a unique state irrespective of the initial condition $\{P_i(t_{\text{init}})\}$ we cannot call it the equilibrium state of the master equation by analogy with (the first law of) macroscopic thermodynamics.

This restriction is required by the compatibility with thermodynamic laws. Below is its demonstration: if a system has a steady state which does not satisfy (3.38), the probability fluxes $\{J_{i \to j}\}$ can be written as a nontrivial superposition of *circulations of probability flux*. Each of the circulations consists of at least three discrete

---

[25] We use the formula $\sum_{j' \in I_i (j' \neq j)} \frac{w_{i \to j'}}{w_{i \to j}} + 1 = \frac{1}{w_{i \to j}} \sum_{j' \in I_i} w_{i \to j'}$.

states, say $S_2 \rightarrow S_3 \rightarrow S_7 \rightarrow S_2$, and we can assign on it a constant flux, e.g., $J_{2\rightarrow 3} = J_{3\rightarrow 7} = J_{7\rightarrow 2} \neq 0$. From such steady-state circulation, we can conceive a hypothetical machine that interacts selectively with the states, $\{S_2, S_3, S_7\}$. This machine could "rotate" in one direction because of the broken symmetry of the flux on these three states. Thus we can extract some systematic work, for example, to wind a string under a load using the rotation of this machine. Now if this steady state were the thermodynamic equilibrium, the machine would be a perpetual machine of the second kind (see Sect. 2.1.2), and this is contradiction to the second law of thermodynamics. Therefore, a steady state violating the detailed balance cannot correspond to a thermal equilibrium state.

* Reconstruction of Equilibrium Probabilities

If a unique steady state of the master equation satisfies detailed balance, we can find easily the equilibrium probabilities, $\{P_j^{\text{eq}}\}$. Below is the protocol:

1. We choose arbitrarily a state, S_i, as a reference.
2. We derive all the P_j^{eq} using (3.39), i.e., $P_j^{\text{eq}} = \frac{w_{i\rightarrow j}}{w_{j\rightarrow i}} P_i^{\text{eq}}$. In case that $w_{j\rightarrow i} = 0$ for some j, we can determine P_j^{eq} indirectly by applying the chain of equalities of (3.39). The presence of a unique steady state assures that it is feasible.
3. Once all the values of P_j^{eq} are determined relative to P_i^{eq}, the normalization condition, $\sum_j P_j^{\text{eq}} = 1$, determines the value of P_i^{eq}.

* Representation of Transition Rates

Suppose that a system with n states, $\{S_1, \ldots, S_n\}$, has a unique state of detailed balance. Let us establish the balance sheet for the number of unknown parameters and the number of conditions. There are $n(n-1)$ parameters of the transition rates and n parameters of the equilibrium probabilities. The constraints of the type (3.39) impose $n(n-1)/2$ conditions. An additional constraint is the normalization of the probabilities. There remains, therefore, $[n(n+1)/2 - 1]$ degrees of freedom.[26] We can write these degrees of freedom in a physically appealing manner [21–24]:

$F_j/k_{\text{B}}T$: "free-energy levels (per $k_{\text{B}}T$)." They amount to $(n-1)$ degrees of freedom, where (-1) is due to the arbitrariness of an additive constant.[27]

$\Delta_{i,j}/k_{\text{B}}T$: "activation free energies (per $k_{\text{B}}T$)" having the symmetry, $\Delta_{i,j} = \Delta_{j,i}$. They amount to $[n(n-1)/2 - 1]$ degrees of freedom, where (-1) is due to the arbitrariness of an additive constant.[28]

ν: the unique "attempting frequency." One degree of freedom.

---

[26] The calculation is $[n(n-1) + n] - [n(n-1)/2 + 1] = n(n+1)/2 - 1$.

[27] In chemistry, one can regard F_j as the Gibbs free energy under isobaric condition, rather than Helmholtz free energy.

[28] The last additive constant can be chosen independently of the one for $\{F_j/k_{\text{B}}T\}$, see the remark after (3.40).

The above parameterization allows us to represent both $\{P_j^{\text{eq}}\}$ and $\{w_{i \to j}\}$ in the context of the thermal activation-assisted transition in a free-energy landscape.

$$P_j^{\text{eq}} = \frac{e^{-F_j/k_B T}}{\sum_i e^{-F_i/k_B T}}, \qquad w_{i \to j} = \nu \exp\left[-\frac{\Delta_{i,j} - F_i}{k_B T}\right]. \tag{3.40}$$

The arbitrariness in both $\Delta_{i,j}$ and F_i can be absorbed by the pre-exponential factor, ν.[29]

The above expression of the transition rates, $\{w_{i \to j}\}$, provides the following intuitive picture [21–24] (see Fig. 3.4): A system in the state S_i has (free) energy F_i. This system attempts to make a state transition to any other state with a common frequency, ν. Because of the (free) energy barrier of activation $(\Delta_{i,j} - F_i)$, the probability of a successful attempt for the transition $S_i \to S_j$ is the Boltzmann factor, $e^{-(\Delta_{i,j} - F_i)/k_B T}$. Note that the purely kinetic parameters $\Delta_{i,j}$ do not appear in the equilibrium probabilities.

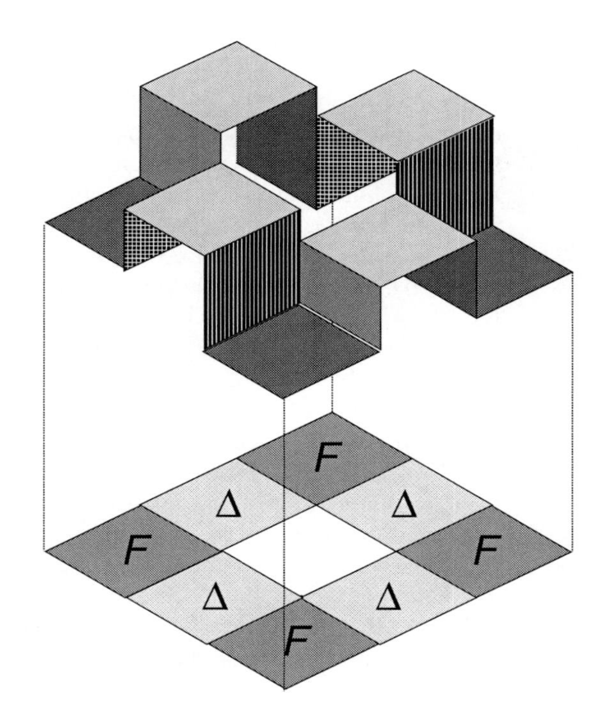

Fig. 3.4 Interpretation of (3.40). The squares on the corners represent the free-energy levels, F_i, while the plateaus between the nearest states represent the barriers, $\Delta_{i,j}$

[29] Once we fix the ν's value, the difference $\Delta_{i,j} - F_i$ has a physical meaning of the activation barrier and contains no arbitrary additive constants.

Nonequilibrium Processes

Once the transition rates are fixed using the detailed balance condition of equilibrium, we can proceed to study the master equation under nonequilibrium conditions. This implies several different things:

Transient nonequilibrium states: Keeping the transition rates of the form (3.40) unchanged, we study the relaxation of the probabilities $P_j(t)$ starting from nonequilibrium initial ones. Also we solve the discrete Langevin equation with these transition rates and look for stochastic processes on the state space.

Nonequilibrium settings I: We may vary F_j's, $\Delta_{i,j}$'s, or $k_B T$ as function of time *in* the transition rates of (3.40). In this case, the instantaneous equilibrium distribution and the landscape picture are still valid according to (3.40). The $P_j(t)$ evolves toward a temporary equilibrium state, although there can be a lag of relaxation.

Nonequilibrium settings II: We modify each *activation barrier*, $\Delta^{\text{act}}_{i \to j} \equiv \Delta_{i,j} - F_i$, disregarding the DB conditions, based on physical arguments.
In fact each activation barrier often has a physical justification of its own, independent of the compatibility with the global equilibrium states. It is like that the macroscopic rate constant k for the reaction A+B→AB can be used either in (3.3) or in (3.1), which is far from equilibrium. Therefore, we can combine these transition rates to build up a reaction network having nonequilibrium steady states. A simple example is given in Sect. 3.3.3.3
Since the number of combinations of $[i \to j]$, i.e., $n(n-1)$, is more than the degrees of freedom left by the DB condition, $n(n+1)/2 - 1$, the modified transition rates can no more be represented by a single-valued landscape like Fig. 3.3. The system now allows steady-state circulations of probability flux.[30]

Relation between the transient nonequilibrium states and the nonequilibrium settings II: When we model the chemical coupling schematized by Fig. 2.7, we can model either the whole closed system, i.e., the chemical engine plus the four particle reservoirs, or the chemical engine as an open system.
The former viewpoint is not very practical but formally simple: we assume the transition rates satisfying detailed balance. The global landscape is fixed like in Fig. 2.8. The nonequilibrium process is regarded as a transient process toward the equilibrium of the whole closed system.
The latter viewpoint is more practical but we must use the nonequilibrium settings of type II. The reservoirs' states are no more taken into account, and only the chemical potentials enter as a parameter for the chemical engine.
In the former formal point of view, the system's evolution is relaxing and downhill on the average. In the latter point of view, the chemical engine makes a stochastic cyclic transitions in its (reduced) state space.

---

[30] One may imagine the Escher staircase.

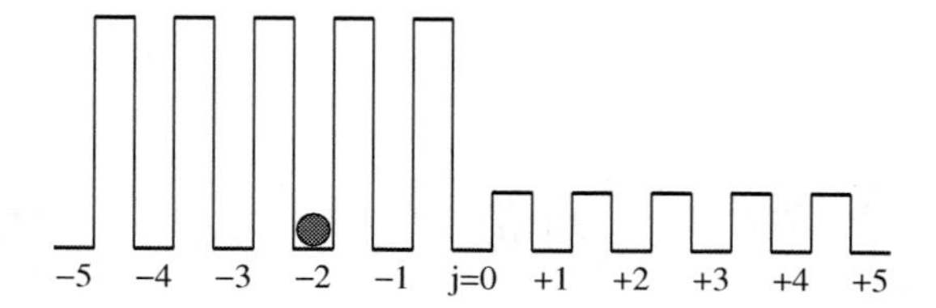

Fig. 3.5 Landscape with inhomogeneous barrier heights. See the text

Simple Example – Will Particles Be Stagnant in the Region of Small Diffusion Coefficient?

Suppose that a system can take the states S_j ($j = 0, \pm 1, \ldots, \pm N$) and that the nonzero transition rates only between j and $j+1$ ($-N \leq j < N$). Figure 3.5 shows the case that $F_j = 0$ for all j, $\Delta_{j,j+1} = W$ for $-N \leq j < 0$, and $\Delta_{j,j+1} = w$ for $0 \leq j < N$ with $w < W$. The general argument above tells that this system has an equilibrium state with detailed balance, and the equilibrium probability is uniform, $P_j^{\mathrm{eq}} = (2N+1)^{-1}$. Nevertheless, might we not expect that the system spends more time in the left region ($j < 0$), where the diffusion takes more time? The key to avoid this trap is to be aware of the opposing effects to this argument. The state $S_{j=0}$ situated between the high barrier W to the left and the low barrier w to the right. If the system is in this state, the (conditional) probability of the transition $S_0 \to S_{-1}$ is smaller than that of the transition $S_0 \to S_{+1}$. Therefore, the chance that the system enters into the $j < 0$ states is relatively small, though the residence time in $j < 0$ states are relatively large. In equilibrium, these two effects exactly cancel.

3.3.1.6 Langevin Equation as a Limit of Discrete Process

Transition Rate of Langevin Equation

The Langevin equation can be regarded as the limit of a discrete process, where the states are infinitely finely distinguished and the transitions are allowed to occur only among the "nearby" states. (The Fokker–Planck equation is, therefore, a limit of the master equation.[31]) Below we demonstrate how the transition rate is obtained for the overdamped Langevin equation.

Let us write the Langevin equation in the form of SDE, see Sect. 1.2.1.1:

$$dx_t = -\frac{1}{\gamma}\frac{dU(x_t)}{dx_t}dt + \sqrt{2D}dB_t, \tag{3.41}$$

where we denote by x_t the value of x at time t and D is the diffusion coefficient. The probability density for x_t is $\langle \delta(x - x_t) \rangle$. Then the conditional probability density for $x_{t+dt} = x_t + dx_t$, given that $x = x_t$ at t, is $\langle \delta(x - x_t)\delta(x' - x_{t+dt}) \rangle / \langle \delta(x - x_t) \rangle$. For $x \neq x'$, this conditional density gives the flow of probability to x' within the time dt. Therefore, the transition rate $w_{x \to x'}$ should be related to this density through

[31] The derivation uses the technique called the Kramers–Moyal expansion. See, for example, [3, 25, 26, 7, 8].

$$\frac{\langle\delta(x-x_t)\delta(x'-x_{t+dt})\rangle}{\langle\delta(x-x_t)\rangle}=\delta(x-x')+w_{x\to x'}dt. \tag{3.42}$$

To evaluate the left-hand side, we develop $\delta(x'-x_{t+dt})$ around $\delta(x'-x_t)$ (Sect. 1.2.2.4) using (3.41) and the Itô formula (1.58).[32] The result is

$$w_{x\to x'}=\frac{1}{\gamma}\frac{dU(x)}{dx}\delta'(x'-x)+D\delta''(x'-x). \tag{3.43}$$

Detailed Balance Condition and Einstein Relation

We can verify that the equilibrium state, $\mathcal{P}^{\rm eq}(x)\propto e^{-U(x)/k_{\rm B}T}$, is the state of detailed balance *if* the diffusion coefficient satisfies the Einstein relation, $D=k_{\rm B}T/\gamma$. As (3.43) includes the derivatives of the δ-function, what corresponds to (3.39) should be expressed by the integral form:

$$\int dx\int dx' f(x)\left[\mathcal{P}^{\rm eq}(x)w_{x\to x'}-\mathcal{P}^{\rm eq}(x')w_{x'\to x}\right]g(x')=0, \tag{3.44}$$

where $f(x)$ and $g(x')$ are arbitrary functions of good properties.[33] A straightforward integrations by parts of (3.44) with (3.43) leads to $D=k_{\rm B}T/\gamma$.

Detailed Balance and Fluctuation–Dissipation (FD) Relation

In Sect. 1.3.1.2 we have seen that the Einstein relation is related also to the fluctuation–dissipation relation. In general, stochastic processes satisfying the detailed balance condition have equilibrium state distribution, and their fluctuations and linear responses obey the fluctuation–dissipation (FD) relation.

For the processes breaking the detailed balance condition, several generalization of the fluctuation–dissipation relation have been formulated [27, 28], and also the discrepancy from the (true) fluctuation–dissipation relation has been related to the heat generation described [29–31]. The latter points will be discussed in the part II.

3.3.2 Stochasticity of Molecule Numbers in the Chemical Reaction Can Be Described by Discrete Master Equation

The method described in the previous section is used to describe the stochastic aspect of chemical reaction. We shall take up the same examples as before, $A+B \rightleftharpoons AB$. The object here is to know the relation between macroscopic and

[32] Note that $\delta(x'-x_t)$ has nonanticipating property with respect to dB_t.

[33] For example, we assume that they are derivable arbitrary many times and are decaying faster than any power of x for $|x|\to\infty$ (Schwartz space).

stochastic parameters, and see how the number of molecular species are distributed at equilibrium.

3.3.2.1 Number State Representation

Fluctuations are unavoidable if we describe a chemical reaction with the resolution of the (integer) number of molecular species. Suppose that the reaction, A+B $\rightleftharpoons$ AB, occurs in a closed container at the temperature T, where the total number of atoms A and atoms B is fixed at $N_{\mathrm{A}}^{\mathrm{tot}}$ and $N_{\mathrm{B}}^{\mathrm{tot}}$, respectively. The number of AB molecule, N_{AB}, is then sufficient to characterize the state of the system, because the numbers of the other molecules, A or B, are given as $N_{\mathrm{A}} = N_{\mathrm{A}}^{\mathrm{tot}} - N_{\mathrm{AB}}$ or $N_{\mathrm{B}} = N_{\mathrm{B}}^{\mathrm{tot}} - N_{\mathrm{AB}}$, respectively. The system has, therefore, $\min\{N_{\mathrm{A}}^{\mathrm{tot}}, N_{\mathrm{B}}^{\mathrm{tot}}\} + 1$ discrete states. Below we shall use N_{AB} to represent the state, $S_{N_{\mathrm{AB}}}$, unless confusions arise. On this level of description, we do not distinguish the individuality of the atoms to define the states, nor their orientations and other internal degrees of freedom. But still the description is more detailed than the macroscopic description of Sect. 3.2. The (number) concentrations, [A], etc., in macroscopic description are readily given as $\langle N_{\mathrm{A}} \rangle / V$, etc., where V (in liter) is the volume of the system.

3.3.2.2 Transition Rates and the Rate Constant

Transition Rates

We consider for a moment only the forward reaction, A + B $\to$ AB, or in stochastic term, the state transition of $N_{\mathrm{AB}} \to (N_{\mathrm{AB}} + 1)$. In the stochastic description, we assume the following simple model for the transition rate, $w_{N_{\mathrm{AB}} \to N_{\mathrm{AB}}+1}$:

$$w_{N_{\mathrm{AB}} \to N_{\mathrm{AB}}+1} = k \frac{N_{\mathrm{A}} N_{\mathrm{B}}}{V}, \tag{3.45}$$

where k is a constant independent of the number of molecules or volume. The approximation leading to (3.45) is that every A molecule and B molecule is distributed randomly in the volume V and that the chance to find a pair of A and B molecule within an atomic distance is $\propto \frac{N_{\mathrm{A}} N_{\mathrm{B}}}{V}$ up to the relative error of $\mathcal{O}(N_{\mathrm{A}}/V, N_{\mathrm{B}}/V)$. The factor k should include the activation factor, or the probability that a collision between an A and a B molecules leads to the formation of an AB molecule.

Rate Constant

We will relate the transition rate $w_{N_{\mathrm{AB}} \to N_{\mathrm{AB}}+1}$ to the rate constant of the macroscopic description of the reaction, described in Sect. 3.2.1.1. If the state transition, $N_{\mathrm{AB}} \to (N_{\mathrm{AB}} + 1)$, occurs the probability $w_{N_{\mathrm{AB}} \to N_{\mathrm{AB}}+1} dt$ for an infinitesimal time dt, N_{AB} should increase approximately by $w_{N_{\mathrm{AB}} \to N_{\mathrm{AB}}+1}\, dt$, that is $dN_{\mathrm{AB}}/dt \simeq w_{N_{\mathrm{AB}} \to N_{\mathrm{AB}}+1} = k\, N_{\mathrm{A}} N_{\mathrm{B}}/V$, where we have used (3.45). Dividing each part of this equation by V, we have

$$\frac{d[\mathrm{AB}]}{dt} \simeq \frac{w_{N_{\mathrm{AB}} \to N_{\mathrm{AB}}+1}}{V} = k[\mathrm{A}][\mathrm{B}]. \tag{3.46}$$

The rightmost the equation is of the same form as the formula for the macroscopic reaction. Therefore, we identify the coefficient k in (3.45) with the rate constant of the reaction, $\mathrm{A} + \mathrm{B} \to \mathrm{AB}$.

Next we take into account the backward state transition, $(N_{\mathrm{AB}} + 1) \to N_{\mathrm{AB}}$, also. This corresponds to the reaction $\mathrm{AB} \to \mathrm{A} + \mathrm{B}$. We assume that the transition rate, $w_{N_{\mathrm{AB}}+1 \to N_{\mathrm{AB}}}$, is proportional to the number of AB molecules *before* the transition, therefore,

$$w_{N_{\mathrm{AB}}+1 \to N_{\mathrm{AB}}} = k'(N_{\mathrm{AB}} + 1), \tag{3.47}$$

where a positive constant, k', is independent of the number of molecules or volume. It is justified if the dissociation occurs in individual AB molecules. By approximately identifying this transition rate with a contribution to $-dN_{\mathrm{AB}}/dt$, we obtain the

$$\frac{d[\mathrm{AB}]}{dt} \simeq k[\mathrm{A}][\mathrm{B}] - k'[\mathrm{AB}]. \tag{3.48}$$

We, therefore, identify the coefficient k' in (3.47) with the rate constant of the reaction, $\mathrm{AB} \to \mathrm{A} + \mathrm{B}$.

Remark: Extensivity of the Transition Rate

The frequency of the transition $N_{\mathrm{AB}} \to N_{\mathrm{AB}} + 1$ has an extensive character that is proportional to the size of the system. Therefore, the time defined, for example, by $(w_{N_{\mathrm{AB}} \to N_{\mathrm{AB}}+1})^{-1}$ has nothing to do with the reaction mechanism of individual molecules.

In order that the modeling as a Markov process be a good approximation, the subsequent transitions, for example, $N_{\mathrm{AB}} \to N_{\mathrm{AB}} + 1$ and $N_{\mathrm{AB}} + 1 \to N_{\mathrm{AB}} + 2$ should be uncorrelated. If the reacting solution is dilute, successive transitions occur at distant spatial locations in the volume, and the Markov approximation is justified, however, large is the transition rate, $w_{N_{\mathrm{AB}} \to N_{\mathrm{AB}}+1}$. That the inverse transition rates scale with V^{-1} is the basis of the van Kampen's expansion method of the master equation [32, 8].

If the nature of chemical reaction is such that two consecutive transitions, e.g., $N_{\mathrm{AB}} \to N_{\mathrm{AB}} + 1$ and $N_{\mathrm{AB}} + 1 \to N_{\mathrm{AB}} + 2$, are strongly correlated on the molecular level, then we can better define the reaction as $2\mathrm{A} + 2\mathrm{B} \rightleftharpoons 2\mathrm{AB}$ and define the state transition as $N_{\mathrm{AB}} \to N_{\mathrm{AB}} + 2$.

3.3.2.3 Probability Fluxes and the Master Equation

Probability Flux of Stochastic Chemical Reaction

We denote by $P(N_{\rm AB}, t)$ $(0 \leq N_{\rm AB} \leq \min\{N_{\rm A}^{\rm tot}, N_{\rm B}^{\rm tot}\})$ the probability that the systems in the state $N_{\rm AB}$ at time t. Applying the definition of probability flux, (3.23), the probability flux of the reaction is given as follows:

$$\begin{aligned} J_{N_{\rm AB}\to N_{\rm AB}+1} &\equiv P(N_{\rm AB}, t)w_{N_{\rm AB}\to N_{\rm AB}+1} - P(N_{\rm AB}+1, t)w_{N_{\rm AB}+1\to N_{\rm AB}} \\ &= -J_{N_{\rm AB}+1\to N_{\rm AB}}. \end{aligned} \tag{3.49}$$

Master Equation of Stochastic Chemical Reaction

With the above probability flux, the master equation for $P(N_{\rm AB}, t)$ is written as

$$\frac{dP(N_{\rm AB}, t)}{dt} = J_{N_{\rm AB}-1\to N_{\rm AB}} - J_{N_{\rm AB}\to N_{\rm AB}+1}. \tag{3.50}$$

The normalization of the probabilities:

$$\sum_{n=0}^{\min\{N_{\rm A}^{\rm tot}, N_{\rm B}^{\rm tot}\}} P(n, t) = \text{const.} \tag{3.51}$$

can be verified by (3.50) at all time t.

Substitution of the expressions of the transition rates, (3.45) and (3.47), we have

$$\begin{aligned} \frac{\partial P(N_{\rm AB})}{\partial t} &= k\frac{(N_{\rm A}+1)(N_{\rm B}+1)}{V}P(N_{\rm AB}-1, t) \\ &- \left\{k'N_{\rm AB} + k\frac{N_{\rm A}N_{\rm B}}{V}\right\} P(N_{\rm AB}, t) + k'(N_{\rm AB}+1)P(N_{\rm AB}+1, t). \end{aligned} \tag{3.52}$$

Equation (3.52) includes macroscopic reaction equation for $[AB]$, i.e., for $\langle N_{\rm AB}\rangle = \sum_{n=0}^{\min\{N_{\rm A}^{\rm tot}, N_{\rm B}^{\rm tot}\}} n\, P(n, t)$. See, for example, [7–9] for the methods to derive the macroscopic equation from (3.52).

3.3.2.4 Equilibrium Properties

State of Detailed Balance

The steady state of (3.52) satisfying the detailed balance (DB) imposes the condition

$$J_{N_{\rm AB}\to N_{\rm AB}+1} = 0, \qquad 0 \leq N_{\rm AB} < \min\{N_{\rm A}^{\rm tot}, N_{\rm B}^{\rm tot}\}$$

under fixed values of $N_{\rm A}^{\rm tot}$ and $N_{\rm B}^{\rm tot}$. This condition gives the probabilities of the equilibrium state:

$$P(N_{AB}, t = \infty) = \mathcal{N} \frac{\tilde{N}_A^{N_A}}{N_A!} \frac{\tilde{N}_B^{N_B}}{N_B!} \frac{\tilde{N}_{AB}^{N_{AB}}}{N_{AB}!} \equiv P^{eq}(N_{AB}), \tag{3.53}$$

where $\mathcal{N}$ is the normalization constant, and the three parameters, $(\tilde{N}_A, \tilde{N}_B, \tilde{N}_{AB})$ are solutions to the following three equations:

$$\frac{(\tilde{N}_A v_0/V)(\tilde{N}_B v_0/V)}{(\tilde{N}_{AB} v_0/V)} = \frac{k'}{k}, \quad \tilde{N}_A + \tilde{N}_{AB} = N_A^{tot}, \quad \tilde{N}_B + \tilde{N}_{AB} = N_B^{tot}, \tag{3.54}$$

where v_0 is a constant with the dimension of volume. The formula (3.53) looks apparently the product of three independent Poisson distributions. In fact, these distributions are not independent for the closed system, because of the constraints $N_A = N_A^{tot} - N_{AB}$ and $N_B = N_B^{tot} - N_{AB}$. Equation (3.54) gives a function of the single variable, N_{AB}.

Nevertheless the form of the product of Poisson distributions is the consequence of only the DB condition, whether or not the system is closed. When the number of N_A and N_B is adjustable by their particle environments, (3.53) describes the true product of Poisson distributions.

Law of Mass Action

The constraint (3.54) suggests already a law of mass action at equilibrium. In fact, when the system is macroscopic, the equilibrium state of master equation (3.53) includes the law of mass action (3.6), that is, $[A][B]/[AB] = k'/k$.

Since (3.54) has a form of law of mass action, we need only to show that the parameters $\tilde{N}_A$, $\tilde{N}_B$, and $\tilde{N}_{AB}$ are the values of N_A, N_B, and N_{AB}, respectively, when $P_{eq}(N_{AB})$ is the maximum respect to N_{AB}. It can be done using the technique of logarithmic derivative.[34] Thus, through the relation (3.54) the equilibrium state (3.53) includes the law of mass action (3.6).

Chemical Potential

In order to see how the chemical potential is interpreted in this simple reaction scheme, $A + B \leftrightarrow AB$, we will compare the result of $P^{eq}(N_{AB})$ with the prediction from statistical mechanics.

We define μ_A^0, μ_B^0, and μ_{AB}^0 as the free energies of *individual* molecules, A, B, and AB, respectively, which reflect the effect of the kinetic energies and the internal degrees of freedom of each molecule in the form of the free energy:

$$\mu_M^0 = -k_B T \ln z_M \qquad (M = A, B, AB), \tag{3.55}$$

[34] We require $\partial[\ln P_{eq}(N_{AB})]/\partial N_{AB} = 0$. Denoting by $\{N_A^*, N_B^*, N_{AB}^*\}$ the numbers $\{N_A, N_B, N_{AB}\}$ at the maximum of the probability, and using the Stirling formula, $n! \sim n^n e^{-n}$, we have the equality $(N_A^*/\tilde{N}_A)(N_B^*/\tilde{N}_B)\,(N_{AB}^*/\tilde{N}_{AB})^{-1} = 1$. The last equation implies $N_A^* = \tilde{N}_A$, etc.

where z_{M} are the partition functions of a molecular species M.[35] These free energies can depend on the temperature. We then assign to each state of the system, N_{AB} ($\leq \min\{N_{\mathrm{A}}^{\mathrm{tot}}, N_{\mathrm{B}}^{\mathrm{tot}}\}$), the following "potential energy":

$$U(N_{\mathrm{AB}}) = N_{\mathrm{A}}\mu_{\mathrm{A}}^{0} + N_{\mathrm{B}}\mu_{\mathrm{B}}^{0} + N_{\mathrm{AB}}\mu_{\mathrm{AB}}^{0},$$

where $N_{\mathrm{A}} \equiv N_{\mathrm{A}}^{\mathrm{tot}} - N_{\mathrm{AB}}$ and $N_{\mathrm{B}} \equiv N_{\mathrm{B}}^{\mathrm{tot}} - N_{\mathrm{AB}}$.[36] We then write down the canonical (configurational) partition function of the form of $\sum e^{-U(N_{\mathrm{AB}})/k_{\mathrm{B}}T}$. We have

$$\frac{Z(T, V, N_{\mathrm{A}}^{\mathrm{tot}}, N_{\mathrm{B}}^{\mathrm{tot}})}{N_{\mathrm{A}}^{\mathrm{tot}}!\, N_{\mathrm{B}}^{\mathrm{tot}}!} = \sum_{N_{\mathrm{AB}}=0}^{\min\{N_{\mathrm{A}}^{\mathrm{tot}}, N_{\mathrm{B}}^{\mathrm{tot}}\}} \frac{\alpha_{\mathrm{A}}^{N_{\mathrm{A}}}\, \alpha_{\mathrm{B}}^{N_{\mathrm{B}}}\, \alpha_{\mathrm{AB}}^{N_{\mathrm{AB}}}}{N_{\mathrm{A}}!\, N_{\mathrm{B}}!\, N_{\mathrm{AB}}!}, \tag{3.56}$$

where $\alpha_{\mathrm{M}} \equiv (V/v_{\mathrm{l}})e^{-\frac{\mu_{\mathrm{M}}^{0}}{k_{\mathrm{B}}T}}$, and v_{l} is a constant. The factorials come from the combinatorial number, ${}_{N_{\mathrm{A}}^{\mathrm{tot}}}C_{N_{\mathrm{A}}} \times {}_{N_{\mathrm{B}}^{\mathrm{tot}}}C_{N_{\mathrm{B}}} \times N_{\mathrm{AB}}!$, for forming N_{AB} of AB molecules out of $N_{\mathrm{A}}^{\mathrm{tot}}$ of A molecules and $N_{\mathrm{B}}^{\mathrm{tot}}$ of B molecules. According to the statistical mechanics of Gibbs, we can identify each term in the sum on the right-hand side of (3.56) with the relative probability. This probability has exactly the form of (3.53), except for a factor of normalization. Therefore, we find

$$\alpha_{\mathrm{M}} = \tilde{N}_{\mathrm{M}}.$$

Combining this result with (3.54) we have[37]

$$\exp\left\{\frac{\mu_{\mathrm{AB}}^{0} - \mu_{\mathrm{A}}^{0} - \mu_{\mathrm{B}}^{0}}{k_{\mathrm{B}}T}\right\} = \frac{k'}{k}. \tag{3.57}$$

We have thus recovered the result of macroscopic thermodynamics, (3.9), and (3.55) gives the microscopic meaning for μ_{A}^{0}, μ_{B}^{0}, and μ_{AB}^{0}.

Remark : Origin of the Combinatorial Factor

Looking back to the origin of the formula (3.53), we see that the combinatorial factor $N_{\mathrm{A}}!$, etc., has nothing to do with the particle-wave duality of quantum physics. A statistical mechanical description of the open system also gives a similar combinatorial factor. We will come back to this point in Sect. 5.2.1.5.

---

[35] Strictly speaking, z_{M} is made nondimensional by a factor related to the atomic specific volumes.

[36] The relation between the "potential energy" and the landscape representation (Sect. 3.3.1.5) will be discussed in the next Sections.

[37] We put $v_0 = v_1$.

3.3.3 Stochastic Open System Is a Class of Stochastic Chemical Reaction System

3.3.3.1 Chemical Potential of the Reservoir

The formula (3.40) gives a landscape representation of the transition rates that satisfy detailed balance. We apply this schema to the system which consists of an open subsystem (open system, for short) *and* a particle reservoir between which molecules of species A are exchanged. See Fig. 3.6 (A). For simplicity, no chemical reactions are assumed to occur within the open system, although the generalization is easy. The only "reaction" is, therefore, the migration between the system (Sys) and the reservoir (Res):

$$\mathrm{A(Sys)} \rightleftharpoons \mathrm{A(Res)}. \tag{3.58}$$

This is a simple model of physical adsorption of molecule A, where the system (S) is a 2D substrate with a "volume" V.

Following the protocol of Sect.3.3.2, we build up the master equation of this model.

1. The state of the system is distinguished by the number of particles in the open system, $N_{\mathrm{A}} \equiv N$.
2. The transition rate $w_{N\to N+1}$ represents the average rate at which a particle enters the open system, while $w_{N+1\to N}$ reflects the rate at which any one particle in the open system leaves for the environment. We will not assume the properties of dilute solutions, but leave the transition rates very general [33].
3. The probability flux is $J_{N\to N+1} = P_N w_{N\to N+1} - P_{N-1} w_{N-1\to N}$.
4. The master equation for P_N is then

$$\frac{dP_N}{dt} = -J_{N\to N+1} - J_{N\to N-1}. \tag{3.59}$$

5. We assume that the system is in equilibrium. We then impose the detailed balance condition (3.40). It relates $w_{N\to N+1}$ with $w_{N+1\to N}$:

$$w_{N+1\to N} = w_{N\to N+1} \exp\left[\frac{F_{N+1} - F_N}{k_{\mathrm{B}}T}\right]. \tag{3.60}$$

6. In order to relate the stochastic description with the macroscopic description, we introduce $\langle N\rangle \equiv \sum_N N P_N$ and $\langle w_{N\to N+1}\rangle \equiv \sum_N w_{N\to N+1} P_N$, etc.

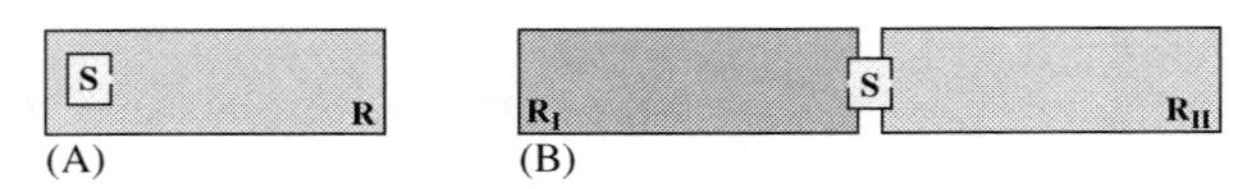

Fig. 3.6 (**A**) A system consisting of an open subsystem, S, and a reservoir, R. (**B**) A system consisting of an open subsystem, S, and two reservoirs, R_{I} and R_{II}

Then by the direct calculation, (3.59) gives[38]

$$\frac{d\langle N\rangle}{dt} = \langle w_{N\to N+1}\rangle - \langle w_{N+1\to N}\rangle. \tag{3.61}$$

The relation (3.60) is not very convenient because F_N concerns the whole system. By a physical argument we will rewrite $(F_{N+1} - F_N)$ in terms of chemical potential. We assume that the coupling between the open system and the reservoir is short-ranged so that the free energy is additive: $F_N = F_N^{\text{sys}} + F_{N^{\text{tot}}-N}^{\text{res}} + (\text{indep. of } N)$, where N^{tot} is the total number of particles in the whole system (S∪R), and the last term represents the interface between the open system and the environment. The chemical potentials of the open system and the reservoir are defined, respectively, by $\mu \equiv \partial F_N^{\text{sys}}/\partial N$ and $\mu^{\text{res}} \equiv \partial F_{N^{\text{res}}}^{\text{res}}/\partial N^{\text{res}}$. Then we have

$$F_{N+1} - F_N = \mu - \mu^{\text{res}}. \tag{3.62}$$

Then we find the following relation between the transition rates:

$$w_{N+1\to N} = w_{N\to N+1}\exp\left[\frac{\mu - \mu^{\text{res}}}{k_B T}\right]. \tag{3.63}$$

In combining with (3.61) we have the following formula:

$$\frac{d\langle N\rangle}{dt} = \langle w_{N\to N+1}\rangle \left(1 - \exp\left[\frac{\mu - \mu^{\text{res}}}{k_B T}\right]\right).$$

Or using $[A] = \langle N\rangle/V$,

$$\frac{d[A]}{dt} = k_{\text{in}}([A])\left(1 - \exp\left[\frac{\mu([A]) - \mu^{\text{res}}}{k_B T}\right]\right), \tag{3.64}$$

where

$$k_{\text{in}}([A]) = \langle w_{N\to N+1}\rangle/V. \tag{3.65}$$

Unlike the law of mass action, (3.64) allows the rate $(d[A]/dt)$ to depend on $[A]$ in nonlinear manner. In the van der Waals model of fluids, μ depends on $[A]$ in a sigmoidal way. As a result, there can be several equilibrium concentrations $[A]$ satisfying $\mu([A]) = \mu^{\text{res}}$ for certain value(s) of μ^{res}. This leads to the phase changes between the cooperative adsorption phase and the dilute adsorption phase.[39] In

[38] We supposed that the distribution of P_N vs. N is sharply peaked around $N = \langle N\rangle (\gg 1)$ and ignored the errors of order $o(\langle N\rangle)$.

[39] The stable equilibria should satisfy $\partial(\mu([A]) - \mu^{\text{res}})/\partial[A] > 0$.

addition to the multi-equilibria, formula (3.64) describes the relaxation and hysteresis of the concentration $[A]$.

3.3.3.2 Butler–Volmer Equation

The above result includes the so-called Butler–Volmer equation in electrochemistry [34] as a special case. Knowing that the transition rates are extensive (see Remark of Sect. 3.3.2.2) and that ν in (3.40) can absorb the global additive constants in $\Delta_{N,N+1}$'s and in F_N's, we choose $\nu = \tilde{\nu}\, V$. Then we can assume that $\Delta_{N,N+1} - F_N$ and $\Delta_{N,N+1} - F_{N+1}$ as well as $\tilde{\nu}$ are nonextensive, i.e., of $\mathcal{O}(V^0)$. Therefore, the weighed average of these two barrier height

$$\psi \equiv \Delta_{N,N+1} - (\alpha_+ F_N + \alpha_- F_{N+1}), \qquad \alpha_+ + \alpha_- = 1, \tag{3.66}$$

is also of $\mathcal{O}(V^0)$ for arbitrary choice of $\alpha_+ (= 1 - \alpha_-)$. Then (3.61) with (3.40) can be rewritten using this ψ and the chemical potentials of (3.62):

$$\frac{d}{dt}\left(\frac{\langle N\rangle}{V}\right) = \tilde{\nu}\exp\left[-\frac{\psi + \alpha_-(\mu - \mu^{\mathrm{res}})}{k_{\mathrm{B}}T}\right] - \tilde{\nu}\exp\left[-\frac{\psi - \alpha_+(\mu - \mu^{\mathrm{res}})}{k_{\mathrm{B}}T}\right]. \tag{3.67}$$

This form (3.67) is called Butler–Volmer equation.

The Butler–Volmer equation has been proposed by geometrical considerations (see textbook, e.g., [34] (p.1048), or recent articles [35, 36]), which is summarized in Fig. 3.7. Recently [37] proposed to use μ derived from a Cahn-Hilliard-type chemical free energy functional [38] so as to include the effect of inhomogeneous concentration. However, the above derivation implies that the experimental fitting with Butler-Volmer equation per se justifies no particular geometrical models of the adsorption because (3.67) is general: α_+ can be arbitrary, even negative, and this ψ can depend on the concentrations $\langle N\rangle / V$ or on the temperature. $\alpha_\pm$ can even depend on the kinetic parameters of the system. Therefore, it is only when $\alpha_\pm$ have separate justification that the geometrical representation Fig. 3.7 is meaningful.

3.3.3.3 Nonequilibrium Open System

When we constructed the Langevin equation, we have used the equilibrium state of the system to determine the relation (Einstein relation) between the viscous friction coefficient, γ, and the strength of random thermal force, b. Once the relation,

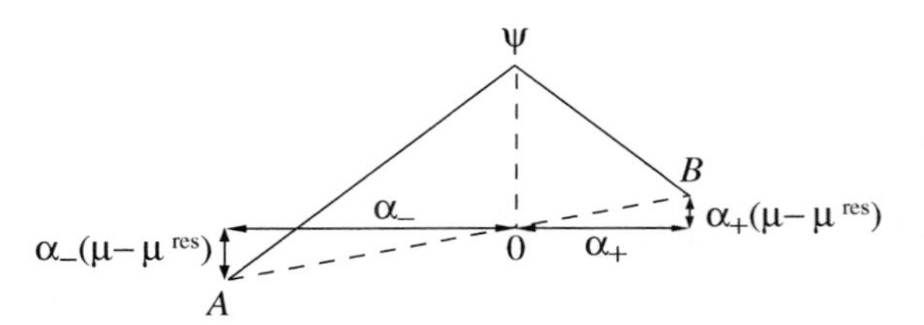

Fig. 3.7 Graphical representation of the "activation free energies," $\psi \pm \alpha_\mp (\mu - \mu^{\mathrm{res}})$ The levels of the "valleys" A and B are different by $\mu - \mu^{\mathrm{res}}$

$b = \gamma k_B T$, is obtained, we use the Langevin equation for the various situations including nonequilibrium conditions.

In the case of discrete stochastic processes, what we first determined using the equilibrium state, or the conditions of detailed balance (DB), was the *relation* between the transition rates, $w_{i\to j}$, and the physical parameters. In Sect. 3.3.1.5 we saw that if the transition rates are consistent with an equilibrium state, they can be formally represented in terms of ν, $\Delta_{i,j}$, F_j, and $k_B T$ (see (3.40)). After that, we related this formal result to the physical parameters, such as μ, μ^{Res}, or k_{in}, in a concrete case (see (3.63) and (3.65)).

As we discussed in Sect. 3.3.1.5,[40] we can then develop physical arguments and modify each transition rate so that those rates are applicable also in nonequilibrium conditions. Below we show how it is practically done in a modeling of nonequilibrium steady states: We consider an open system, S, and the two particle reservoirs, R_I and R_{II}, with which the system can exchange molecules, A. See Fig. 3.6 (B). We generalize the schema (3.58) to

$$\text{A(Res-I)} \rightleftharpoons \text{A(Sys)} \rightleftharpoons \text{A(Res-II)} \tag{3.68}$$

and denote the state of the whole system by the number of particles in the system, $N \equiv N_A$, and that in the reservoir R_I, $N' \equiv N_{\text{A}}^{\text{Res-I}}$. (Note that the number of particles in reservoir R_{II}, $N_A^{\text{Res-II}}$ depends on N and N').

The transitions we consider are
$(N, N') \to (N \pm 1, N' \mp 1)$: migration of an A molecule between S and R_I.
$(N, N') \to (N \pm 1, N')$: migration of an A molecule between S and R_{II}.

One of the easiest ways of modeling the transition rates for these transitions is to consider the special case where one of the two reservoirs is practically inaccessible due to an extremely high energy barrier between the system and the reservoir. We can then use the previous results (3.63) and (3.65) for the transition rates for the migration with the reservoir which is not blocked. A physical hypothesis that we take here is that the transition rates between the system and a reservoir are unchanged whether or not the migration with the other reservoir is blocked.[41] If this is a good approximation, we can repeat the same argument by exchanging the blocked reservoir and unblocked one, we obtain the following model:

$$w_{(N,N')\to(N+1,N'-1)} \simeq k_{\text{in}}^{\text{Res-I}}\, V,$$

$$w_{(N,N')\to(N-1,N'+1)} = w_{(N,N')\to(N+1,N'-1)} \exp\left[\frac{\mu - \mu^{\text{Res-I}}}{k_B T}\right],$$

---

[40] See *Nonequilibrium settings II*.

[41] Such a hypothesis is *not* always plausible. For example, some proteins might function so that the accessibility to the system is exclusive, called *alternative access model* [39]. (cf. *Exchange of binding* in Sect. 7.2.1.4.)

$$w_{(N,N')\to(N+1,N')} \simeq k_{\text{in}}^{\text{Res-II}}\, V,$$
$$w_{(N,N')\to(N-1,N')} = w_{(N,N')\to(N+1,N')} \exp\left[\frac{\mu - \mu^{\text{Res-II}}}{k_{\text{B}}T}\right]. \tag{3.69}$$

We recall the note on the *Relation between the transient nonequilibrium states and the nonequilibrium settings II* (Sect. 3.3.1.5). If we describe the process in the whole (N, N') space, we may construct a landscape that matches the above transition rates. (In fact the state space is separated into slices according to the total number of particles.) But it is not a practical description. We rather use a (reduced) representation where we look at only the open system and regard the particle environments as stationary reservoirs. Then each line of (3.69) represents the rate of entrance from R_{I}, of departure to R_{I}, entrance from R_{II}, and of departure to R_{II}, respectively. Here the chemical potential μ depends on N.

On the macroscopic level, the equation for [A] corresponding to (3.64) is

$$\frac{d[A]}{dt} = k_{\text{in}}^{\text{Res-I}}\left[1 - e^{\frac{\mu-\mu^{\text{Res-I}}}{k_{\text{B}}T}}\right] + k_{\text{in}}^{\text{Res-II}}\left[1 - e^{\frac{\mu-\mu^{\text{Res-II}}}{k_{\text{B}}T}}\right], \tag{3.70}$$

where μ, $k_{\text{in}}^{\text{Res-I}}$, and $k_{\text{in}}^{\text{Res-II}}$ are functions of the concentration [A] in the system. Under nonequilibrium condition, $\mu^{\text{Res-I}} > \mu^{\text{Res-II}}$, the value of [A] in the steady-state solution is determined by solving (3.70) with $d[A]/dt = 0$.

3.3.3.4 Stochastic Michaelis–Menten Kinetics

The merit and at the same time demerit of the Poisson distribution is that it is governed by only one parameter. Its variance, therefore, adds no new information to the knowledge of the average value. It is not the case when the discrete states are, for example, binary {0, 1}. In that case the stochastic level observation and description of chemical reactions bring more information than the rate constants. As an example, we will again take up the Michaelis–Menten kinetics (3.17), i.e., $\text{E} + \text{S} \rightleftharpoons \text{ES} \to \text{E} + \text{P}$ (see Sect. 3.2.2.2). We will show that, from the stochastic data of the slow process, $\text{ES} \to \text{E} + \text{P}$, we can extract the parameter of the fast process, $\text{E} + \text{S} \rightleftharpoons \text{ES}$.

Parameters of the Single-Enzyme Michaelis–Menten Kinetics [40, 41]

We introduce several stochastic parameters characterizing a particular event of the release of the product P from a single enzyme, E (see Fig. 3.8).

$t_v^{(k)}$: The kth $(1 \le k \le n)$ time interval of the free enzyme E.
$t_r^{(k)}$: The kth $(1 \le k \le n)$ time interval of the complex ES.
n: The times of the reaction, $\text{E} + \text{S} \to \text{ES}$, at which the product release, $\text{ES} \to \text{E} + \text{P}$, occurs.
t_{P}: The period between the last product release and the present product release:

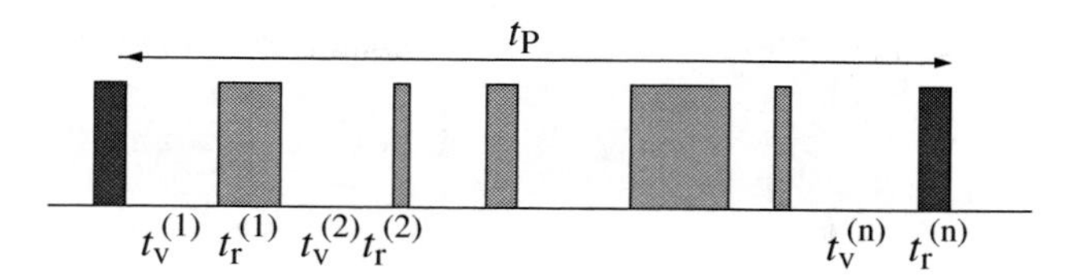

Fig. 3.8 Temporal "data" of single-enzyme reaction of Michaelis–Menten type. Each black [gray] bar denotes the complex ES which dissociates into E+P [E+S], respectively. The gaps between the bars denote the free enzyme E

$$t_P = \sum_{k=1}^{n} (t_r^{(k)} + t_v^{(k)}). \tag{3.71}$$

Setup of Problem

We suppose that one can only observe the events of the product release, that is the sequence of t_P. Also we assume that one can do the observation with different values of the substrate concentration, [S]. We introduce a Markov model which contains the following statistical parameters:

$T_v{}^{-1}$: The transition rate of the formation of the complex ES.
κ: The coefficient such that $T_v{}^{-1} = \kappa[S]$.
$T_r{}^{-1}$: The transition rate of the termination of the complex ES.
q: The probability of the production, ES $\to$ E + P, from the state ES.

Our goal is to determine as many parameters as possible from the observation.

Result: Informations Obtained from Stochastic Data

By the analysis using the probability of individual events, we have the following relations:

$$\frac{1}{\langle t_P \rangle} = \frac{\frac{1-q}{T_r}[S]}{\frac{1}{\kappa T_r} + [S]} \tag{3.72}$$

and

$$\frac{\langle t_P \rangle}{\langle t_P \rangle^2 - 2\langle t_P{}^2 \rangle} = \kappa[S] + \frac{1}{T_r}. \tag{3.73}$$

For the derivation, see Appendix A.3.2. The first result (3.72) is the rate of production per enzyme. This could have also been obtained from the macroscopic production rate, v, of (3.21) if we knew the total molar concentration of the enzyme,

$[E]^{tot}$. By contrast, the second result, (3.73), contains intrinsically the information of the stochastic data. From the data of $\langle t_P \rangle$ and $\langle t_P{}^2 \rangle$ for various concentrations of the substrate, [S], the formulas (3.72) and (3.73) allow us to obtain the three constants, κ, and $T_r{}^{-1}$, and q. Especially, $\langle n \rangle = (1-q)^{-1}$ is obtained.

3.4 Discussion

We have analyzed chemical reactions using two levels of descriptions, macroscopic deterministic and continuous description and mesoscopic stochastic and discrete description. The latter description includes the former as a result of coarse graining. But if the total number of a molecular species is not large, the coarse-grained description is not valid.

Even the discrete stochastic description may not be valid if the transition rate depends strongly on the parameters which are not represented by the chemical formula, such as the orientations or configurations of participating molecules and the internal states of solvent molecules. For example, a rapid water exchange between the hydration shell of a molecule and the surrounding fluid water is beyond the description of the previous sections.

Depending on what spatiotemporal scale is decisively important for the reaction rate, we should choose different methods, such as the Langevin equation, molecular dynamics simulation, or density functional description. Except for the full quantum descriptions of whole atoms and electrons, we always ignore some details as rapidly (often said "adiabatically") following degrees of freedom, but the justification of such hypothesis – separation of fast and slow degrees of freedom – is often done a posteriori by the comparison of model results with experimental observations.

References

1. R. Messina, C. Holm, K. Kremer, Phys. Rev. E **64**, 021405 (2001)
2. P.W. Atkins, *Physical Chemistry*, 4th edn. (Oxford University Press, Oxford, 1990)
3. H.A. Kramers, Physica **7**, 284 (1940)
4. P. Faccioli, M. Sega, F. Pederiva, H. Orland, Phys. Rev. Lett. **97**, 108101 (2006)
5. P. Hänggi, P. Talkner, M. Borkovec, Rev. Mod. Phys. **62**, 251 (1990)
6. C.W. Gardiner, *Handbook of Stochastic Methods for Physics, Chemistry for Natural Sciences*, 3rd edn. (Springer, 2004)
7. N.G. van Kampen, *Stochastic Processes in Physics and Chemistry*, Revised edn. (Elsevier Science, 2001)
8. R. Kubo, K. Matsuo, K. Kitahara, J. Stat. Phys. **9**, 51 (1973)
9. K. Kitahara, Adv. Chem. Phys. **29**, 85 (1973)
10. L.D. Landau, E.M. Lifshitz, *Statistical Physics (Part 1): Course of Theoretical Physics*, Vol 5, 3rd edn. (Butterworth-Heinemann, 1980), sects. 7
11. T.M. Cover, J.A. Thomas, *Elements of Information Theory* (John Wiley & Sons, Inc., New York, 1991)
12. M. Paniconi, Y. Oono, Phys. Rev. E **55**, 176 (1997)
13. Y. Oono, M. Paniconi, Prog. Theor. Phys. Suppl. **130**, 29 (1998)

14. S. Yukawa, cond-mat/0108421 (2001)
15. S. Sasa, H. Tasaki, J. Stat. Phys. **125**, 125 (2006)
16. K. Kawasaki, in *Phase Transitions and Critical Phenomena*, vol. 2, ed. by C. Domb, M.S. Green (Academic, New York, 1972)
17. D.T. Gillespie, J. Chem. Phys. **113**, 297 (2000)
18. D. Gillespie, J. Comput. Phys. **22**, 403 (1976)
19. D.T. Gillespie, J. Phys. Chem. **81**, 2340 (1977)
20. D.T. Gillespie, Ann. Rev. Phys. Chem. **58**, 35 (2007)
21. P.G. Bergmann, J.L. Lebowitz, Phys. Rev. **99**, 578 (1955)
22. J.L. Lebowitz, P.G. Bergmann, Ann. Phys. **1**, 1 (1957)
23. H. Spohn, J.L. Lebowitz, Adv. Chem. Phys. **38**, 109 (1978)
24. T.L. Hill, *Free Energy Transduction and Biochemical Cycle Kinetics*, dummy edn. (Springer-Verlag, New York, 1989)
25. J.E. Moyal, J. Roy. Stat. Soc (London) B **11**, 150 (1949)
26. M. Lax, Rev. Mod. Phys. **32**, 25 (1960)
27. H. Risken, *Fokker-Planck Equation*, 2nd edn. (Springer-Verlag, Berlin, 1989)
28. T. Speck, U. Seifert, Europhys. Lett. **74**, 391 (2006)
29. T. Harada, Europhys. Lett. **70**, 49 (2005)
30. T. Harada, S. Sasa, Phys. Rev. Lett. **95**, 130602 (2005)
31. J. Deutsch, O. Narayan, Phys. Rev. E **74**, 026112 (2006)
32. N.G. van Kampen, Can. J. Phys. **39**, 551 (1961)
33. K. Sekimoto, *Stochastic Energetics* (Iwanami Book Ltd., 2004, in Japanese)
34. J. Bockris, A. Reddy, M.Gamboa-Aldeco, *Modern Electrochemistry, Second Edition, Fundamentals of Electrodics,* vol 2A (Kluwer Acad., New York, 2000)
35. J.M. Rubi, S. Kjelstrup, J. Phys. Chem. B **107**, 13471 (2003)
36. J.M. Rubi, D.B.S. Kjelstrup, J. Phys. Chem. B **111**, 9598 (2007)
37. G.K. Singh, G. Ceder, M.Z. Bazant, Electrochim. Acta **53**, 7599 (2008)
38. J.W. Cahn, J.E. Hilliard, J. Chem. Phys. **28**, 258 (1958)
39. C. Tanford, Annu. Rev. Biochem. **52**, 379 (1983)
40. R. Gentry, L. Ye, Y. Nemerson, Biophys. J. **69**, 356 (1995)
41. R. Gentry, L. Ye, Y. Nemerson, Biophys. J. **70**, 2033 (1996)

Part II
Basics of Stochastic Energetics

Chapter 4
Concept of Heat on Mesoscopic Scales

In this chapter we introduce the concept of heat in the physics of the Langevin equation, as well as its discrete version.

In the fluctuating world, some important aspects of the system cannot be described in terms of the ensemble average of physical observables. Einstein was among the first to study the connection between fluctuations and energy in the fluctuating world [1]. Then Kramers described this connection in the context of stochastic processes. In his seminal paper [2], he represented a chemical reaction as a process in which the state of molecule(s) undergoes fluctuations and eventually changes its state qualitatively. That is, the molecules' state (reaction coordinate) which was originally found near the bottom of a potential ($U(x)$) can overcome a potential barrier, see Fig. 4.1. Kramers analyzed this situation by constructing a Fokker–Planck equation, called the Kramers equation, and derived the reaction rate (or transition rate).

Now, our interest is how much energy is exchanged between the system and the thermal environment along an individual realization of such process. Does the energy exchanged vary from one realization to other, or not? Knowing the conservation law of energy, we may argue that the energy absorbed from the thermal environment equals to the height of the potential energy barrier. In order to *show* explicitly the transferred energy and answer to the above question, the Fokker–Planck (Kramers) equation is not appropriate, because this equation deals with an ensemble of stochastic processes, but not an individual one. Therefore, our study of energetics should be based on the Langevin equations[1]:

$$\frac{dp}{dt} = -\frac{\partial U(x,a)}{\partial x} - \gamma \frac{p}{m} + \xi(t), \quad \frac{dx}{dt} = \frac{p}{m} \quad \text{(underdamped)} \tag{4.1}$$

or

$$0 = -\frac{\partial U(x,a)}{\partial x} - \gamma \frac{dx}{dt} + \xi(t) \quad \text{(overdamped)}, \tag{4.2}$$

if the inertial term is negligible.

[1] In the following, we shall omit the symbol ^ of $\hat{X}$, etc., denoting a random variable or a stochastic process, unless some confusion is possible.

Sekimoto, K.: *Concept of Heat on Mesoscopic Scales*. Lect. Notes Phys. **799**, 135–174 (2010)
DOI 10.1007/978-3-642-05411-2_4

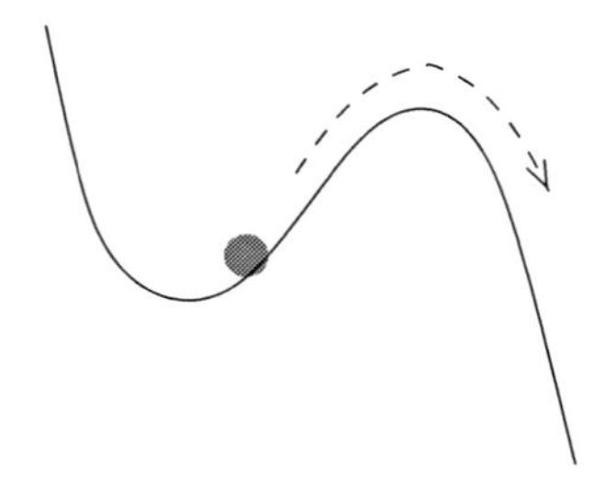

Fig. 4.1 Kramers' picture of chemical reaction: A new state is reached once a potential barrier is overcome (*dashed arrow*) with the help of the thermal fluctuations

The fluctuating world described by a Langevin equation supposes a minimal triad consisting of a system, a thermal environment, and an external system. Along a single realization of the stochastic process, energy is exchanged among these three components. Heat is defined as the energy exchanged between the system and the thermal environment. The law of energy balance is derived for each realization. Many theoretical models have been studied by this framework, and several experimental demonstrations of this law have been done. For numerical simulations, the framework of energetics turned out to require more precision than a simple convergence of the solution of the Langevin equation. The framework of energetics is also formulated for the discrete Langevin equation. In both continuous and discrete cases, the ensemble properties of energetics can be reduced once the energetics for an individual realization is formulated.

When a fluctuating system is in contact with two thermal environments, the energetics shows some aspects which are not found in the macroscopic case. Of special interest is the energetics of autonomous heat engines: (i) Feynman pawl and ratchet wheel and (ii) Büttiker–Landauer ratchet.

In this chapter we focus mainly on heat. The aspects related to work will be discussed in the next chapter.

4.1 Framework

4.1.1 * The Similarity of Setup Between the Fluctuating World and the Thermodynamics Leads to a Natural Definition of Mesoscopic Heat

4.1.1.1 Triad: System, Thermal Environment, and External System

The Langevin equations (4.1) or (4.2) involve three parts, i.e., system, thermal environment, and external system as mentioned in Chap. 1. The summary below emphasizes the parallelism with thermodynamics described in Chap. 2.

System: Of the whole world, a part which is properly cut out is called the system. Its state is represented by x (and p) in (4.1) or (4.2). These equations describe the evolution of the state of the system. We consider that the potential energy and the kinetic energy belong to the system.[2]

Thermal environment: The background system to which the system is connected is called the thermal environment. This environment is characterized in itself by a single parameter, the temperature T. The conservation laws for the total energy, mass or volume, and momentum have no effect on the state of the thermal environment. The environment returns instantaneously to its equilibrium state and keeps no memories of the system's action in the past. The interaction between the system and the thermal environment is characterized by the friction coefficient γ as well as temperature. The strength of the thermal random force, $\xi(t)$, is specified by these two parameters.[3]

External system: It is an agent which is capable of controlling macroscopically the system through the parameter a of the potential energy $U(x, a)$. The term "external" indicates that the evolution of the parameter a is not determined by Eqs. (4.1) or (4.2). In the case of coupled Langevin equations, e.g., of the system described by the state variables (x_1, x_2), the x_2 could take the position of a, that is, the external system with respect to the subsystem, x_1. For the moment, however, we reserve the concept of the external system for the variables whose dynamics are not determined by stochastic equations.[4]

The above triad is our starting point. In terms of characteristic timescale, we normally suppose that they represent the intermediate (system), micro (thermal environment), and macro (external system) scales, respectively. Not all composite fluctuating systems may be decomposed into this type of triad. But we start from an elementary prototype.

4.1.1.2 Definition of Heat in the Fluctuating World

In the macroscopic world, we do not observe the vigorous molecular motions of a hot object. We can detect it by touch, by the melting of ice in contact, etc., through some amplifying mechanisms of microscopic thermal motions. In the microscopic world, on the contrary, we have no notion of "hotness," but everything is represented

---

[2] This point will be discussed later in more detail in Sect. 4.1.2.

[3] Unlike the macroscopic thermodynamics, the definition of thermal environment does not exclude the interaction energy between the system and the thermal environment. This point has not been well discussed.

[4] More discussion is given in Chap. 6.

We do not allow the external system to control the interaction between the system and the thermal environment. In other words, γ should not depend on a. The reason will be described later (Sect. 4.1.2.2).

as motion. The fluctuating world is between these two limits. Let us try to imagine the mechanical processes undergone by a Brownian particle.

The law of action and reaction: Depending on the velocity of a Brownian particle dx/dt, this particle should receive the unbalanced number of collisions with environment, e.g., solvent molecules, more from forward than from backward. The total transfer of momentum per unit time is the restoring force $-\gamma dx/dt$ and the random thermal force $\xi(t)$ with zero mean. We suppose that the *law of action and reaction* holds always between the Brownian particle and its environment: when the environment exerts a force $-\gamma dx/dt + \xi(t)$ on the particle, the particle exerts the reaction force, $-(-\gamma dx/dt + \xi(t))$, on the environment.

Concept of heat: Let us denote by $dx(t)$ the evolution of x over a time interval dt, according to the Langevin equation. It is determined for each realization. The work done by the particle on the environment upon this change $dx(t)$ is equal to the product of this variation and the above mentioned reaction force, that is, $-(-\gamma dx/dt + \xi(t)) \circ dx(t)$. We adopt the Stratonovich-type product $\circ$ in the above. (Justification is given below.) This work can be either positive or negative. When positive, it represents energy lost by the system. From the standpoint of the environment, energy $(-\gamma dx/dt + \xi(t)) \circ dx(t)$ is lost as work to the Brownian particle. In any case, $(-\gamma dx/dt + \xi(t)) \circ dx(t)$ is the energy transferred from the environment to the system. We then define this energy transfer as heat [3]. Although the microscopic motions in the thermal environment are not explicitly represented in the Langevin equation, the law of action and reaction allows us to identify how much energy has been transferred.

Sign convention and the formula of heat: As in macroscopic thermodynamics (Chap. 2), we assign the positive sign for the energy received by the system. For instance, when a positive amount of work is done on the system by the thermal environment, we say that the system receives a positive amount of heat $d'Q > 0$. The equation defining $d'Q$ then reads

$$d'Q \equiv \left(-\gamma \frac{dx}{dt} + \xi(t)\right) \circ dx(t). \tag{4.3}$$

When inertia is taken into account, we replace dx/dt above by p/m. We shall use the same sign convention for the work done by the external system on the system (see below). In (4.3) and hereafter, we add "d dash (d')" to the heat Q or the work W when it concerns the process during an infinitesimal time, dt. We distinguish this from "d" and reserve the latter for the differential.[5] In general $d'Q$ is not differentials.

[5] That is, the differentials of a monovalent function, like $dx(t)$, or the total differentials of multi-valued function, like $dU(x, a)$.

4.1.2 * Energy Balance Along a Single Realization Follows from the Definition of Heat

4.1.2.1 The Law of Energy Balance

The definition of heat (4.3) and the Langevin equations (4.1) or (4.2) are sufficient to establish the relation of energy balance. See Fig. 4.2.

Energy balance with inertia: The reaction force by the thermal environment in (4.3) is rewritten using (4.1). Then $d'Q$ becomes $d'Q = (dp/dt + \partial U/\partial x) \circ dx(t)$. We then use the identities for the kinetic energy and the potential energy:

$$\frac{dp}{dt} \circ dx(t) = \frac{dp}{dt} \circ \frac{p}{m} dt = d\left(\frac{p^2}{2m}\right),$$

$$\frac{\partial U}{\partial x} \circ dx(t) = dU(x(t), a(t)) - \frac{\partial U}{\partial a} \circ da.$$

We recall that Stratonovich calculus allows us to use the usual rules of calculus about the differentials. After substituting these two expressions in the last expression of $d'Q$, we obtain

$$d\left(\frac{p^2}{2m} + U(x, a)\right) = d'Q + \frac{\partial U}{\partial a} \circ da. \tag{4.4}$$

Now we identify the total energy E of the system

$$E \equiv \frac{p^2}{2m} + U(x, a) \tag{4.5}$$

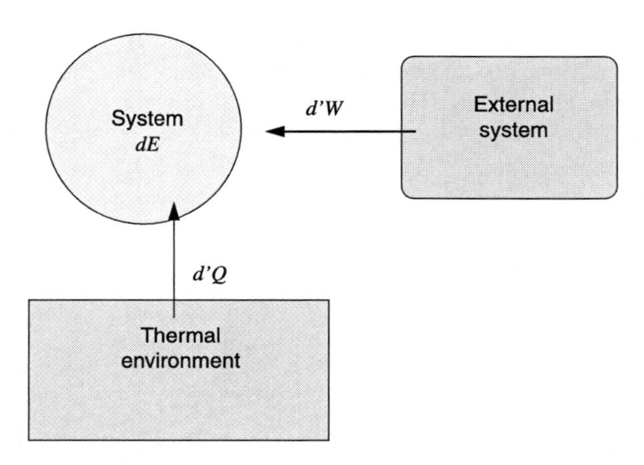

Fig. 4.2 Energy balance among dE, $d'Q$, and $d'W$

and the work done by the external system on the system

$$d'W \equiv \frac{\partial U}{\partial a} \circ da. \tag{4.6}$$

Then (4.4) is expressed in the form shown in Fig. 4.2:

$$dE = d'Q + d'W. \tag{4.7}$$

Equation (4.7) expresses the balance of energy concerning the system. It is analogous to the first law of thermodynamics; both are based on the principle of energy conservation in nonrelativistic system.[6] We would stress that (4.7) holds for each realization of the stochastic process. A purely mechanical energy balance, $dE = d'W$, will hold in case that a can be changed "suddenly," that is, very fast with respect to the characteristic relaxation time of x, but slow enough so as not to violate the time resolution of the Langevin model.

Energy balance without inertia: The definition of heat (4.3) and the Langevin equation (4.2) leads to the following result:

$$dU(x,a) = d'Q + \frac{\partial U}{\partial a} \circ da. \tag{4.8}$$

The calculation is very similar to and even simpler than the previous case. The energy balance relation is

$$dU = d'Q + d'W \quad \text{(when the inertia can be neglected).} \tag{4.9}$$

Compared to (4.7), the kinetic energy term is missing from the complete differential form. However, if the temperature T depends on the variable x, the evolution equation (1.94) in Sect. 1.3.2.1 leads to the general form of energy balance (4.7).

The above is the basis of stochastic energetics, i.e., the energetics of a single realization of stochastic process associated with the Langevin equation. The assignments of E, $d'Q$, and $d'W$ are all consistent with the law of mechanics and the usual definition of work. And the law of energy balance is not postulated but is derived from the Langevin equation.

The choice of Stratonivich-type calculus for $d'Q$ (4.3) was crucially important for the derivation of the law of energy balance. In the overdamped case, the temporal integration of (4.3) requires some care. The separate integral of $-\gamma dx/dt \circ dx(t)$ or of $\xi(t) \circ dx(t)$ does not assure finite results.

---

[6] Can we conceive the similar problem about a relativistic system? We should remember that the thermal environment imposes a particular reference frame to define $dx/dt = 0$, i.e., the Langevin equation does not support the Galilean invariance.

4.1.2.2 Remarks

General

1. Energetics about a single realization is more detailed than energetics based on the Fokker–Planck equation. The latter describes an ensemble statistics at each instant of time. For the purpose of studying the energy exchange associated with a particular fluctuation event the latter method is not adequate. For example, an exceptionally large heat absorption should be associated to a rare event of climbing over an energy barrier. This characteristics will be masked if we used the ensemble statistics because such rare fluctuations in the ensemble are not synchronized in time (See also Sect. 4.2.2 below).
We know, however, that the Fokker–Planck equation contains the same information as the Langevin equation. There is no contradiction between this statement and the previous one. The above mentioned difference concerns their solutions, i.e., the difference between the instantaneous probability distributions and the single realization of stochastic process.[7]
2. In some literatures after [3] the heat for the overdamped system was defined using the following identity:

$$dU(x,a) = \frac{\partial U}{\partial x}dx + \frac{\partial U}{\partial a}da. \tag{4.10}$$

They argued that the term, $\frac{\partial U}{\partial x}dx$, must be the heat since it is the total change of energy dU *less* the work, $\frac{\partial U}{\partial a}da$. For the underdamped case, the corresponding identity for $E = p^2/(2m) + U$ reads

$$\begin{aligned} dE(p,x,a) &= \frac{p}{m}dp + \frac{\partial U}{\partial x}dx + \frac{\partial U}{\partial a}da \\ &= \left[\frac{dp}{dt} + \frac{\partial U}{\partial x}\right]dx + \frac{\partial U}{\partial a}da. \end{aligned} \tag{4.11}$$

The last expression gives the definition of heat such as $d'Q = \left[\frac{dp}{dt} + \frac{\partial U}{\partial x}\right]dx$, with the quantities in [ ] being the negative of inertial and potential forces. These are mathematically correct as an identity. Einstein treated heat in the above mentioned way when he formulated statistical mechanics.[8] Also this way will be adopted for the energetics of the discrete Langevin equation (Sect. 4.1.2.6).
The definition of heat (4.3) is, however, more general for the system with inhomogeneous temperature $T(x)$ [4]: The overdamped Langevin equation, interpreted in the Stratonovich sense, is (see (1.94))

---

[7] Roughly speaking, the correspondence between $(\partial P(x,t)/\partial t)dt$ and $\{\langle dE\rangle, \langle d'W\rangle, \langle d'Q\rangle\}$ is far from 1 to 1, while the mapping $\{dx(t), da(t)\} \mapsto \{dE, d'W, d'Q\}$ in the overdamped case is almost bijection.

[8] Y. Oono pointed this out to the author.

$$\gamma \frac{dx}{dt} = -\frac{\partial}{\partial x}\left[U(x,a) + \frac{k_{\rm B}T(x)}{2} \right] + \xi(t), \tag{4.12}$$

where the random thermal force, $\xi(t)$, is multiplicative. In this case, the definition of heat, (4.3), leads to energy balance (4.7), where the energy E is redefined as $E \equiv k_{\rm B}T(x)/2 + U(x,a)$, not as (4.5). Also the definition based on the law of action reaction, (4.3), provides a clearer view of heat when stochastic energetics is generalized for the fluctuating hydrodynamics (Sect. A.4.7.1) and for the suspension of hard spheres (Sect. A.4.7.2).

About Heat

1. Long before the introduction of the concept of heat [3], an attempt to relate the Langevin equation to the thermal physics had been made [5]. They derived the following expression about the energy, $dE = d\,[p^2/2m + U(x,a)]$:

$$dE = -\frac{2\gamma}{m}\left(\frac{p^2}{2m} - \frac{k_{\rm B}T}{2}\right) dt + d'W + \sqrt{2\gamma k_{\rm B}T}\,\frac{p}{m}\cdot dB_t. \tag{4.13}$$

Here "·" before dB_t denotes the product of Itô type. See Appendix A.4.1 for the derivation. Equation (4.13) is mathematically equivalent to (4.7), but the nonanticipating term ($\sim \cdot dB_t$) was explicitly sorted out. As mentioned above, Itô-type calculus is preadapted to the ensemble average, and the first term on the right-hand side has an allusion that the Langevin dynamics generates only the relaxation processes to equilibrium, like the equipartition. By contrast, (4.7) emphasizes balance of energy for individual processes.
2. Can heat defined in the above be derived by the projection method, the method which transformed the microscopic Hamiltonian dynamics to the Langevin equation (see Sect. 1.2.1.5)? There are simple cases where we can identify energies or heat on two levels: In the model from which Zwanzig demonstrated the nonlinear Langevin equation (Sect. 1.2.1.3), the potential energy $U_0(X)$ in the resulting Langevin equation (1.41) was nothing but the proper energy of X in the starting Hamiltonian dynamics, apart from an additive constant that reflects the eliminated microscopic degrees of freedom. A similar situation will be realized when a Brownian particle in a fluid moves under an optical trapping potential [6].
However, it is not the case in general. While the law of energy conservation holds in both levels of description, what are meant by energy and by heat differ generally from one level to the other.[9] The full analysis of energy and heat at different levels of description will be done in Chap. 6 [7]. Here we note only two points:

(1) The potential energy $U_{\rm eq}(\boldsymbol{A})$ in the Langevin equation (1.42) obtained by projection method is in fact a (constrained) free energy (see (1.45)).

[9] "Each level has its own thermodynamics."

(2) The heat [energy] of different levels of description can be related quantitatively.

We will, therefore, continue to use the definitions of the heat and energy in this section. Whenever necessary, we can translate those quantities to their counterpart on the microscale.[10]

About Work

1. In Sect. 4.1.1.1 we prohibited the external system to control the interaction between the system and the thermal environment. One reason is now clear from the above formalism: our definition of work is based on the potential force. Related but more fundamental reason is that the stochastic energetics is conscious about the scales of description: For example, let us consider the rotational Brownian motion of a vane in a fluid medium (cf. Fig. 4.6). The friction coefficient γ can be modified by changing the shape of the vanes.[11] In order to assess the work to change the friction coefficient γ for this motion, we need structural and mechanical information about the interface between the system and the thermal environment. Such a requirement conflicts with our assumption about the thermal environment: the latter should be memory-free, nonstructured, and uncorrelated. A solution for the above example would be to include some region of the thermal environment near the interface as a part of the extended system. See Sect. 7.1.1.5 for a concrete procedure of redefining the system.
2. The work $d'W$ has been defined such that it is 0 if the control parameter a is constant. For example, suppose that a represents a constant external force $g_0 > 0$ applied to a Brownian particle (position $x(t)$) in the direction of $x > 0$. According to our definition, the external system applying this force does not do work while the particle drifts toward $x(t) \to \infty$. We count $(-g_0\, dx)$ as a part of the change of the internal energy of the *system*. Therefore, we regard that the particle dissipates its potential energy $-g_0 x$. In contrast, if an external system increases the force strength from $g = 0$ to g_0, then $\int_0^{g_0}(-x(t)) \circ dg(t)$ should be counted as the work on the system, $d'W$.[12]

4.1.2.3 Examples

Deformation of an Ideal Chain

Let us consider an ideal chain consisting of many microscale rigid rods joined together by completely flexible joints (Fig. 4.3). The only interactions are steric repulsion among rods. Therefore, the system's potential energy, $U(\{\boldsymbol{x}_i\})$, is constant

---

[10] The impatient reader could give a quick glance at Sect. 6.2 (especially the Eq. (6.23)) of Chap. 6, where $\tilde{F}$ stands for $U(x, a)$ here.

[11] This example has been brought by T. S. Komatsu.

[12] Jarzynski discussed this issue [8] in the context of Jarzynski nonequilibrium work relation (Sect. 5.4.1).

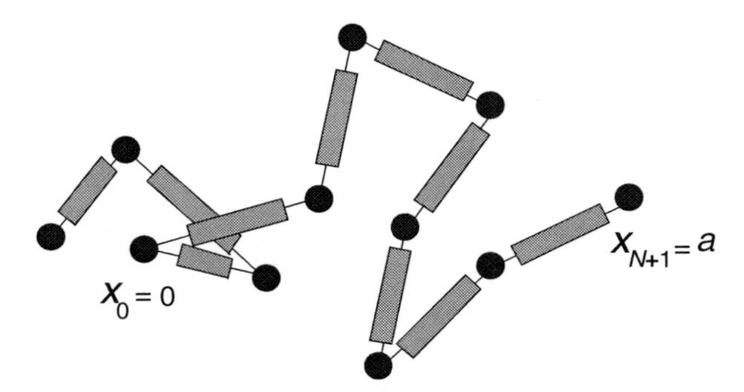

Fig. 4.3 Ideal chain consisting of rigid rods (*gray rectangles*) and free joints (*filled discs*) immersed in a thermal environment. One end, $\boldsymbol{x}_{N+1} \equiv \boldsymbol{a}$, can be externally controlled, while the other end, $\boldsymbol{x}_0$, is fixed at $\boldsymbol{x} = 0$

as function of the positions of joints, $\{\boldsymbol{x}_i\}$ $(0 \leq i \leq N+1)$. Equilibrium statistical mechanics tells that this system shows (entropic) elasticity with respect to the end-to-end distance $|\boldsymbol{x}_{N+1} - \boldsymbol{x}_0|$, and the average elastic force for a given end-to-end distance is proportional to $k_{\mathrm{B}}T$. Now we are interested in the energetics during the free shrinkage of this chain after the end point $\boldsymbol{x}_{N+1} \equiv \boldsymbol{a}$ is released while the other end point $\boldsymbol{x}_0$ is fixed at $\boldsymbol{x} = 0$. Above all, no work is done by the external system; $d'W = 0$. If the overdamped Langevin equation describes well the conformational changes of this system, the balance of energy (4.7) tells immediately that $d'Q = dU = 0$. If the inertia of the rods and joints are not negligible, (4.9) tells that the heat should balance with the kinetic energy, through $d'Q = dE$. Using this framework of energetics, energy exchange can be studied for different initial conditions. One situation of special interest might be the relaxation of a stretched long chain, either with or without Brownian particle at the movable end, $\boldsymbol{x}_{N+1}$. This issue provokes many problems about the heat on mesoscopic scales. We will discuss this in depth in Chaps. 5 and 6.

"Jump-and-catch" mechanism of binding

In proteins, often a large conformational change is caused by the interaction between very localized and specific binding sites, see Fig. 4.4 (top). We might ask

(i) How does a short-ranged bonding interaction (of some Å range) cause a large-scale conformational change of proteins (of 10-nm range)?
(ii) What supplies the energy for the proteins to deform prior to the release of the bonding energy?

There is no work in (4.7), $d'W = 0$. The balance of energy is then $dU = d'Q$. As for the potential energy U, Fig. 4.4 (bottom) gives the decomposition of the energy into the short-range binding interaction, U_{bond}, and the energy of protein's

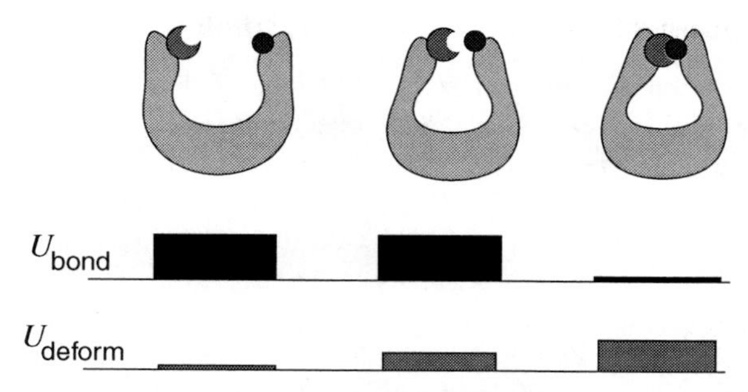

Fig. 4.4 Schematic presentation of the "jump-and-catch" transition (*top*) and its energetics (*bottom*). A protein (*shaded object*) has an intramolecular binding pair (*filled disc* and *filled crescent*). The height of rectangles shows the energies $U_{\rm bond}$ (upper law) and $U_{\rm deform}$ (lower law), respectively. The process of jump (from *left* to *right*) requires to borrow energy from the thermal environment

conformation change, $U_{\rm deform}$. In order to achieve the transition from the "open" state (left) to the "closed" state (right), the thermal environment can transfer heat $d'Q$ to the protein.[13] The supplied energy is stocked as the potential energy for deformation, $dU_{\rm deform} = d'Q > 0$. If bond formation is unsuccessful, the protein conformation will return to the relaxed state, and the stocked energy is returned to the thermal environment as heat. If bond formation occurs, the binding energy of the bond $-dU_{\rm bond} > 0$ is released to the thermal environment, while the deformation energy $dU_{\rm bond}$ is retained in the proteins. If the gain of the bond energy $|dU_{\rm bond}|$ overcomes the energy cost of deformation, $dU_{\rm deform}$, we have $(dU_{\rm deform} + dU_{\rm bond}) < 0$, and the transition is stabilized.

In the above, the thermal environment worked as a bank of (free) energy. It allowed the protein to explore large deformations and find a short-ranged binding pair which were originally far apart. While the average energy of fluctuations is $\sim k_{\rm B}T \simeq 4$pN nm per a degree of freedom, rare fluctuations may attain the heat transfer of more than $10k_{\rm B}T$. If the protein waits long enough time, e.g., $\sim$ ms, it can attain large conformational fluctuations.[14]

Borrowing energy from the thermal environment is a characteristic feature of the fluctuating world. This principle is fully used to extract work from the thermal or chemical energy. Energetics of mesoscopic thermal and chemical engines are discussed in Sects. 8.1.1 and 8.1.2. Stochastic energetics brings much more informations with respect to the ensemble theory of statistical mechanics.

---

[13] Here we abuse the notations $d'Q$, etc., to represent finite variations.

[14] A rough order estimation: if a binding pair diffuses with diffusion coefficient $D = 10^{-7}{\rm cm}^2/{\rm s}$ from the initial separation $\ell = 5$ nm, the pair can meet in $\simeq \ell^2/(2D) \simeq 10^{-6}$ s times the Boltzmann factor $e^{\epsilon/k_{\rm B}T}$, where ϵ is the barrier energy of protein deformation. Then the waiting for 1 ms may allow to cross a barrier ϵ of $\tau e^{\epsilon/k_{\rm B}T} = 1$ ms, or $\epsilon \simeq 7k_{\rm B}T$.

4.1.2.4 Experimental Demonstration of Energy Balance

What does the measurement of heat mean if heat is directly associated with the fluctuating motions? Is $d'Q$ neither readily observable in experiments nor calculated from observable quantities? The first question concerns the very question of the heat, and it is discussed in Chap. 6. For the second question, the brief answer is yes. Recent experimental technique [6, 9] enabled to measure both the energy change dU and the work $d'W$ at high accuracy.[15] The heat $d'Q$ is, therefore, assessed using the balance of energy (4.7) i.e., $dU = d'Q + d'W$. They used either optically trapped bead [6] or brass wire-held pendulum [9]. See Fig. 4.5. They observed, on the one hand, the time series of the fluctuating variable $x(t)$, i.e., the position of the trapped bead or the tiny rotation angle of the pendulum, respectively, at a high temporal resolution (faster than ms). On the other hand, the potential function $U(x, a)$, and therefore the (reaction) force $\partial U/\partial x$, are identified through separate calibrating experiment. With these data of $x(t)$ together with the protocol of the external parameter, $a(t)$, the energy change $dU(x, a)$ and the work $d'W = \partial U/\partial x\, da(t)$ were deduced.

The heat $d'Q$ was not measured independently by these experiments, nor was the law of energy balance proven. However, these experiments also demonstrated the Jarzynski's nonequilibrium work theorem and the fluctuation theorem (see Sects. 5.4.1 and 5.4.2). The latter verification implies that the process $x(t)$ is a Markovian process. Therefore, the overdamped Langevin equation is a good model for these experiments.[16]

We should note that, in the experiments cited above, the potential energy is a true energy, i.e., without entropic part. Therefore, the molecular heating up of the local environment should be observed if a very sensitive thermometer is developed and if experiments are conducted at very low temperature. For the moment ultralow temperature experiments involving the energy balance have not been done. But there are already the measurements of thermal conduction by phonons [10] and that by photons [11].

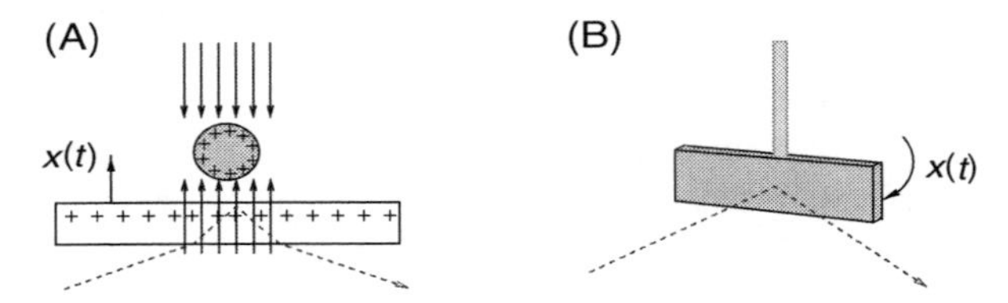

Fig. 4.5 Schematic setups for measuring dU and $d'W$. (a) Optically trapped bead [6] and (b) brass wire-held pendulum [9]. The dashed lines are light for detecting $x(t)$. The parallel *arrows* (a) and *rectangle* (b) denote the optical+electrostatic and mechanical mechanisms, respectively, for the restoring forces

[15] For example, [6] had an error of only $\sim k_B T/4$ vs. about $10 k_B T$ of total energy variations.

[16] The effect of inertia is too short lived (~ 10 ns) to be captured.

4.1.2.5 Numerical Precision Required for the Energetics

Energetics of stochastic process brings a new notion in the numerical calculation of the Langevin equation or of the SDE. We have to distinguish between the convergence of solution and the convergence of the associated energetics. In fact the latter requires a higher precision of the solution than its mere convergence.

It concerns the "overdamped" case, where we neglect the inertia. Let us take a simple Langevin equation:

$$0 = -\frac{dU(x)}{dx} - \gamma\frac{dx}{dt} + \xi(t), \tag{4.14}$$

where $\langle\xi(t)\rangle = 0$ and $\langle\xi(t)\xi(t')\rangle = \sqrt{2\gamma k_{\rm B}T}\,\delta(t-t')$. Our object is to solve this equation by using discretization in time with a finite time step, h. We fix the total time interval of integration. The total number of steps N is, therefore, $N \propto \frac{1}{h}$.

The discretized solution with an arbitrary $\xi()$ is said to converge to the true solution of (4.14) with the same $\xi()$ if the difference between the discretized (rectilinear) $x(t)$ approaches indefinitely to the true $x(t)$ in the limit of $h \to 0$. Let us take the simplest convergent scheme of discretization:

$$0 = -\frac{dU(\tilde{x}_t)}{dx}h - \gamma(x_{t+h} - x_t) + w_{t,t+h}, \tag{4.15}$$

where[17] $w_{t,t+h} \equiv \sqrt{2\gamma k_{\rm B}T}\,(B_{t+h} - B_t)$ with B_t being the Wiener process, and $\tilde{x}_t \equiv \theta x_t + (1-\theta)x_{t+h}$ with θ being arbitrary for the moment except that it is normal to choose such that $0 \le \theta \le 1$. In fact (4.15) is an integral of (4.14) from t to $t+h$. There is, therefore, a value $\theta \in [0, 1]$ for which (4.15) is exact.[18] We can show that, with any choice of θ, the error in $x(t)$ by this scheme is $\mathcal{O}(h^{\frac{1}{2}})$. (See Appendix A.4.2.) In other words, the difference between the approximate $x(t)$ of (4.14) and the true $x(t)$ of (4.15) decreases as $\mathcal{O}(N^{-\frac{1}{2}})$ for $N \to \infty$. The explicit and lowest order Euler scheme uses $\theta = 0$. Therefore, this scheme is convergent with the error in $x(t)$ being $\mathcal{O}(N^{-\frac{1}{2}})$.

In order to numerically assess heat, we need to integrate the heat $d'Q$ of (4.3). The unambiguous choice to discretize $d'Q$ in the lowest order is

$$\Delta Q \equiv \frac{1}{h}\big[-\gamma(x_{t+h} - x_t) + w_{t,t+h}\big](x_{t+h} - x_t). \tag{4.16}$$

Similarly, the change in the potential energy should be expressed as

$$\Delta U \equiv U(x_{t+h}) - U(x_t). \tag{4.17}$$

---

[17] If we abuse the integral and derivative, $w_{t,t+h} = \int_t^{t+h} \xi(s)ds$ and $\xi(t) = dB_t/dt$.

[18] But such θ depends on x, $\xi()$, and t.

The convergence of energetics, therefore, requires the error of each step, $\Delta r \equiv \Delta U - \Delta Q$, to be less than $\sim h$ so that the cumulated error $\sim N\Delta r$ vanishes for $N \to \infty$. In Appendix A.4.2 we show that this condition requires taking $\theta = \frac{1}{2}$ in (4.15). With this choice, the error in the solution of $x(t)$ itself is better than otherwise, and the convergence in $x(t)$ is of $\mathcal{O}(N^{-\frac{3}{2}})$ instead of $\mathcal{O}(N^{-\frac{1}{2}})$.

What occurs if we took $\theta \neq \frac{1}{2}$ in (4.15)? Suppose, for example, the case of $U(x,a) = U(x)$ with a single minimum. Apparently the time-discretized $x(t)$ behaves normally, since this is a good approximate of the true solution. The residence time distribution of $x(t)$ will approach the canonical distribution. Therefore, the time average of $U(x_t)$ approaches its canonical average. At the same time, the energy balance of (4.9) should be $dU = d'Q$, because $d'W = 0$ here. Numerically, however, the simple scheme predicts $\Delta Q = \mathcal{O}(h)$ at each step, that is, heat is generated or absorbed steadily, in contradiction to the conservation of energy.

4.1.2.6 Energy Balance for Discrete States

For a particular realization of a discrete Markov process, the system undergoes a transition from the state $i_{\alpha-1}$ to the state i_α at time t_α with $1 \leq \alpha \leq n$ and $0 < t_1 < \cdots < t_n < t$. The energy level of the system changes accordingly. We denote by $E_i(a)$ the energy of state i, which may depend on time through the external parameter a. In the context of master equation, the energy balance of this process has long been discussed: along the process between $t = 0 (\equiv t_0)$ and $t = t(\equiv t_{n+1})$ it is written as follows:

$$\Delta E = \Delta' W + \Delta' Q, \tag{4.18}$$

where the change of energy ΔE,

$$\Delta E = E_{i_n}(a(t)) - E_{i_0}(a(0)), \tag{4.19}$$

is decomposed into work which is assigned as

$$\Delta' W \equiv \sum_{\alpha=0}^{n} \left[E_{i_\alpha}(a(t_{\alpha+1})) - E_{i_\alpha}(a(t_\alpha)) \right] \tag{4.20}$$

and heat defined by

$$\Delta' Q \equiv \sum_{\alpha=1}^{n} \left[E_{i_\alpha}(a(t_\alpha)) - E_{i_{\alpha-1}}(a(t_\alpha)) \right]. \tag{4.21}$$

In order to deal with individual processes, we have introduced the discrete "Langevin equation" (3.32), i.e.,

$$\frac{d}{dt}|\psi_t\rangle = \sum_j (|j\rangle - |\psi_t\rangle) \cdot \zeta_{\psi_t,j}(t).$$

Through a process, the energy of the system changes as $E_{\psi_t}(a(t))$. Using the discrete Langevin equation, the energetics defined above can be described simply.

We first express the energy $E_{\psi_t}(a(t))$ using an energy operator, $\mathsf{E}(a)$, defined by

$$\mathsf{E}(a) \equiv \sum_i E_i(a)|i\rangle\langle i|. \tag{4.22}$$

Then the energy for the state $|\psi_t\rangle$ is

$$E_{\psi_t}(a(t)) = \langle\, |\mathsf{E}(a(t))|\psi_t\rangle, \tag{4.23}$$

where $\langle\, | \equiv \sum_j \langle j|$ [12]. With these notations, the energy balance, which corresponds to $dE = d'W + d'Q$ for the Langevin equation, is as follows:[19]

$$d[\langle\, |\mathsf{E}(a(t))|\psi_t\rangle] = \langle\, | d[\mathsf{E}(a(t))]\, |\psi_t\rangle + \langle\, |\mathsf{E}(a(t))\, d[|\psi_t\rangle]. \tag{4.24}$$

4.1.3 *The Ensemble Average Heat Flux has Several Different Expressions*

In this section we will derive the formula for the average heat transferred per unit time to the system from thermal environments.

4.1.3.1 Case with Inertia

We consider the Langevin equation (4.1). We will derive a formula for the average of heat $d'Q$ over the stochastic processes between t and $t+dt$. For this purpose, the Itô-type representation of the energy balance (4.13), that is,

$$\begin{aligned} d'Q &= dE - d'W \\ &= -\frac{2\gamma}{m}\left(\frac{p^2}{2m} - \frac{k_B T}{2}\right) dt + \sqrt{2\gamma k_B T}\,\frac{p}{m}\cdot dB_t \end{aligned} \tag{4.25}$$

is the most convenient. The result is

$$\langle d'Q\rangle = -\frac{2\gamma}{m}\left[\left\langle\frac{p^2}{2m}\right\rangle - \frac{k_B T}{2}\right] dt. \tag{4.26}$$

The result indicates that, independent of the form of the potential function $U(x, a)$, the exchange of heat with the thermal environment proceeds through the kinetic energy of the system.[20]

---

[19] We remind that the vectors $|j\rangle$ or $\langle j|$ are nominative, and their time derivatives are always 0, even if the physical state "j" changes through the parameter $a(t)$.

[20] For the Langevin equation without inertia, the kinetic energies exchanged between the system and the thermal environment are hidden, but should exist. That $\langle d'Q\rangle$ is linear in the imbalance

This simple formula gives an illusion of stability, but may mask what occurs in the system. For example, suppose that the system starts from a metastable state of a potential $U(x)$ as in Fig. 4.1. Until the system escapes from the metastable valley, $\frac{p^2}{2m}$ remains $\simeq \frac{k_B T}{2}$, but the net release of the heat will occur from the escaping particle beyond the potential barrier.

4.1.3.2 Case Without Inertia

If the temporal resolution is not enough to observe, we assume that the kinetic energy $\frac{p^2}{2m}$ is mostly equal to $\frac{k_B T}{2}$. However, the substitution of this hypothesis into (4.26) leads to $\langle d'Q\rangle = 0$ for any process. This is apparently wrong. We must, therefore, reformulate the average heat for the case without inertia from the Langevin equation (4.2), and the definition of heat, $d'Q = (-\gamma dx/dt + \xi(t)) \circ dx$. The results is

$$\langle d'Q\rangle = -\frac{1}{\gamma}\left[\left\langle\left(\frac{\partial U}{\partial x}\right)^2\right\rangle - k_B T\left\langle\frac{\partial^2 U}{\partial x^2}\right\rangle\right]dt. \tag{4.27}$$

The derivation is given in Appendix A.4.3. [21]

As example of (4.27), suppose that the potential energy is harmonic, $U(x) = \frac{K}{2}x^2$. The above formula then gives

$$\langle d'Q\rangle = -\frac{2K}{\gamma}\left[\frac{K}{2}\langle x^2\rangle - \frac{k_B T}{2}\right]dt. \tag{4.28}$$

4.1.3.3 Expression of Average Heat Flow in Terms of the Probability Current

The above results can be cast into a common form, using the probability current of the Fokker–Planck equation. We calculate the average heat $\langle d'Q\rangle$ as $\langle d'Q\rangle = \langle dE\rangle - \langle d'W\rangle$, where the energy function $E(X, P, a)$ is $P^2/(2m) + U(X, a)$ in the case with inertia (Sect. 4.1.3.1) and $E(X, a) = k_B T/2 + U(X, a)$ in the case without inertia (Sect. 4.1.3.2). Using the probability density $\mathcal{P}$ obeying the Fokker–Planck equation, the above relation is

$$\begin{aligned}\frac{\langle d'Q\rangle}{dt} &= \frac{d}{dt}\int E\mathcal{P}d\Gamma - \frac{da}{dt}\int \frac{\partial E}{\partial a}\mathcal{P}d\Gamma \\ &= \int E\frac{\partial \mathcal{P}}{\partial t}d\Gamma,\end{aligned} \tag{4.29}$$

of the kinetic energy comes from the linearity of the Langevin equation in γ. The Navier–Stokes equation of hydrodynamics is also the linear theory from this point of view.

[21] A byproduct of (4.27) is the equilibrium relationship, when the parameter a is fixed: $\left\langle\left(\frac{\partial U}{\partial x}\right)^2\right\rangle_{eq} = k_B T\left\langle\frac{\partial^2 U}{\partial x^2}\right\rangle_{eq}$. This equality can also be obtained using the canonical equilibrium distribution $\propto e^{-U/k_B T}$.

where $d\Gamma \equiv dXdP$ (with inertia) and $d\Gamma \equiv dX$ (without inertia), respectively. We then substitute for $\frac{\partial \mathcal{P}}{\partial t}$ the Fokker–Planck equation, $\frac{\partial \mathcal{P}}{\partial t} = -\nabla \cdot \boldsymbol{J}$, see (1.76) and (1.78) in Sect. 1.2.3.2. with $\nabla \cdot \boldsymbol{J} = \partial J_x/\partial X + \partial J_p/\partial P$ (with inertia) and $\nabla \cdot \boldsymbol{J} = \partial J_x/\partial X$ (without inertia), respectively. The probability fluxes (J_x, J_p) in case with inertia and J_x in case without inertia are defined, respectively, by (1.79) and by (1.77). Performing the integral by parts, and assuming that the boundary terms vanish, we obtain the following formulas[22]:

$$\begin{aligned} \frac{\langle d'Q \rangle}{dt} &= \int \left[\frac{\partial E}{\partial X} J_x + \frac{\partial E}{\partial P} J_p \right] dXdP \qquad \text{(with inertia effect)}, \\ \frac{\langle d'Q \rangle}{dt} &= \int \frac{\partial U}{\partial X} J_x \, dX \qquad \text{(without inertia effect)}. \end{aligned} \tag{4.30}$$

The expressions of the average heat (4.30) have long been found in the context of Fokker–Planck equation [13].

The first equation of (4.30) includes, as special case, the result for the purely mechanical system. In that case, $E = H$, $(J_x, J_p) = \left(\frac{dX}{dt}\mathcal{P}, \frac{dP}{dt}\mathcal{P}\right)$, and (X, P) obey the Hamiltonian equation, $\left(\frac{dX}{dt}, \frac{dP}{dt}\right) = (-\partial H/\partial P, \partial H/\partial X)$. We then have $\frac{\partial E}{\partial X} J_x + \frac{\partial E}{\partial P} J_p \equiv 0$, that is, the heat transfer in the first equation of (4.30) vanishes identically.

4.2 Generalization

4.2.1 Heat on the Mesoscopic Scale Can Be Generalized to the System in Contact with More Than One Thermal Environments

4.2.1.1 Formal Results

When more than one thermal environment interacts with a system, the analysis of average heat flux in the previous section can be generalized. In order to avoid direct interactions *among* the thermal environments, we assume that the thermal environments do not couple with the same degree(s) of freedom of the system. We assume a simple model energy of the system:

$$E(\{x_i, p_i\}, a) = \sum_i \frac{{p_i}^2}{2m_i} + U(\{x_i\}, a), \qquad (i = 1, \ldots, n), \tag{4.31}$$

[22] An alternative proof of (4.30) will be given in the next section.

where m_i is the mass of the ith degrees of freedom, and a is external control parameter(s). The Langevin equation for the ith degree of freedom is

$$\frac{dx_i}{dt} = \frac{p_i}{m_i}$$
$$\frac{dp_i}{dt} = -\gamma_i \frac{p_i}{m_i} - \frac{\partial U}{\partial x_i} + \sqrt{2\gamma_i k_B T_i}\, \theta_i(t), \tag{4.32}$$

where γ_i are the friction constants associated with the coupling between the system and the ith thermal environment of the temperature T_i.[23] $\theta_i(t)$ are white Gaussian random noises with zero mean and $\langle \theta_i(t)\theta_j(t')\rangle = \delta_{i,j}\delta(t-t')$.

The heat from the ith thermal environment to the system, $d'Q_i$ is defined as

$$d'Q_i = \left(-\gamma \frac{p_i}{m_i} + \xi_i(t) \right) \circ dx_i, \tag{4.33}$$

while the work $d'W$ is defined as before,

$$d'W = \frac{\partial U}{\partial a} \circ da. \tag{4.34}$$

It is easy then to verify that the balance of energy:

$$dE = d'W + \sum_i d'Q_i, \tag{4.35}$$

Following exactly the same procedure as we did in Sect. 4.1.3.1 for a single thermal environment, we can derive the following formula of the average heat flux from ith thermal environment to the system: (cf. (4.30))[24]:

$$\frac{\langle d'Q_i \rangle}{dt} = \int \left[\frac{\partial E}{\partial X_i} J_{i,x} + \frac{\partial E}{\partial P_i} J_{i,p} \right] dX dP \qquad \text{(with inertia effect)},$$
$$\frac{\langle d'Q_i \rangle}{dt} = \int \frac{\partial U}{\partial X_i} J_{i,x}\, dX \qquad \text{(without inertia effect)}, \tag{4.36}$$

where $J_{i,x} \equiv \frac{P_i}{m_i}\mathcal{P}$ and $J_{i,p} \equiv -\left(\frac{\partial U}{\partial X_i} + \gamma_i \frac{P_i}{m_i} \right)\mathcal{P} - \gamma_i k_B T_i \frac{\partial}{\partial P_i}\mathcal{P}$ are the ith flux components of the Fokker–Planck equation associated with X_i and P_i, in the case with inertia, and also $J_{i,x} \equiv -\frac{1}{\gamma_i}\left[\frac{\partial U}{\partial X_i}\mathcal{P} + k_B T_i \frac{\partial}{\partial X_i}\mathcal{P} \right]$ is the ith probability flux associated with X_i in the case without inertia. The derivation is given in Appendix A.4.4.

The interpretation of terms like $\frac{\partial E}{\partial X_i} J_{i,x} + \frac{\partial E}{\partial P_i} J_{i,p}$ as an analog of $d'Q = (dp/dt + \partial U/\partial x) \circ dx(t)$ (see the paragraph above (4.4)) would be far fetched, because the derivation of (4.36) uses integral by parts. The former, therefore, is not adapted to a local interpretation.

---

[23] If some "internal" degrees of freedom do not directly interact with the thermal environments, we assign $\gamma_i = 0$ for those degrees of freedom.

[24] Here $dX \equiv \prod_i dX_i$ and $dP \equiv \prod_i dP_i$.

About the average heat fluxes, a type of "H-theorem" has been derived [13]. It concerns the Shannon entropy S of the system:

$$S \equiv -\int \mathcal{P} \log \mathcal{P} d\Gamma. \tag{4.37}$$

Direct reorganization of terms shows the following equalities:[25]

$$\frac{dS}{dt} - \sum_i \frac{1}{k_B T_i} \frac{\langle d'Q_i \rangle}{dt} = \int \frac{1}{\mathcal{P}} \sum_i \frac{\gamma_i}{k_B T_i} \left[\frac{P_i}{m_i} \mathcal{P} + k_B T_i \frac{\partial \mathcal{P}}{\partial P_i} \right]^2 dX dP$$
(underdamped),

$$\frac{dS}{dt} - \sum_i \frac{1}{k_B T_i} \frac{\langle d'Q_i \rangle}{dt} = \int \frac{1}{\mathcal{P}} \sum_i \frac{\gamma_i (J_{i,x})^2}{k_B T_i} dX \qquad \text{(overdamped).}$$

The right-hand sides of these identities are nonnegative. Therefore,

$$\frac{dS}{dt} - \sum_i \frac{1}{k_B T_i} \frac{\langle d'Q_i \rangle}{dt} \geq 0 \tag{4.38}$$

in both cases. This inequality is interpreted [13] as that the Shannon entropy of the whole system is nondecreasing, $-\frac{1}{k_B T_i} \frac{\langle d'Q_i \rangle}{dt}$ is regarded as the increment rate of the Shannon entropy of ith thermal environment. In the steady state, where $dS/dt = 0$, (4.38) implies the inequality:

$$\sum_i \frac{1}{k_B T_i} \frac{\langle d'Q_i \rangle}{dt} \leq 0 \qquad \text{(stationary state).} \tag{4.39}$$

This relation imposes a constraint among the steady-state heat currents.

4.2.1.2 Conduction, Partition, and Diffusion of Heat Between Two Thermal Environments

Apart from the formal results, we will take below a simple model system that interacts with two thermal environments (Fig. 4.6). We will see several different aspects: heat conduction through mechanical motion of system, the partition of heat into two environments upon the external work on the system, and the diffusion of heat between two environments through the system.

[25] We use $\int (\partial \mathcal{P}/\partial t) \log \mathcal{P} d\Gamma = \int \sum_i \boldsymbol{J}_i \cdot \nabla_i \log \mathcal{P}$, where ∇_i denotes the derivatives with respect to ith degree of freedom. We use also $-\sum_i \int \left(\frac{P_i}{m_i} \frac{\partial}{\partial X_i} - \frac{\partial U}{\partial X_i} \frac{\partial}{\partial P_i} \right) \mathcal{P} d\Gamma = 0$, assuming that the surface integral vanishes.

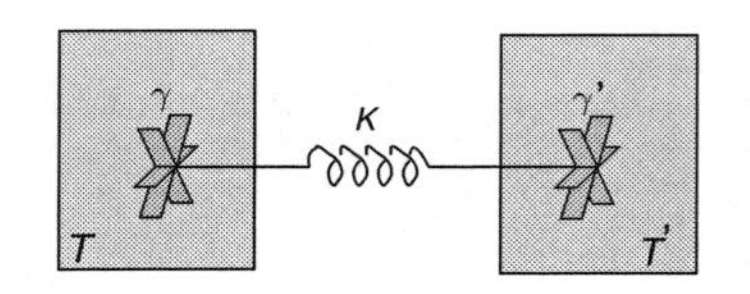

Fig. 4.6 Two vanes immersed in their respective thermal environments. These vanes are mechanically connected by a spring. See the text

Model

In Fig. 4.6 two thermal environments are at the temperatures T and T', respectively. In each of these a vane is immersed. The coupling of these vanes to their thermal environments is characterized by the friction coefficients γ and γ', respectively. The two vanes are joined through a harmonic torsional spring with the elastic constant, K. We will neglect the inertia effect for a moment. Let us denote by x and x' the rotation angles of the vanes in the environment of temperature T and T', respectively. The Langevin equations for the vanes are

$$0 = -\gamma\dot{x} + \xi(t) - K(x - x'), \qquad 0 = -\gamma'\dot{x}' + \xi'(t) - K(x' - x), \tag{4.40}$$

where $\xi(t)$ and $\xi'(t)$ are two independent Gaussian white noisees with zero mean and $\langle\xi(t_1)\xi(t_2)\rangle = 2\gamma k_B T\,\delta(t_1 - t_2)$ and $\langle\xi'(t_1)\xi'(t_2)\rangle = 2\gamma' k_B T'\,\delta(t_1 - t_2)$. We will use the "dot" ($\dot{x}$ etc.) to mean the time derivative.

We define heat by $d'Q = (-\gamma\dot{x} + \xi) \circ d'x$ and $d'Q' = (-\gamma'\dot{x}' + \xi') \circ d'x'$. The balance of energy is in a time differential form,

$$\frac{d'Q}{dt} + \frac{d'Q'}{dt} = \frac{d}{dt}\left[\frac{K}{2}(x - x')^2\right] - \frac{(x - x')^2}{2}\frac{dK}{dt}. \tag{4.41}$$

The first and second terms on the right-hand side of (4.41) shows dE/dt and $d'W/dt$, respectively.

We can also show the following relation:

$$\gamma\frac{d'Q}{dt} - \gamma'\frac{d'Q'}{dt} = K(x - x') \circ (\xi(t) + \xi'(t)). \tag{4.42}$$

Moreover, we can derive the Langevin equation for the relative rotation angle, $\mu \equiv x - x'$, from the original coupled Langevin equation:

$$\frac{d\mu}{dt} = -\left(\frac{1}{\gamma} + \frac{1}{\gamma'}\right)K\,\mu + \left(\frac{\xi(t)}{\gamma} - \frac{\xi'(t)}{\gamma'}\right). \tag{4.43}$$

These three equations give the heat currents $\frac{d'Q}{dt}$ and $\frac{d'Q'}{dt}$ for each realization of $(\xi(), \xi'())$. Since $Q(t) \equiv \int_0^t (d'Q/dt)dt$ and $Q'(t) \equiv \int_0^t (d'Q'/dt)dt$ are quadratic functionals in the Gaussian stochastic processes, ξ and ξ', detailed statistics of $Q(t)$ and $Q'(t)$ are available through the generating function $\langle e^{-\lambda Q(t)}\rangle$, etc. Using this fact, for example, the theorem of heat fluctuation [14] has been tested [15]. This theorem is a development of the so-called fluctuation theorem (FT) [16, 17]. See also [18] and a recent review to date [19]. We will not go into details of these statistics. Below we will discuss the average and variance of heat currents.

Note: The original coupled Langevin equation is decoupled by using the relative angle, μ, and the "center of diffusion" $X \equiv (\gamma x + \gamma' x')/(\gamma + \gamma')$. The latter obeys a free rotative Brownian motion, $(\gamma+\gamma')\dot{X} = \xi(t)+\xi'(t)$.[26] Therefore, the diffusion of X does not contribute to the heat currents.

Heat Conduction Between Two Thermal Environments

Using the solution of (4.43) for $\mu(t)$, the averages of (4.42) and (4.41)[27]:

$$\frac{\langle d'Q\rangle}{dt} + \frac{\langle d'Q'\rangle}{dt} = \frac{K(t)}{2}\frac{d}{dt}\langle (x - x')^2\rangle, \tag{4.44}$$

$$\gamma\frac{\langle d'Q\rangle}{dt} - \gamma'\frac{\langle d'Q'\rangle}{dt} = K(t)(k_{\mathrm{B}}T - k_{\mathrm{B}}T'). \tag{4.45}$$

In the steady state with K constant, the right-hand side of (4.44) vanishes. Then we have the average heat currents:

$$\frac{\langle d'Q\rangle}{dt} = -\frac{\langle d'Q'\rangle}{dt} = \frac{K}{\gamma + \gamma'}(k_{\mathrm{B}}T - k_{\mathrm{B}}T'). \tag{4.46}$$

The above result describes the heat conduction mediated by a mechanical spring. When $T = T'$ the average heat current vanishes. Also if γ or γ' goes to infinity, there is no conduction. The average heat current increases with K. It means that the stiffer spring can transmit energy more efficiently from one thermal environment to the other. According to (4.46) the heat current diverges in the limit $K \to \infty$. This apparently unphysical result is a sign of the abuse of the Langevin equation without inertia: In the limit of high stiffness of the spring, the timescale $\sim (\gamma + \gamma')/K$ characterizing the variation of $\mu = (x - x')$ is beyond the time resolution of the modeling by such Langevin equation. In order to avoid the spurious divergence, the model has to take into account the effects of inertia, for example, due to the moment of inertia of the vanes. In the limit of $K = \infty$, appropriate Langevin equations are as follows:

$$I\ddot{x} = -\gamma\dot{x} + \xi(t) + f(t), \qquad I'\ddot{x}' = -\gamma'\dot{x}' + \xi'(t) - f(t), \tag{4.47}$$

where I and I' are the moments of inertia of the vanes, $\pm f(t)$ is the force of action/reaction through the rigid spring, determined so that $x(t) - x'(t) = 0$ at

[26] The "apparent temperature" for this diffusion is $(\gamma T + \gamma' T')/(\gamma + \gamma')$.

[27] Technical detail: We use a rule $\int_0^t \delta(s)ds = \frac{1}{2}$ to be consistent with the Stratonovich type calculus

any t.[28] A new energy balance relation is written as

$$\frac{d'Q}{dt} + \frac{d'Q'}{dt} = \frac{d}{dt}\left(I\frac{\dot{x}^2}{2} + I'\frac{\dot{x}'^2}{2} \right) \tag{4.48}$$

and the average heat current in the steady state is

$$\frac{\langle d'Q \rangle}{dt} = -\frac{\langle d'Q' \rangle}{dt} = \frac{\gamma\gamma'}{(\gamma+\gamma')(I+I')}(k_B T - k_B T'). \tag{4.49}$$

Nanomachine to Absorb the Heat vs. Maxwell's Demon

It will be instructive to look the above model from the standpoint of the hot reservoir. Let us suppose that $T > T'$. How do things look like if we observe the microscopic events around the vane in the high-temperature environment of temperature T?

The rotational motion of the vane $x(t)$ is apparently random, and the thermal environment (T) is isolated except for the mechanical link through the vane immersed therein. However, this random motion of the vane results in extracting energy from this environment. Moreover, the "Feynman pawl and ratchet" (Sect. 1.3.4) replaces the spring by more intelligent attachments and enables work to be done using the energy extracted through this random motion.

If we compare the above situations with the experiment of agitating a "spoon" in water, often mentioned after J. P. Joule (around 1850), the spoon could only warm it up. Therefore, the random motion of the vane in the high-temperature environment T is essentially different from the agitation of the spoon. What is the difference? We note that the extraction of energy by the vane does not contradict the second law of thermodynamics. What the second law excludes is the extraction of energy from a system with a single thermal environment. Our system of vanes would indeed cease to extract the energy as soon as we remove the second thermal environment (temperature $T'(< T)$).

The key to the answer is related to the well-known "paradox of Maxwell's demon": The demon D sits by a gate between two containers of gas, which are initially at the same temperature. D and the two containers are isolated from the rest of the world. D can recognize the velocity of the gas particles coming toward the gate. Now D operates the gate so that only the hot particles go from the left container to the right one, and that only the cool particles go from the right container to the left one. Then D can create the temperature difference between the two containers. This contradicts with the zeroth and second laws of thermodynamics. The fault of logic in this paradox is that, in fact, D must keep contact with a low-temperature environment so that D can recognize the velocities of gas particles.

Coming back to our vane that absorbs heat, it recognizes the fluctuations in the hot environment T and moves so that it absorbs energy. For this action, contact with

[28] K can be neglected for $(\gamma^{-1} + \gamma'^{-1})^{-1}/(I + I') \gg (\gamma + \gamma')/K$. See (4.49) below.

the cool environment T' is indispensable. It is the hot environment that exerts the random force on the vane, but the motion of the vane is not incoherent like Joule's spoon. To see this point clearer, let us analyze the behavior of $x(t)$ of the system (4.40) putting $T' = 0$. Under this assumption we can solve the second equation of (4.40) for $x'(t)$:

$$x'(t) = \frac{K}{\gamma'} \int_0^{+\infty} e^{-\frac{K}{\gamma'}s} x(t-s) ds. \tag{4.50}$$

Substituting this result into the first equation of (4.40), we have the following equation for $x(t)$:

$$\gamma \dot{x}(t) = -K\left[x(t) - \frac{K}{\gamma'} \int_0^{+\infty} e^{-\frac{K}{\gamma'}s} x(t-s) ds\right] + \xi(t). \tag{4.51}$$

The integral on the right-hand side of (4.51), i.e., $x'(t)$ of (4.50), shows a feedback control in the restoring force. As x' follows adaptively to the motion of x in the past, the restoring force, $-K(x(t) - x'(t))$, is weakened as compared with a spring with fixed end ($x'(t)$ =const.). As the result, the vane $x(t)$ returns less energy to the environment T than what was injected from this environment. A related model is found in [20].

Based on (4.51) we can design a cooling device in the fluctuating world: we somehow trap a Brownian particle (position $x(t)$) around the trapping center (the position $x'(t)$). If $x(t)$ can be observed and if the feedback circuit can change $x'(t)$ quickly enough to realize (4.50), then we can extract energy around the Brownian particle as if the particle is in contact with another environment with $T' = 0$.

Partition of Heat

Now we suppose that $T = T'$ for the two thermal environments while they have no direct contact with each other. The object is to find how much heat the spring absorbs from respective thermal environments when the spring constant $K(t)$ is slowly lowered. Analogous situation was mentioned in Sect. 2.2, see, Fig. 4.7. There, the volume of the cylinder $V(t)$ played a role of the inverse square root of the spring constant, $K(t)^{-1/2}$. If we pull up the piston quasistatically, what will be the partitioning of the heat supplied from these environments?

Conventional thermodynamics did not answer to this question.[29]

The average heat currents are found from the relation (4.45) with $T = T'$ and (4.44). The result is

$$\gamma \frac{\langle d'Q \rangle}{dt} = \gamma' \frac{\langle d'Q' \rangle}{dt} = \frac{\gamma\gamma'}{\gamma + \gamma'} \frac{K(t)}{2} \frac{d}{dt} \langle (x - x')^2 \rangle. \tag{4.52}$$

---

[29] Linear nonequilibrium thermodynamics may devise a model with introducing phenomenological thermal resistances. In the present model, γ and γ' are sufficient.

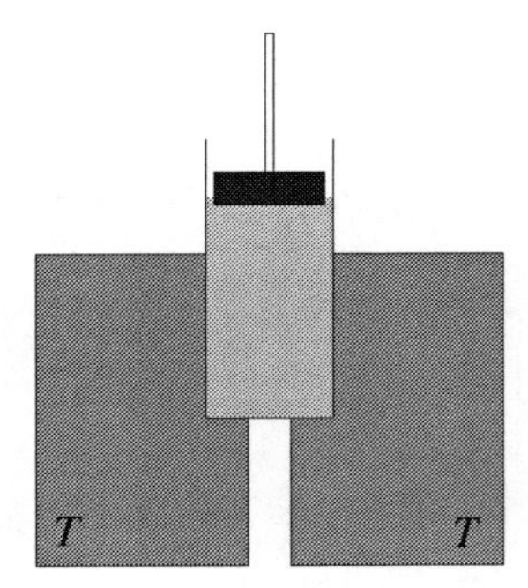

Fig. 4.7 (The same figure as Fig. 2.4.) Partition of heat into two thermal environments of the same temperature T (*left* and *right* boxes) through the change of the volume of a gas (*center*)

This shows that whatever the change of $K(t)$ is, the heat currents are partitioned according to the inverse ratio of γ and γ'. More heat is exchanged with the environment of smaller value of γ. In particular, for the slow change of $K(t)$, we may have a concrete result: Using the equipartition law, $K(t)\langle(x-x')^2\rangle/2 = k_\mathrm{B}T/2$, we integrate (4.52) to obtain the average cumulate heat from respective environments, $\langle \Delta Q\rangle$ and $\langle \Delta Q'\rangle$:[30]

$$\gamma\langle \Delta Q\rangle = \gamma'\langle \Delta Q'\rangle = \frac{\gamma\gamma'}{\gamma+\gamma'}k_\mathrm{B}T\,\ln\sqrt{\frac{K_\mathrm{init}}{K_\mathrm{fin}}} \qquad \text{(quasistatic)}, \tag{4.53}$$

where K_init and K_fin are the initial and final values of $K(t)$. If the spring is softened indefinitely ($K(t) \to +0$), the external system extract arbitrarily large amount of energy from the thermal environments, in the proportion of γ^{-1} vs. γ'^{-1}.[31]

Diffusion of Heat

Even in the case of $T = T'$ and $K =$ const., heat flows randomly between the two thermal environments. The cumulated heat undergoes a Brownian motion around the average steady change. For the case of $K =$ const., this fluctuation is characterized by the "thermal diffusion constant" $\mathcal{D}$ through the following relation:

$$\left\langle \left[\int_0^t \frac{d'Q}{dt}dt - \int_0^t \frac{\langle d'Q\rangle}{dt}dt\right]^2\right\rangle = 2\mathcal{D}t. \tag{4.54}$$

Some calculation shows[32]

$$\mathcal{D} = \frac{K}{\gamma+\gamma'}\left(\frac{\gamma k_\mathrm{B}T + \gamma' k_\mathrm{B}T'}{\gamma+\gamma'}\right)\left(\frac{\gamma' k_\mathrm{B}T + \gamma k_\mathrm{B}T'}{\gamma+\gamma'}\right). \tag{4.55}$$

---

[30] We integrate $\frac{K(t)}{2}\frac{d}{dt}\frac{k_\mathrm{B}T}{K(t)} = \frac{k_\mathrm{B}T}{2}\frac{d}{dt}\ln\frac{1}{K(t)}$.

[31] The energy in the spring is $\sim k_\mathrm{B}T/2$, independent of the stiffness.

[32] The expression of $\mathcal{D}$ in the Japanese Version [21] should be corrected as that of (4.55).

See Appendix A.4.5 for the derivation. This formula is invariant under the exchange of T and T'. At high temperature and/or for very slow change of $K(t)$, the diffusion of heat may mask the prediction about the partition of heat, (4.53).

4.2.2 *Energetics of Thermal Ratchet Motors*

We revisit the thermal ratchet motors (Sect. 1.3.4) and discuss several aspects of energetics. Figure 4.8 recapitulates the Feynman pawl and ratchet and Büttiker and Landauer ratchet.

4.2.2.1 Heat Leak and Onsager Coefficient in the Feynman Pawl and Ratchet Wheel

Introduction

As mentioned in Sect. 1.3.4 the Feynman ratchet model played an important role in the construction of the stochastic energetics.

We recall below the Langevin equation of this model:

$$0 = -\gamma\frac{dx}{dt} - \frac{\partial U}{\partial x} + \sqrt{2\gamma k_{\mathrm{B}}T}\,\theta(t), \qquad 0 = -\gamma'\frac{dy}{dt} - \frac{\partial U}{\partial y} + \sqrt{2\gamma' k_{\mathrm{B}}T'}\,\theta'(t), \tag{4.56}$$

where $U(x,y) = U_1(x-\phi(y)) + U_2(y) - fx$, and $\theta(t)$ and $\theta'(t)$ are mutually independent white Gaussian random noises with zero mean and $\langle\theta(t)\theta(t')\rangle = \langle\theta'(t)\theta'(t')\rangle = \delta(t-t')$. For $T \neq T'$ this model can rotate on average the ratchet wheel x in one direction. Here the load f is regarded as a part of the system, and its

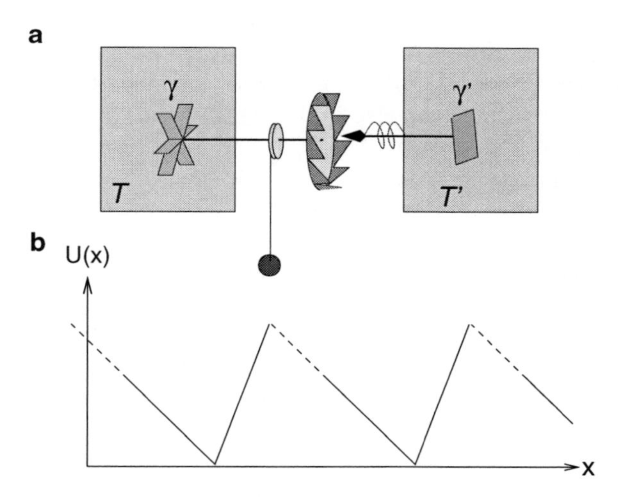

Fig. 4.8 (**a**) Feynman ratchet and pawl and (**b**) Büttiker and Landauer ratchet. (The same figures as Figs. 1.9 and 1.10. See those legends.)

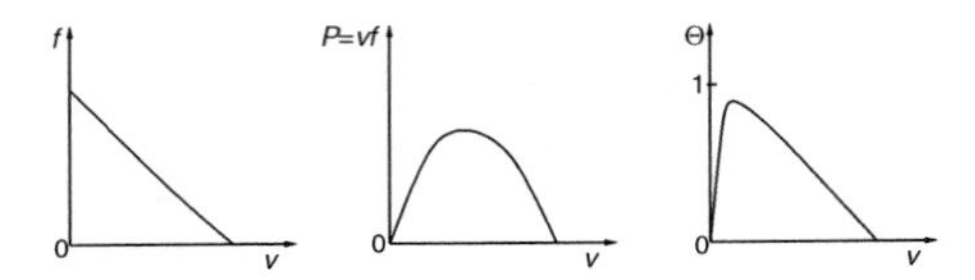

Fig. 4.9 Schematic presentation of the load f, power $P = vf$, and the energy efficiency Θ of motors as function of its velocity v. The force at $v = 0$ is the *stall force*

potential energy $(-fx)$ is counted in the system's potential energy. This system is, therefore, autonomous with no external parameters.

After Feynman, there have been many variations of his model, and we will not go into details of those model. We only summarize in Fig. 4.9 general qualitative features of this model, in terms of the average working velocity $v = \langle \dot{x} \rangle$, load f, average power output $P = vf$, and the efficiency Θ, i.e., the ratio of P to input energy from high temperature bath (T) *normalized* by the Carnot maximum value, $|\Delta T|/T$ (see (2.24)).

We will discuss mainly two aspects of energetics: the stalled state where the load is just strong enough that the motor moves neither forward nor backward, and around the equilibrium state where the efficiency by linear nonequilibrium thermodynamics becomes undetermined (of 0/0 type).

Stalled State and Efficiency

One of the main questions was if Feynman's model can attain $\Theta = 1$, i.e., the efficiency of the ideal Carnot engine.[33] Feynman himself argued affirmatively in the original textbook [22], since he focused on the critical and rare events when a pawl finds a new space between the ratchet tooth. Later his argument was represented more formally in terms of some stochastic boundary condition at the peaks of the ratchet profile [23].

While these critical events are essential for the function of this motor, the other "nonsuccessful" fluctuations are also important from the energetics point of view. Especially, at the stalled state Parrondo [24] and then the author [3] noticed that the efficiency should be $\Theta = 0$. While the pawl jiggles around a valley between neighboring ratchet tooth, the fluctuations of the ratchet wheel and those of the pawl are mechanically coupled. Heat conduction from the high-temperature environment to the low-temperature one then takes place, as we have analyzed a simple case in Sect. 4.2.1.2.

The same reasoning can also be described in terms of the probability flux of the Fokker–Planck equation at the stalled state. The stationary probability flux (J_x, J_y) exhibits vorticity in the phase space (xy-plane) [25, 26], which can then be related to the dissipation of heat without contributing to work.

---

[33] That is, the ratio η of the available work to the supplied heat from the high-temperature (T) thermal environment is $\eta_{\max} \equiv 1 - T'/T$.

In conclusion, there is a leakage of heat at the finite rate at the stalled state. The efficiency of energy conversion then vanishes at the stalled state. Experience shows that, if the parameters of the model are well tuned, efficiency can attain almost 1 (theoretical maximum) just off this stalled state while the efficiency *at* the stalled is 0. The efficiency has, therefore, a dip at this state. From practical viewpoint, if the size of the system is increased, the effect of jiggling may become negligible, and the dip in the efficiency may be too narrow to be observed. Discrete modeling may also suppress or lessen this jiggling.

Near Equilibrium State and Onsager Coefficients

Near the equilibrium, $T \simeq T'$, and weak load, $f \sim 0$, Onsager coefficients characterize the function of Feynman ratchet on macroscopic (ensemble average) level.

In this framework, the average velocity of wheel rotation $\mathcal{J}_p = \langle \frac{dx}{dt} \rangle$ (see (1.114)) and the heat flux $\mathcal{J}_q$ from the hot thermal environment are linearly related to the thermodynamic force concerning the load, f/T, and the temperature gradient, $\Delta T/T^2$ ($\Delta T \equiv T' - T$) with the symmetric Onsager coefficients [27, 28]:

$$\begin{pmatrix} \mathcal{J}_p \\ \mathcal{J}_q \end{pmatrix} = \begin{pmatrix} \mu T & L_{pq} \\ L_{pq} & \lambda T^2 \end{pmatrix} \cdot \begin{pmatrix} f/T \\ \Delta T/T^2 \end{pmatrix}, \tag{4.57}$$

where $\mu = \mathcal{J}_p/f$ for $\Delta T = 0$ and $\lambda = \mathcal{J}_q/\Delta T$ for $f = 0$ are, respectively, the mobility coefficient and the heat conductivity coefficient. Onsager theory says that the off-diagonal coefficients L_{pq} are the same. This symmetry in the context of ratchet motors has been discussed in [29]. Exactly at equilibrium ($f = \Delta T = 0$), the above phenomenology is not valid. For example, the efficiencies of energy conversion become singular, simply because (4.57) yields 0/0, nondeterminate. In fact, in the very vicinity of equilibrium, the fluctuations in the fluxes dominate the average fluxes, and the fluctuations are not singular.

The modeling by the Langevin equation tells the underlying mechanism of the relation (4.56), especially, of the cross-coupling coefficient L_{pq}.

First, we notice that there is no *kinetic* cross-coupling in the Langevin equation (4.56), that is, no forces on x directly drives dy/dt, or vice versa. It is through the potential coupling $U(x, y)$ that the cross-coupling L_{pq} is realized. Through the nonlinearity of this potential coupling, the model also predicts the nonlinear force–flux relation. There are models of heat engine that assume the cross-coupling as a kinetic mechanism. (See below.)

Second, the modeling by the Langevin equation allows the analysis of energetics along an individual process. Even at equilibrium, a *cycle* of fluctuations in (x, y)-space can accompany the transfer of heat coupled with the space translocation (wheel rotation).[34] In fact this analysis gives the physical interpretation of the

---

[34] Because of detailed balance at equilibrium, forward cycles and backward cycles occur equally likely. Due to this cancellation this phenomena cannot be captured by macroscopic phenomenology.

cross-coupling constant L_{pq} almost quantitatively [30]. Below we summarize their analysis.

Using a slightly modified model of Feynman pawl and ratchet wheel, the "quantum" of the transferred heat, $q_{\rm step}$, is measured upon a directed, e.g., forward, step of ratchet potential (in equilibrium). Assuming that this quantum step is not singular at $f = 0$, the relation between the $\mathcal{J}_q$ and the (small) load f is written as $\mathcal{J}_q = q_{\rm step}(\mu f)/\ell_{\rm step}$, where $\ell_{\rm step}$ is known step size of the ratchet potential. Comparing this relation with (4.57), the coefficient L_{pq} is $L_{pq} = q_{\rm step}\mu T/\ell_{\rm step}$. This result is valid also at $f = 0$. On the other hand the Einstein relation tells that μT is the diffusion coefficient, $\ell_{\rm step}^2/\tau_{\rm step}$, where $\tau_{\rm step}$ is the characteristic time of spontaneous step in equilibrium. $\tau_{\rm step}$ is also measurable from the model analysis. Combining the above results the cross-coupling coefficient is directly correlated with the energetics of a single-step event:

$$L_{pq} = q_{\rm step}\left(\frac{\ell_{\rm step}}{\tau_{\rm step}}\right). \tag{4.58}$$

This argument is confirmed numerically and is also justified from the calculation of Green–Kubo formula [28].[35]

4.2.2.2 Inertia as a Singular Perturbation: Case of Büttiker and Landauer Ratchet

Introduction

Stochastic energetics has been applied to the Büttiker and Landauer ratchet (see Fig. 4.8 (b) and Sect. 1.3.4.3).

For the model without inertia, (1.112), i.e.,

$$\gamma\, dx = \sqrt{2\gamma k_{\rm B}T(x)} \circ dB_t - \frac{\partial}{\partial x}\left[U(x,a) + fx + \frac{k_{\rm B}T(x)}{2}\right]dt, \tag{4.59}$$

the analysis showed the efficiency Θ of energy conversion up to 1[36] [4, 31]. The maximum is attained in the stall state.

If the inertia is taken into account, the model is (1.111), i.e.,

$$\frac{dp}{dt} = -\frac{\partial U(x)}{\partial x} - f - \gamma\frac{p}{m} + \xi(t), \quad \frac{dx}{dt} = \frac{p}{m}, \tag{4.60}$$

[35] Note that the potential energy of the system $U(x, y)$ does not contribute to the "quantum" of the transferred heat, $q_{\rm step}$, except for the load's potential energy, $f\ell_{\rm step}$. The remaining part of $U(x, y)$ returns to the original value, on the average, after a "quantum" cycle is completed.

[36] i.e., the Carnot efficiency η up to $\eta_{\rm max} = 1 - T'/T$.

where $\xi(t)$ is the white Gaussian noise with $\langle \xi(t)\xi(t') \rangle = 2\gamma k_B T(x)\delta(t - t')$. The analysis of the latter model concluded that the efficiency should drop to 0 at the stalled state [32–34], in contradiction to the model without inertia.

This discrepancy implies that the limit of $m \to 0$ is not equivalent to the model with $m = 0$. If we put $m = 0$ in the Langevin equation, the equation changes the order of differential. Such a change may drastically affect the behavior of $x(t)$, and the energetics of Büttiker and Landauer ratchet is sensitive to this change.

Scale of Description

To better understand the cause of discrepancy and to know the true efficiency, we first remember that we should respect the time resolution of each Langevin equation (see, Sect. 1.3.2). In fact, both models abuse the Langevin equations beyond their validity range. When a particle moves across the discontinuity of the temperature, the environment for the particle changes within a infinitesimal time, shorter than any finite time resolution that is associated with a Langevin model.

Next we note the role of the velocity relaxation time. This time is finite for the model with inertia, $\tau_p \equiv m/\gamma$, while it is assumed to be infinitesimal for the model without inertia. We have a physical argument (see below) that, if we keep track of this timescale τ_p, the sudden jump of temperature is "buffered" and the model with inertia gives a meaningful result. We should not take the overdamped limit first.

Physical argument

When a particle switches its thermal environment, a thermalization of the kinetic energy of the particle occurs. On the average, the particle's kinetic energy will relax toward its (new) equipartition value. Through this process, an irreversible heat exchange takes place with a new thermal environment [35, 32]. This is the cause of heat leakage.

To discuss more quantitatively, suppose that a particle moves from the temperature T to the temperature T'. When the particle cross the border, its motion is almost ballistic during the time $\sim \tau_{\rm p} \equiv m/\gamma$. During this period, the particle keeps its original velocity $\sim v_{\rm th} \equiv \sqrt{k_B T/m}$. Then thermalization occurs, and the particle exchanges energy, which is about the difference of the equipartition kinetic energy $|k_B T - k_B T'|/2$. This energy does not depend on m.

In summary, as far as the particle's kinetic energy is concerned, the temperature border is smoothed. The effective "temperature gradient" is $\sim |T - T'|/(v_{\rm th}\tau_{\rm p}) \propto m^{-1/2}$.[36, 32]. The heat leakage through the thermalization of the kinetic energy occurs with this temperature gradient. At the stalled state this heat leakage continues at a finite rate. The efficiency is, therefore, 0 at the stalled state.

The conclusion is parallel to the Feynman pawl and ratchet wheel. But the mechanism of the leakage in the two models is very different. The maximum of the efficiency is realized near but off the stalled state, and less than the Carnot limit.

Kinetic Argument

Is there a remedy for the model without inertia to recover the above mentioned leakage [33, 34]? Below is the argument against this possibility: for a Wiener process, or a overdamped free Brownian motion ($0 = -\gamma dx/dt + \xi$), there is no characteristic timescale. The implication is that, once it visits a point, it revisits the same point indefinitely many times within a short period. Therefore, the to-and-fro of the particle at the temperature boundary can occur indefinitely many times, before the bias field drives the particle off the boundary. It is, therefore, impossible to incorporate the thermalization of kinetic energy in the overdamped model. If the space is discretized [37], this problem disappears apparently. But justification is needed for the discretization.

Proof by Numerics

The above physical argument supporting the Langevin model with inertia was finally verified by a careful molecular dynamic simulation [38]. Their analysis of energy transfer confirmed the divergence $\propto m^{-1/2}$ of the heat leakage ($\dot{Q}_{\rm kin}$) due to the kinetic energy carried by the particle. See, Fig. 4.10. Their simulation used 2D hard-core gas as thermal environments of two different temperatures. The gas is dense enough that the Knudsen number (mean-free path/system size) is < 1. The $\dot{Q}_{\rm kin} \propto m^{-1/2}$ behavior holds well even for $\tau_{\rm p} = m/\gamma$ as small as the microscopic time, i.e., the inverse collision frequency of gas particles against the Brownian particle.[37]

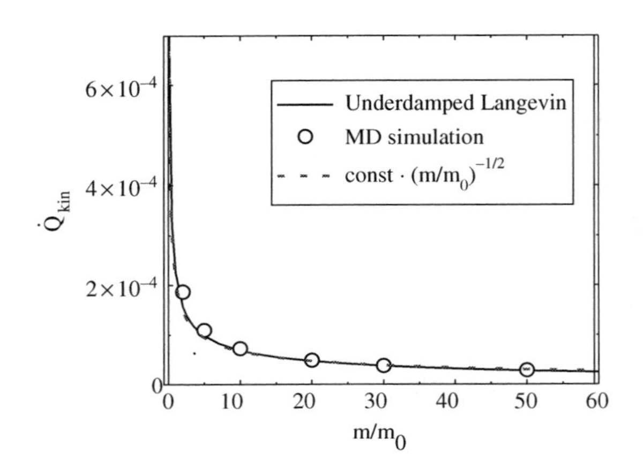

Fig. 4.10 Kinetic energy contribution of heat transfer, $\dot{Q}_{\rm kin}$. It diverges as $(m/m_0)^{-1/2}$, where m and m_0 are, respectively, the mass of the Brownian particle and the mass of the gas particle constituting the model thermal environment (Figure by the courtesy of R. Kawai)

[37] This result also justifies the Langevin equation with inertia down to the microscopic timescale. The validity of this equation at the intermediate scale, ~ 10–$10^2\tau_{\rm p}$, will be scrutinized in Sect. 6.3.1.

In conclusion, the study of Büttiker and Landau ratchet gives a lesson that an apparent reasonable result (Carnot's limit) can be wrong, and apparently singular result can be true.

4.2.3 Fluctuating Open System Does Not Exchange Chemical Potential as Energy of Particle

4.2.3.1 * Definition of the System

On the level of description of the Langevin equation, we always identify the positions of particles. We, therefore, take into account explicitly the space coordinates. We define as the *open system* a spatial region Ω within an entire space which the particles can explore. For example, Ω can be a volume confined by a piston and a cylinder with a valve (cf. Fig. 7.18), or a focused region of laser light under optical tweezers. Or Ω can also represent binding potential wells on a membrane (cf. Fig. 8.3).

We call "particle environment" the complementary region, Ω^c. We assume that the entire system, $\Omega \cup \Omega^c$, is closed and immersed in a thermal environment of temperature T. In this chapter, the particle environment need not be very large. See Fig. 4.11 (*top*). The stochastic energetics described up to the previous sections applies to this entire system. For simplicity we will consider only one species of particles.

We suppose that each particle (position $x_j(t) \in \Omega \cup \Omega^c$) obeys the Langevin equation without inertia term:

$$0 = -\gamma \frac{dx_j}{dt} + \xi_j(t) - \frac{\partial E_{\text{tot}}}{\partial x_j} \qquad (j = 1, \ldots, N), \tag{4.61}$$

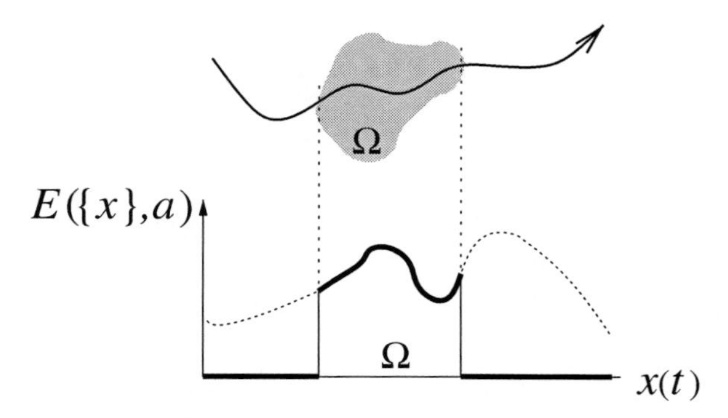

Fig. 4.11 *Top:* Open system Ω and its "particle environment" Ω_c, which is outside of Ω. Thick curve is a trajectory of the particle which passes through the system. *Bottom:* The energy of the system, $E_{(\{x\},a)}$ (*thick curve* and *thick line* segments). In the absence of particles, the energy of open system is zero, $E(\{\},a) = 0$

with $\{\xi_j(t)\}$ being mutually independent white Gaussian random forces with zero mean and correlations $\langle \xi_j(t)\xi_k(t')\rangle = 2\gamma k_B T \delta_{j,k}\delta(t-t')$.[38]

However, we explicitly describe only those particles in the system Ω. Once a particle exits from the system, we lose track of it, i.e., we lose its degree of freedom. In summary, the open system is a region, and the particles belong to this system while they are in this region.

4.2.3.2 * States of Open System

It is useless to distinguish *all* the particles in the entire system, first because any particular particle can spend most of the time in the environment, Ω^c, second because we need not to distinguish which particular particles are in the system, whether or not this particle is the one which has left the system before.

We, therefore, define the state of an open system as follows:

$$
\begin{aligned}
\{\} &: \text{“Null” state, where there is no particle in } \Omega. \\
\{x_1\} &: \text{the state with one particle at } x_1. \\
\{x_1, x_2\} &: \text{the state with two particles with their positions being } x_1 \text{ and } x_2. \\
&\cdots \\
\{x_1, \ldots, x_n\} &: \text{the state with } n \text{ particles with their positions being } x_1, \ldots, x_n. \\
&\cdots
\end{aligned}
\tag{4.62}
$$

Therefore, the number of particles in the system, $\hat{n}(t)$, is also a random variable.[39] If there are more than one species of particles, we will distinguish them and denote $x_1^{(\alpha)}$, etc.

4.2.3.3 Energy of Open System

As the energy of an open system, we count only the energies of those particles in the region Ω. Interactions among the particles require a refinement to this definition. This complication is unavoidable since energy of small system does not have the extensive character, in general.

One-Particle Energy of the Open System

For a moment we shall ignore all the interactions among the particles. Let us denote by $U_1(x,a)$ the energy of a particle at the position x. We assume that $U_1(x,a)$ is defined on the entire system, $x \in \Omega \cup \Omega^c$. As for the dependence on the external

[38] The generalization to the case with inertia is straightforward, but we will not discuss here.

[39] For those who know second quantization, this state space may be reminiscent of the Fock space. In the Fock space, however, the identity of individual particles is lost because of the symmetrization or antisymmetrization of the product states of single particle. In the present state space, the particle's identity is preserved while particle remains in the system.

parameter a, we require that $U_1(x, a)$ depends on a only for $x \in \Omega$.[40] The reason is that we have excluded the direct interactions between the external system and the system's environment. The one-particle energy U_1 has an arbitrariness of an additive constant.

If only one particle passes through the system Ω, the system's energy will change as shown in Fig. 4.11 (*bottom*). If there are N particles in the entire system, the energy of the system, E, is represented as (see (4.62))

$$E = \sum_{j=1}^{N} U_1(x_j, a)\theta_\Omega(x_j), \tag{4.63}$$

where we have introduced the (single-particle) characteristic function $\theta_\Omega(x)$ for the zone Ω:

$$\theta_\Omega(x) = \begin{cases} 1 \ (x \in \Omega), \\ 0 \ (x \notin \Omega). \end{cases} \tag{4.64}$$

Remark on the choice of Ω*:* Often Ω may be chosen as a region surrounded by potential barriers. So as to describe the external control of these barriers, it is practical to include the barrier region as a part of Ω and count the barrier potential in U_1.[41] If the system's periphery is an electric double layer, then the potential energy $U_1(x, a)$ has a physical discontinuity across this layer. In this case it is practical to define Ω so that the double layer is found in the interior of Ω. The stochastic energetics of the transmembrane diffusion of ions may explain that, if diffusion is caused by the difference of ion concentrations, it will release no heat, while if it is due to the electrostatic double layer, it accompanies a heat release. The electrochemical potential alone does not distinguish these two cases.

Inclusion of the Interactions Among Particles

Things become suddenly complicated if the particles interact among themselves [39]. For example, let us consider the situation of Fig. 4.12; the two particles are

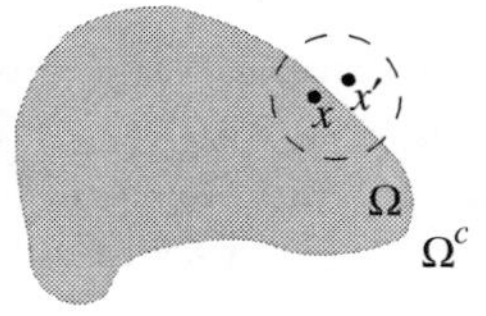

Fig. 4.12 The situation where a particle in the environment (x') is within the interaction distance (illustrated by the *dashed circle*) of another particle in the system (x) across the system boundary $\partial\Omega$. We will then count the interaction energy as a part of energy of the system, E (see the text)

---

[40] That is, $\partial U_1/\partial a = 0$ if $x \notin \Omega$.

[41] Ω is considered to be a "closed set."

close to each other across the boundary of Ω, and that interactions between them are not negligible. There is no a priori reason to include or ignore this interaction energy as a part of energy of the system, E. The same is true for the interaction energy involving more than two particles. For macroscopic thermodynamics, the premise of the extensive property of thermodynamic quantities justified to exclude such ambiguity.[42] In the study of mesoscopic systems, however, we cannot avoid this boundary effect.

We will then take up the following definition for the energy of the system. If a cluster of particles are interacting with each other, and if at least one of those particles is in Ω, we count this interaction energy in the energy of the system, E.

Leaving the details of analysis in Appendix A.4.6, the result is

$$E = \sum_{i=1}^{N} U_1(x_i, a)\theta_\Omega^{(1)}(x_i) + \sum_{j=1}^{N} \sum_{k=j+1}^{N} U_2(x_j, x_k)\theta_\Omega^{(2)}(x_j, x_k) + \sum_{j=1}^{N} \sum_{k=j+1}^{N} \sum_{l=k+1}^{N} U_3(x_j, x_k, x_l)\theta_\Omega^{(3)}(x_j, x_k, x_l) + \dots, \quad (4.65)$$

where $U_2(x_j, x_k)$ and $U_3(x_j, x_k, x_l)$ are properly defined two and three body interactions, etc., and $\theta_\Omega^{(2)}(x_j, x_k)$, $\theta_\Omega^{(3)}(x_j, x_k, x_l)$, ..., takes the value 1 if at least one of their argument takes the value in Ω and 0 otherwise. Because of the sharp boundary of the system Ω, the energy of the system E can vary discontinuously.

The above definition of the system's energy is not the unique choice. Nor this choice is proven to be the best one. Apparently, this definition needs modifications when there are two open systems Ω_1 and Ω_2 which share a part of their boundaries.

For later use, we write down also the total energy E_{tot} of the entire system $\Omega \cup \Omega^c$:

$$E_{\text{tot}} = \sum_{i=1}^{N} U_1(x_i, a) + \sum_{j=1}^{N} \sum_{k=j+1}^{N} U_2(x_j, x_k) + \sum_{j=1}^{N} \sum_{k=j+1}^{N} \sum_{l=k+1}^{N} U_3(x_j, x_k, x_l) + \dots . \quad (4.66)$$

Remark : Necessity of steric repulsion for a single-particle binding site. On the level of description by the Langevin equations, the hard-core repulsive interaction between particles should be explicitly accounted for as a part of the term U_2, even if the single-particle binding potential, $U_1(x, a)$, affords only room for single particle. Otherwise, more than one particles, e.g., $\{x_1, x_2\}$, can enter the same binding site at the same time. As is the case with usual hard-core repulsion, the last interaction has no direct contribution to the energy of the system: it only restricts the available phase space of x_1, x_2, etc.

[42] In all cases, we exclude the long-range interactions: we exclude unscreened electrostatic interaction and gravitational interaction.

4.2.3.4 Energy Balance of Open System

During dt all the particles in the entire system undergo small steps. For those particles which continue to belong to the system Ω, we will apply the previous definition of the heat. We call this heat internal heat and denote it by $d'Q^{(\mathrm{in})}$. Using $\theta_\Omega(x)$, internal heat is

$$d'Q^{(\mathrm{in})} \equiv \sum_j \theta_\Omega(x_j)\left(-\gamma\frac{dx_j}{dt} + \xi_j(t)\right) \circ dx_j. \tag{4.67}$$

The work $d'W$ is also defined as before:

$$d'W \equiv \sum_j \theta_\Omega(x_j)\frac{\partial U_1(x_j, a)}{\partial a}da(t). \tag{4.68}$$

We can rewrite this as follows:

$$d'W = \sum_{j=1}^{N} \frac{\partial[\theta_\Omega(x_j)U_1(x_j, a)]}{\partial a}da(t) = \frac{\partial E_{\mathrm{tot}}}{\partial a}da(t). \tag{4.69}$$

Recall that E_{tot} depend on a only through $U_1(x, a)$ and the latter depends on a only when $x \in \Omega$.

Because the particles can migrate during the time dt, and because the energy E should be updated upon the migration of particles, we do not expect the relation $dE = d'Q^{(\mathrm{in})} + d'W$ (wrong). In order to compare dE and $d'Q^{(\mathrm{in})} + d'W$, we use, in addition to the Langevin equation (4.61), the following identity, valid for any functions $f(x, a)$ and $\theta(x)$:

$$d[f(x, a)\theta(x)] \equiv \left[\frac{\partial f(x, a)}{\partial a}da + \frac{\partial f(x, a)}{\partial x}dx\right]\theta(x) + f(x, a)d\theta(x). \tag{4.70}$$

We then find the law of energy balance for the open system [39]:

$$dE = d'W + d'Q^{(\mathrm{in})} + d'Q^{(\mathrm{mig})} + d'Q_{\partial\Omega}. \tag{4.71}$$

Below, we will briefly describe the additional terms, $d'Q^{(\mathrm{mig})} + d'Q_{\partial\Omega}$.[43] We used the notation "Q" for these terms because these energies are not directly controlled by the external system.

$d'Q^{(\mathrm{mig})}$: Heat due to the migration of particles. This term accounts for the energy caused by the migration of particles. In the identity (4.70), this energy comes from the last term on the right-hand side. This energy is as follows:

$$dQ^{(\mathrm{mig})} \equiv \sum_{j=1}^{N} U_1(x_j, a)d\theta_\Omega^{(1)}(x_j) + \sum_{j=1}^{N}\sum_{k=j+1}^{N} U_2(x_j, x_k)d\theta_\Omega^{(2)}(x_j, x_k) + \ldots. \tag{4.72}$$

---

[43] For the details, see [39].

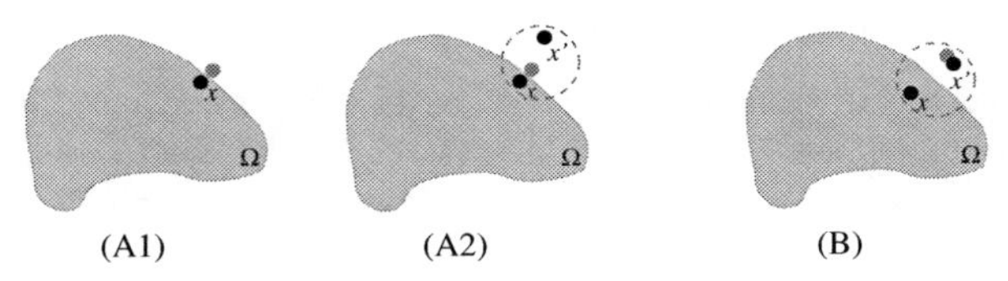

Fig. 4.13 The first (A1) and second (A2) terms on the *right*-hand side of (4.72) for $dQ^{(\mathrm{mig})}$, and the term (4.73) for $dQ_{\partial\Omega}$ (B). These terms appear when a particle changes its position from the gray disc to its adjacent black disc in each figure

Figure 4.13 (A1) and (A2) illustrates the first and second terms, respectively. In (4.72), $d\theta_\Omega^{(p)}(x_{j_1}, \ldots, x_{j_n})$ takes nonzero value when, during the infinitesimal time dt, a particular p-particle cluster comes to participate in the system ($d\theta_\Omega^{(p)} = 1$) or ceases to belong to the system ($d\theta_\Omega^{(p)} = -1$).

$d'Q_{\partial\Omega}$: Heat due to the interaction with the particles just outside the boundary. As illustrated in Fig. 4.13 (B), the system's energy changes also by the displacement of those particles which are outside Ω but interacts with a particle (or particles) inside Ω.For example, the terms in $dQ_{\partial\Omega}$ attributed to the two-particle interaction are

$$\sum_{j=1}^{N}\sum_{k=j+1}^{N} \theta_\Omega^{(2)}\left(\frac{\partial U_2}{\partial x_j}dx_j + \frac{\partial U_2}{\partial x_k}dx_k\right) - \sum_{j=1}^{N}\theta_\Omega^{(1)}(x_j)\sum_{k=1}^{N(k\neq j)} \theta_\Omega^{(2)}\frac{\partial U_2}{\partial x_j}dx_j, \tag{4.73}$$

where $U_2 \equiv U_2(x_j, x_k)$ and $\theta_\Omega^{(2)} \equiv \theta_\Omega^{(2)}(x_j, x_k)$. Only those terms with j and k such that $x_j \in \Omega$ and $x_k \notin \Omega$ [or $x_k \notin \Omega$ and $x_j \in \Omega$] contribute and leave the terms, $(\partial U_2(x_j, x_k)/\partial x_k)dx_k$ [or $(\partial U_2(x_j, x_k)/\partial x_j)dx_j$]. General expression is more complicated but the principle is the same [39].

It is important to notice that, in the present level of description of open system, the chemical potential μ appears nowhere in the energy balance equation.[44] See (2.9) for comparison. The chemical potential, which reflects the density of particles, appears when we move from the scales where we can in principle follow the individual particle to the scales where we view the system through the external system. This is the subject of Sect. 5.2.1.4.

4.3 Discussion

4.3.1 *Applicability of Stochastic Energetics to Different Forms of Langevin Equations*

The method of stochastic energetics can be applied to different forms of Langevin equations. We mention three examples below. These will show how the basic

---

[44] For example, we do not say unconditionally such as "an electron carries the Fermi energy, ε_{F}, (=chemical potential of electrons) when it moves across a junction."

framework is applied, i.e., the assignment of system and its environment, definition of heat, and the relation of energy balance.

Hydrodynamic fluctuations have become more and more accessible through the Brownian particle in a real fluid [40] or through a molecular dynamic simulation of fluid jet [41]. The fluid dynamic equation with spontaneous thermal noise is described by a Langevin equation for field. This subject is called fluctuating hydrodynamics and is developed by Landau and Lifshitz [42]. The energetic aspects associated with these fluctuation phenomena can be formulated along the principle described in Sect. 4.1. In Appendix A.4.7.1 we derive the formula of energy balance for the fluctuation hydrodynamics, with several simplifying restrictions.

The stochastic motion of suspended hard spheres has been described by the Langevin equation. In this Langevin equation, there is cross-coupling between the force on a sphere and the velocities of the other spheres, due to hydrodynamic interactions. In Appendix A.4.7.2we derive the formula of energy balance for the suspension of hard spheres, based on the recent model equation by [43, 44].

Stochastic motion expressed by curvilinear coordinates, or the stochastic motion on a curved manifolds (surface, curve, etc.) are described by Langevin equations with multiplicative noises. That is, the amplitudes of thermal random force depend on the variable of the equation. In Appendix A.4.7.3 we derive the formula of energy balance for the Langevin equation on the manifold.

4.3.2 Applicability to Nonequilibrium Processes and Limitations of the Langevin Description

The Langevin equation has the microscopic basis of the projection technique (Sect. 1.2). But in order that its derivation and the Markov approximation justify the Langevin equation, the eliminated degrees of freedom, i.e., the thermal environment, should behave not far from the equilibrium fluctuations.

A question is how far we can extend the use of the Langevin equation beyond the equilibrium fluctuations. Can we study the transient or steady nonequilibrium phenomena, especially their energetics?

There are many examples where stochastic energetics are applied to the nonequilibrium phenomena, and their studies brought, in most cases, physically sound and interesting insights about the nonequilibrium phenomena. Below is an incomplete list of those studies:

- Theoretical analysis of the energetics of the ratchet models. (There are many papers and we refer only two reviews, [45, 46].)
- Theoretical demonstration of the fluctuation theorem about the heat [47, 48]
- Definition of pathwise entropy [48].
- Numerical analysis of the Feynman pawl and ratchet [26, 49].
- Experimental assessment of the energy balance.

1. Using optically trapped bead [6].
2. Using the brass wire-held pendulum [9]

- A formulation of steady–state thermodynamics. [50, 51].
- The violated fluctuation–dissipation relation in nonequilibrium steady states.

1. The short-time limit of the discrepancy in the fluctuation–dissipation (FD) relation for the velocity in nonequilibrium steady state can be expressed in terms of the average heat flow to the thermal environment [52, 53].[45]
2. [54] extended the relation of [53] to the Langevin equations with memory. An elementary derivation of [53] is also found therein.
3. [55] checked [53] using an optically driven colloid and also demonstrated [54] using a Brownian particle in a viscoelastic fluid [56].

- Fluctuation–dissipation-like relation in nonequilibrium steady states [57]. The authors demonstrated that, if the velocity $\dot{x}$ is replaced by the relative velocity with respect to the "mean velocity," $v_s(x) \equiv j_s/p_s(x)$, the ratio of the steady-state probability flux j_s to the steady-state probability, $p_s(x)$.[46]

Still, we could mention the case where Langevin modeling is *invalid* under nonequilibrium setup. Suppose that we measure the random force on a Brownian particle while we apply a constant force onto it, e.g., in the positive x direction. In this steady non–equilibrium state, the spatial symmetry of $\pm x$ is apparently broken. We, therefore, expect a broken symmetry in the statistics of the random force (i.e., the "skewness" in the force distribution). However, the thermal random force of the Langevin equation, $\xi(t)$, is always symmetric, by definition.

Therefore, we should be conscious about what type of nonequilibrium settings we can study using the Langevin equation and its stochastic energetics. There is no unique criterion for this point, partly because it depends on our exigence, partly because our knowledge of nonequilibrium phenomena is incomplete. Still the reflection on the above successful examples and also on the cases of abuse suggests the following (evident) thing:

> *This method works if the non-equilibrium is in the system, but it does not work if the non-equilibrium is at the interface between the system and the environment.*

In fact, the absence of skewness in the random force mentioned above is apparently the latter case.

[45] For the Langevin equation without inertia, both the velocity response and the velocity correlation diverge in the short-time limit, $|t - t'| \downarrow 0$. But the difference remains finite.

[46] A drawback to this beautiful formula is that $v_s(x)$ is not a local observable as function of x and t.

4.3.3 Comments

Can we associate entropy with each trajectory?

Entropy and its production along a stochastic trajectory has been proposed to derive the "integral fluctuation theorem" [48]. This somehow contradictory notion, entropy vs. individual sample, is in fact well defined in the context of the path ensemble average. In this book we limit to disuss those observables which are determined or measured for an isolated realization.

Is it only the degrees of freedom that vary in an open system?

Open systems concern the creation and annihilation of some object. The open system that we describe in this book is a special and simple case because we have fixed the kind of objects that are created. More generally, the state of the object (and its energy, etc.), as well as its position, may also vary. For example, actin gel is generated and degraded within a cell. There, the newly created gel can vary its elastic moduli as well as its state of deformation. In other words, it is the *functions* which are inserted to or deleted from an open system. We discuss this point in Appendix A.4.8.

References

1. Einstein, *Investigation on the theory of the Brownian Movement*, (original edition, 1926) ed. (Dover Pub. Inc., New York, 1956), chap.V-1, pp. 68–75
2. H.A. Kramers, Physica **7**, 284 (1940)
3. K. Sekimoto, J. Phys. Soc. Jpn. **66**, 1234 (1997)
4. M. Matsuo, S. Sasa, Physica A **276**, 188 (2000)
5. H. Hasegawa, T. Nakagomi, J. Stat. Phys. **21**, 191 (1979).
6. V. Blickle, T. Speck, L. Helden, U. Seifert, C. Bechinger, Phys. Rev. Lett. **96**, 070603 (2006)
7. K. Sekimoto, Phys. Rev. E **76**, 060103(R) (2007)
8. C. Jarzynski, C. R. Physique **8**, 495 (2007)
9. S. Joubaud, N.B. Garnier, S. Ciliberto, J. Stat. Mech. **2007/09/**, P09018 (2007)
10. K. Schwab, E.A. Henriksen, J.M. Worlock, M.L. Roukes, Nature **404**, 974 (2000)
11. D.R. Schmidt, R.J. Schoelkopf, A.N. Cleland, Phys. Rev. Lett. **93**, 045901 (2004)
12. K. Kawasaki, in *Phase Transitions and Critical Phenomena, Vol.2,* ed. by C. Domb and M.S. Green (Academic, New York, 1972)
13. H. Spohn, J.L. Lebowitz, Adv. Chem. Phy. **38**, 109 (1978)
14. C. Jarzynski, D.K. Wójcik, Phys. Rev. Lett. **92**, 230602 (2004)
15. F. van Wijland, Phys. Rev. E **74**, 063101 (2006)
16. D.J. Evans, E.G.D. Cohen, G.P. Morriss, Phys. Rev. Lett. **71**, 2401 (1003)
17. G. Gallavotti, E.G.D. Cohen, Phys. Rev. Lett. **74**, 2694 (1995)
18. J. Kurchan, J. Phys. A **31**, 3719 (1998)
19. J. Kurchan, J. Stat. Mech. P07005 (2007)
20. H. Qian, J. Math. Chem. **27**, 219 (2000)
21. K. Sekimoto, *Stochastic Energetics* (Iwanami Book Ltd., 2004, in Japanese)
22. R.P. Feynman, R.B. Leighton, M. Sands, *The Feynman Lectures on Physics – vol.1* (Addison Wesley, Reading, Massachusetts, 1963), §46.1§–§46.9

23. H. Sakaguchi, J. Phys. Soc. Jpn. **67**, 709 (1998)
24. J. Parrondo, P. Español, Am. J. Phys. **64**, 1125 (1996)
25. T. Hondou, F. Takagi, J. Phys. Soc. Jpn. **67**, 2974 (1998)
26. M.O. Magnasco, G. Stolovitzky, J. Stat. Phys. **93**, 615 (1998)
27. C. Van den Broeck, R. Kawai, Phys. Rev. Lett. **96**, 210601 (2006)
28. N. Nakagawa, T. Komatsu, Europhys. Lett. **75**, 22 (2006)
29. F. Jülicher, A. Ajdari, J. Prost, Rev. Mod. Phys. **69**, 1269 (1997)
30. T. Komatsu, N. Nakagawa, Phys. Rev. E **73**, 065107(R) (2006)
31. M. Asfaw, M. Bekele, Eur. Phys. J. B **38**, 457 (2004)
32. T. Hondou, K. Sekimoto, Phys. Rev. E **62**, 6021 (2000)
33. B.-Q. Ai, H.-Z. Xie, D.-H. Wen, X.-M. Liu, L.-G. Liu, Eur. Phys. J. **B 48**, 101 (2005)
34. B.-Q. Ai, L. Wang, L.-G. Liu, Phys. Lett. A **352**, 286290 (2006)
35. I. Derényi, M. Bier, R.D. Astumian, Phys. Rev. Lett. **83**, 903 (1999)
36. I. Derényi, R.D. Astumian, Phys. Rev. E **59**, R6219 (1999)
37. M. Asfaw, M. Bekele, Phys. Rev. E **72**, 056109 (2005)
38. R. Benjamin, R. Kawai, Phys. Rev. E **77**, 051132 (2008)
39. T. Shibata, K. Sekimoto, J. Phys. Soc. Jpn. **69**, 2455 (2000)
40. B. Lukić et al., Phys. Rev. Lett. **95**, 160601 (2005)
41. M. Moseler, U. Landman, Science **289**, 1165 (2000)
42. L. Landau, E.M. Lifshitz, Sov. Phys.-JETP **5**, 512 (1957)
43. M. Tokuyama, I. Oppenheim, Phys. Rev. E **50**, R16 (1994)
44. M. Tokuyama, I. Oppenheim, Physica A **216**, 85 (1995)
45. J.M.R. Parrondo, B.J. De Cisneros, Appl. Phys. A **75**, 179 (2002)
46. P. Reimann, Phys. Rep. **361**, 57 (2002)
47. C. Maes, K. Netoný, M. Verschuere, J. Stat. Phys. **111**, 1219 (2003)
48. U. Seifert, Phys. Rev. Lett. **95**, 040602 (2005)
49. C. Jarzynski, O. Mazonka, Phys. Rev. E **59**, 6448 (1999)
50. T. Hatano, S. Sasa, Phys. Rev. Lett. **86**, 3463 (2001)
51. T. Komatsu, N. Nakagawa, S. Sasa, H. Tasaki, Phys. Rev. Lett. **100**, 230602 (2008)
52. T. Harada, Europhys. Lett. **70**, 49 (2005)
53. T. Harada, S. Sasa, Phys. Rev. Lett. **95**, 130602 (2005)
54. J. Deutsch, O. Narayan, Phys. Rev. E **74**, 026112 (2006)
55. S. Toyabe, T.N.H.-R. Jiang, Y. Murayama, and M. Sano, Phys. Rev. E **75**, 011122 (2007)
56. S. Toyabe and M. Sano, Phys. Rev. E **77**, 041403 (2008)
57. T. Speck, U. Seifert, Europhys. Lett. **74**, 391 (2006)

Chapter 5
Work on the Mesoscopic Systems

The balance of energy in Chap. 4 has been heuristically introduced by an analogy to mechanics. We have interpreted work to thermal environments as heat, and energy balance took the form of the first law of thermodynamics. However, it was not clear if the structure of energetics on the scale of fluctuations adapts truly to that of thermodynamics. In this chapter we will scrutinize the work done through the change of the external parameter a. The thermodynamic relationship between work and thermodynamic functions will be found. In other words, this chapter deals with the "second law" of thermodynamics. The quotation mark "" is used because we always assume the equilibrium thermodynamic character of the thermal environment.

Before the introduction of stochastic energetics, even the balance of energy on the level of Langevin equation has not been truly realized and used in the studies of fluctuation phenomena. But what is more surprising would be that the Langevin equation can realize the reversible, or quasistatic, processes. From the microscopic viewpoint, the Markov approximation breaks the time-reversal symmetry of (pure) mechanical systems. From the standpoint of the system, however, the thermal environment does not cause by itself irreversibility. The work done is equal to the change of the Helmholtz free energy of the system in the limit of slow variation of the control parameter. This convergence occurs for *each* realization of the stochastic process.

The difference with macroscopic thermodynamics is the absence of the fourth law. We do not use the extensive property with respect to the system size. For the fluctuating system we should think over what is the external system and what is the external parameters.

Once quasistatic process is understood, we should study the processes with finite but slow change of control parameters. Such processes have their own importance because (i) the irreversible work (dissipation) and the slowness of the control obeys a complementarity relation and (ii) the relation between the external system and the system exhibits a parallelism with the relation between the system and the thermal environment.

When the external parameter is changed at a finite rate, the irreversible work is a random variable, and its average is shown to be nonnegative. The demonstration uses the Jarzynski nonequilibrium work relation (for continuous case) or the fluctuation theorem (for discrete case).

Sekimoto, K.: *Work on the Mesoscopic Systems*. Lect. Notes Phys. **799**, 175–201 (2010)
DOI 10.1007/978-3-642-05411-2_5

5.1 * Work Done by External System

What does the parameter "a" stands for?

In case that the potential energy function, $U(x, a)$, represents a "rigid container" (that is, $U = 0$ for $|x| < a$ and $U = \infty$, otherwise), the parameter a is the one that controls the *position* of the hard wall. If, however, $U(x, a)$ takes the form of $U_0(x) - ax$, then a is the parameter to control the uniform external *force* field applied to the state point, x. Thus a can represent either the position or the force, depending on the case.[1] In both cases, a is regarded to be a slow variable to characterize the state of an external system: In case of the rigid container, a is related to the position of the wall as a whole. In the case of the uniform field, a is related to the position of the source of the field.

$U(x, a)$ includes all the interaction energy between the system and the external system. Therefore, $-\partial U(x, a)/\partial a$ is the force which the system exerts on the external system. By the same token as in Sect. 4.1.1.2, the external system exerts the reaction force, $\partial U(x, a)/\partial a$, onto the system. The expression (4.6), that is, $d'W \equiv \frac{\partial U}{\partial a} \circ da$ is the product of this force with the "displacement," da. It is naturally understood to be the work to the system.

Where is the boundary between the system and the external system?

We do not describe the dynamics of the external system. It implies the hypothesis that the external system is not influenced by the system's state and its dynamics.

In reality, however, it is very difficult to impose a fixed position of a rigid wall: first of all, there does not exist microscopically rigid container, and the container is also subject to thermal fluctuations. In order to take into account the fluctuation of the wall surface, one may regard the materials composing the wall surface as a part of a new enlarged system. Then the question is where we separate the system and the external system and how we define a (see Sects. 5.2.3.3 below). To the author's knowledge, no systematic argument is developed about the condition of the external control parameter.

We deal with the cases where x represents a very few, typically a single degree of freedom. We assume that stable control parameter(s) a can be found. Such simplifications allow us to concentrate on the principal things. We must, however, remember all the above mentioned ambiguities when we consider a modeling of experimental setups.

What enables the comparison between different equilibrium states?

If the parameter a is fixed, the system visits its different states so that the cumulated residence time distribution approaches asymptotically the canonical equilibrium dis-

[1] Microscopically, the interaction with the rigid wall is also due to the force field by the wall onto the system's variable.

tribution at a given temperature T and at the fixed value of a. For this system, those equilibrium states specified by other values of a have not particular meaning.[2] It is the external system and its controllability against the system that define the *process* from one equilibrium to the other one. It is the force $\partial U/\partial a$ that tells which equilibrium state is preferred by the system among the candidates parameterized by a.

Remarks

(1) We have excluded direct interactions between the external system and the thermal environment.

Therefore, in $\partial U/\partial a$ the potential energy U can always be replaced by the total energy U^{tot} of the system *plus* environments. When we apply the framework to an experimental setup, the above point should be verified. For example, if we apply an electric field to a protein motor, the influence of the field on the surrounding water as thermal environment is not a priori counted in the original formalism of stochastic energetics.

(2) Keeping the value of a constant is in general not equivalent to keeping constant the force on the system, $-\partial U(x, a)/\partial x$.

The only exception is if $U(x, a)$ can be written in the form of $U_0(x)-\phi(a)\,x$. Otherwise, keeping $\partial U(x, a)/\partial x$ constant needs the adaptation of $a(t)$. Such a feedback control from $\hat{x}(t)$ to $a(t)$ introduces a correlation between the protocol of $a(t)$ and the particular realization.[3]

(3) Force, $\partial U(x, a)/\partial a$, depends on the resolution of a.

For example, if we added small "details" $u(a) = \epsilon \sin(\frac{a}{\epsilon})$ to $U(x, a)$, the resulting force would be changed by $\cos(\frac{a}{\epsilon})$.[4]

5.2 Work Under Infinitely Slow Variation of Parameters

5.2.1 The Quasistatic Work of a Single Trajectory Leads to a Pertinent Free Energy and Is, Therefore, Reversible

5.2.1.1 * Simple Example

Let us consider a Brownian particle trapped by a harmonic potential, $U(x, a) = a\,x^2/2$, see Fig. 5.1 (left). The external system controls the "spring" constant, a. For example, a laser tweezer can trap a Brownian particle, and its light intensity can be controlled, see Fig. 5.1 (right). Neglecting the inertia effect, the Langevin equation for the Brownian particle is written as follows:

---

[2] The system does not know what boundary conditions are variable and what others are not.

[3] cf. the nanomachine to absorb the heat (Sects. 4.2.1.2).

[4] cf. the coarse graining about x, Sects. 1.3.2.2.

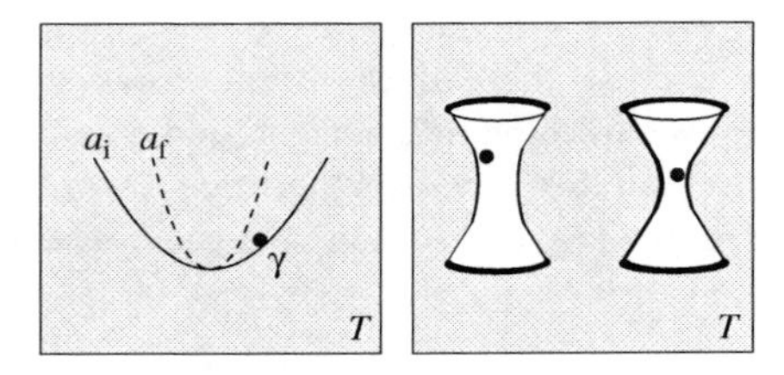

Fig. 5.1 *(Left)* Brownian particle *(thick dot)* in thermal environment (temperature T) is trapped in a harmonic potential (*thick curves*). The external system changes the profile of this potential. (*Right*) Trapping of Brownian particle under laser tweezer. The focusing controls the profile of trapping potential

$$-\frac{\partial U(x,a)}{\partial x} + \left[-\gamma \frac{dx}{dt} + \xi(t)\right] = 0, \tag{5.1}$$

where $\xi(t)$ is the Gaussian white random noise with zero mean and $\langle \xi(t)\xi(t')\rangle = 2\gamma k_B T \delta(t-t')$.

We will calculate the work done to change a from a_i to a_f taking very long operation time, τ_{op}. A similar model has been considered in Sects. 4.2.1.2. But here we do *without* the ensemble averages over realizations. The work W to change the parameter $a(t)$ from a_i to a_f is

$$W = \int_{a(0)=a_i}^{a(\tau_{op})=a_f} \frac{\partial U(x(t),a)}{\partial a} da(t) = \frac{1}{2}\int_{a(0)=a_i}^{a(\tau_{op})=a_f} x(t)^2\, da(t). \tag{5.2}$$

In the above, the value of W is evaluated with a particular realization of $x(t)$ obeying the Langevin equation (5.1). Therefore, W is a random variable. We will see that, in the limit of $\tau_{op} \to \infty$, the random variable W converges to a single value which depends only on a_i and a_f, not the protocol of $a(t)$.

We formulate the limit of slow process as in Sects. 1.3.3.2. That is, first we define a protocol $\tilde{a}(s)$ that takes unit time, i.e., $\tilde{a}(0) = a_i$ and $\tilde{a}(1) = a_f$. And then we "expand" this protocol to the time τ_{op} by $a(t) \equiv \tilde{a}(\frac{t}{\tau_{op}})$.[5] Using this representation of the protocol, the last integral of (5.2) is then written as

$$\begin{aligned} W &= \frac{1}{2}\int_0^1 x(\tau_{op}\, s)^2 \frac{d\tilde{a}(s)}{ds} ds \\ &= \int_{s=0}^{s=1} \frac{1}{\tilde{a}(s)} \frac{d\tilde{a}(s)}{ds} ds \times \underline{\frac{1}{\tau_{op}\, ds} \frac{\tilde{a}(s)\, x(\tau_{op}\, s)^2}{2} d(\tau_{op} s)}. \end{aligned} \tag{5.3}$$

We interpret the second line on the right-hand side as follows: if τ_{op} is large enough, we can imbed a very long history of $x(t)$ in the small element *ds* for which $\tilde{a}(s)$ scarcely changes. More precisely, we compare τ_{op} with the characteristic

[5] $\tilde{a}(s)$ must be a continuous function of s.

timescale of (5.1), i.e., $\gamma/\min(a)$. If $\tau_{op}ds$ is much larger than this timescale, i.e., $\mathcal{N} \equiv \tau_{op}ds/(\gamma/\min(a)) \gg 1$, then $x(\tau_{op}\, s)$ in the integrand experiences large number ($\sim \mathcal{N}$) of uncorrelated, or i.i.d., fluctuations. Then the underlined part of the integral in (5.3) is the longtime average of $ax(t)^2/2$ with a being virtually fixed.[6] By the footnote below (1.105), this temporal average converges to its canonical average, $k_B T/2$. Therefore, we have

$$W \to \frac{k_B T}{2} \int_{s=0}^{s=1} \frac{d\tilde{a}(s)}{\tilde{a}(s)} = k_B T \ln \sqrt{\frac{a_f}{a_i}} \quad (\tau_{op} \to \infty). \tag{5.4}$$

For finite $\mathcal{N}$, the typical error is estimated to be $\sim \mathcal{N}^{-1/2}$ by the central limit theorem (see Sects. 1.1.2.3).[7] The result (5.4) is instructive for two reasons:

1. The result is definite. Although $x(t)$ varies temporally and differs from one realization to another, the work W takes asymptotically the same value.
2. The result corresponds to the statistical mechanical result. In the Gibbs' statistical mechanics, the Helmholtz free energy $F(a, \beta)$ ($\beta \equiv 1/k_B T$) of the present system is

$$F(a, \beta) = -k_B T \ln \frac{1}{\sqrt{a}} + (\text{terms independent of } a). \tag{5.5}$$

   The work W obtained above is equal to the difference of this free energy, $F(a_f, \beta) - F(a_i, \beta)$. This correspondence is what we expect for any quasistatic thermodynamic process.

Gibbs statistical mechanics considers the ensemble of realizations. Any results from that framework is, therefore, the statistical average over the ensemble. The approach of the Fokker–Planck equation also yields results about the ensemble of realizations. The above analysis, however, dealt with only a single realization. For $\mathcal{N} \to \infty$ the convergence is of probability 1 due to the property of *self-averaging* for slow process. In this sense, the thermodynamic structure appeared due to the (strong) law of large numbers (of realizations).[8]

5.2.1.2 * General Theory

When a system is in contact with a single thermal environment of temperature T, the work W done by the external system on the system is given by

---

[6] $(\tau_{op}ds)^{-1}A(s, \tau_{op}s)d(\tau_{op}s)$ is approximated by $(\tau_{op}ds)^{-1} \int_{\tau_{op}s}^{\tau_{op}s+\tau_{op}ds} A(s, T)\, dT$, and the last expression is then approximated by the "long" time average about T. This approach is called the *method of multiple scale*. See, for example, Chap. 6 of [1].

[7] This argument owes to C. Jarzynski in the context of his analysis of ergodic adiabatic invariant [2, 3].

[8] Y. Oono, private communication.

$$W = \int_{a(0)=a_{\rm i}}^{a(\tau_{\rm op})=a_{\rm f}} \left. \frac{\partial U(x,a)}{\partial a} \right|_{(x,a)=(x(t),a(t))} da(t). \tag{5.6}$$

We will show that, in the limit of slow variation of the operation ($\tau_{\rm op} \to \infty$), the work W converges to the difference of the Helmholtz free energy, ΔF, defined by

$$\Delta F \equiv F(a_{\rm f}, \beta) - F(a_{\rm i}, \beta), \qquad e^{-\beta F(a,\beta)} \equiv \int e^{-\beta U(X,a)} dX. \tag{5.7}$$

Derivation : In Sects. 1.3.3.2 we have demonstrated that for the integral

$$\hat{\mathcal{I}} \equiv \int_{a(0)=a_{\rm i}}^{a(\tau_{\rm op})=a_{\rm f}} \Phi(\hat{x}(t), a(t)) da(t) \tag{5.8}$$

converges to the following integral over the parameter a in the limit of slow variation of $a(t)$:

$$\hat{\mathcal{I}} \to \int_{a_{\rm i}}^{a_{\rm f}} \langle \Phi(\cdot, a) \rangle_{\rm eq} \, da \qquad (\tau_{\rm op} \to \infty), \tag{5.9}$$

where $\langle \Phi(\cdot, a) \rangle_{\rm eq}$ is defined in (1.109), that is, $\langle \Phi(\cdot, a) \rangle_{\rm eq} \equiv \int \Phi(X,a) \mathcal{P}^{\rm eq}(X, a; T) dX$, and $\mathcal{P}^{\rm eq}(X, a; T)$ is the canonical probability distribution at $k_{\rm B}T = \beta^{-1}$ with a given value of a

$$\mathcal{P}^{\rm eq}(X, a; T) \equiv \frac{e^{-\beta U(X,a)}}{\int e^{-\beta U(X',a)} dX'}. \tag{5.10}$$

We show next that the limit in (5.9) is unique, independent of the protocol $\tilde{a}(s)$ between $a_{\rm i}$ and $a_{\rm f}$. Applying the above general formula to $\Phi(X,a) = \partial U(x,a)/\partial a$, we have $\langle \partial U(x,a)/\partial a \rangle_{\rm eq}$ in the integrand. This can be rewritten by using the so-called Ehrenfest formula:[9]

$$\left\langle \frac{\partial U(x,a)}{\partial a} \right\rangle_{\rm eq} \equiv \int \frac{\partial U(x,a)}{\partial a} \mathcal{P}^{\rm eq}(X, a; T) dX = \frac{\partial F(a,\beta)}{\partial a}. \tag{5.11}$$

By integrating the rightmost of (5.11) with respect to a, we have ΔF.(*End.*)
In summary, we have shown that the following relation is valid for any individual realization:

$$W \to \Delta F \qquad (\tau_{\rm op} \to \infty). \tag{5.12}$$

[9] We can verify (5.11) by differentiating the normalization condition of the canonical distribution, $\int e^{\beta(F(a,\beta) - U(X,a))} dX = 1 \Rightarrow \int \frac{\partial}{\partial a} e^{\beta(F(a,\beta) - U(X,a))} dX = 0$.

5.2.1.3 Discrete Case

The work W due to the change of parameter $a(t)$ on this level of description has been given:[10]

$$W = \int_0^{\tau_{\rm op}} \left\langle \left| \frac{d\mathsf{E}(a(t))}{dt} \right| \psi_t \right\rangle dt = \int_{a(0)=a_{\rm i}}^{a(\tau_{\rm op})=a_{\rm f}} \left\langle \left| \frac{d\mathsf{E}(a(t))}{da} \right| \psi_t \right\rangle da(t). \tag{5.13}$$

In the limit of slow variation of $a(t)$, we can replace the integrand in the rightmost of (5.13) by the canonical average, $\langle d\mathsf{E}(a)/da\rangle_{\rm eq}$, thus

$$W \to \int_{a_{\rm i}}^{a_{\rm f}} \left\langle \frac{d\mathsf{E}(a)}{da} \right\rangle_{\rm eq} da \qquad (\tau_{\rm op} \to \infty). \tag{5.14}$$

For $\langle d\mathsf{E}(a)/da\rangle_{\rm eq}$ we use the Ehrenfest formula, $\langle d\mathsf{E}(a)/da\rangle_{\rm eq} = \partial F(a,\beta)/\partial a$.[11] In summary, we have the formula for any individual realization:

$$W - \Delta F = \int_{a(0)=a_{\rm i}}^{a(\tau_{\rm op})=a_{\rm f}} \left[\left\langle \left| \frac{d\mathsf{E}(a(t))}{da} \right| \psi_t \right\rangle - \left\langle \frac{d\mathsf{E}(a)}{da} \right\rangle_{\rm eq} \Bigg|_{a=a(t)} \right] da(t)$$
$$\to 0 \qquad (\tau_{\rm op} \to \infty). \tag{5.15}$$

5.2.1.4 Quasistatic Process of Open System

We can immediately apply the result (5.12) to the entire system including the system and the "environment," $\Omega \cup \Omega^c$. Using the expression of the work $d'W$, (4.69), we have

$$W \to \Delta F_{\rm tot} \qquad (\tau_{\rm op} \to \infty), \tag{5.16}$$

where $F_{\rm tot}$ is the Helmholtz free energy of the entire system, defined by [12]

$$e^{-F_{\rm tot}/k_{\rm B}T} = \frac{1}{N_{\rm tot}!} \int e^{-E_{\rm tot}/k_{\rm B}T} d^{N_{\rm tot}}x, \tag{5.17}$$

where $N_{\rm tot}$ is the number of particles in the entire system, and the integral runs over the entire system for each particle.

---

[10] See Sects. 3.3.1.3 and 4.1.2.6. We wrote W instead of $\Delta' W$ for the consistency of notations between (5.12) above and (5.15) below.

[11] We use $\sum_j e^{\beta(F(a,\beta)-E_j(a))} = 1 \Rightarrow \sum_j \frac{\partial}{\partial a} e^{\beta(F(a,\beta)-E_j(a))} = 0$.

[12] For the facility of calculation we put the factor $(N_{\rm tot}!)^{-1}$ and render $F_{\rm tot}$ extensive. See below.

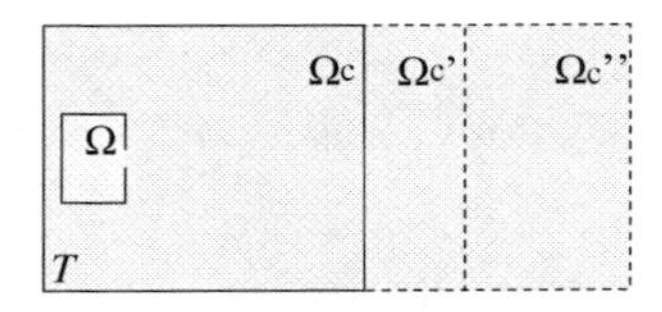

Fig. 5.2 An open system Ω and several "environments," Ω^c, $\Omega^{c\prime}$, and $\Omega^{c\prime\prime}$

We need to relate ΔF_{tot} to the thermodynamical quantity of open system. For this purpose we take the volume of the "environment," $\|\Omega^c\|$, to infinity in F_{tot}, while keeping only the part which is relevant to the open system. See in Fig. 5.2. The result is

$$\lim_{\|\Omega^c\|\to\infty} \Delta F_{\text{tot}} = \Delta J. \tag{5.18}$$

Derivation: First we identify the volume specific free energy $f^c(T, \mu)$ by

$$f^c(T, \mu) = \lim_{\|\Omega^c\|\to\infty} \frac{F_{\text{tot}}}{\|\Omega^c\|}.$$

Then we define J by subtracting from F_{tot} the (asymptotic) free energy of the environment, $\|\Omega^c\| f^c(T, \mu)$:

$$J(a, t, \mu) \equiv \lim_{\|\Omega\cup\Omega^c\|\to\infty} \left[F_{\text{tot}} - \|\Omega^c\| f^c(T, \mu)\right]. \tag{5.19}$$

This $J(a, T, \mu)$ is the thermodynamic potential for the open system Ω (see Sect. 2.1.4). $J(a, T, \mu)$ represents the particle environment only through the temperature T and the chemical potential of the particle in the environment, μ (or the density of the particles in Ω^c). Since $f^c(T, \mu)$ characterizing the particle environment should not depend on the external parameter a, we arrive at the result (5.18). (*End.*)

In conclusion, the quasistatic work done on the open system for a particular realization of stochastic process is given by the change of the thermodynamic potential for the open system [4]:

$$W \to \Delta J \qquad (\tau_{\text{op}} \to \infty). \tag{5.20}$$

In Appendix A.5.1 we recall a statistical mechanical derivation of J in (5.19) with simplifying assumptions. The result writes

$$e^{-J/k_{\text{B}}T} = \sum_{n=0}^{\infty} e^{-(F^{(n)}-\mu n)/k_{\text{B}}T}, \tag{5.21}$$

where $F^{(n)}$ is the Helmholtz free energy of the open system Ω when it contains n particles;

$$e^{-F^{(n)}/k_{\mathrm{B}}T} \equiv \frac{1}{n!}\int_{(\Omega)} e^{-E_n/k_{\mathrm{B}}T} d^n x. \tag{5.22}$$

The relations (5.16) and (5.18) are the key steps through which the chemical potential (in J) enters the energetics based on the Langevin equation.

5.2.1.5 Remark: Gibbs' Paradox and Extensivity

At the end of Sects. 3.3.2.4, we noticed that the combinatorial factor $N_{\mathrm{A}}!$, etc., in the formula (3.53) appeared independently from the particle–wave duality of the quantum physics. This factor came simply from the fact that we do not distinguish any A molecule from other A molecules. In fact these combinatorial factors are independent of whether or not the molecules consist of isotope forms or have internal parameters of long memory *as far as* the molecular reaction $\mathrm{A} + \mathrm{B} \rightleftharpoons \mathrm{AB}$ is not influenced by this variability. Also in (5.22) above, the factor $n!$ came out from purely combinatorial reason, not of quantum mechanics.

The so-called Gibbs' paradox is related to this combinatorial factor. This paradox says that

(i) In order for the thermodynamic functions to be extensive, we need to divide the phase integrals like (5.22) by the combinatorial factor corresponding to the permutation of identical particles.
(ii) Since this operation is not explained by classical mechanics, the factor is ascribed to the particle–wave duality of quantum mechanics.[13]

The resolution of the paradox is that

(i) In the study of the thermodynamic *processes* in the classical regime, increasing *all* the materials by, for example, twice is not a thermodynamic process. It contradicts the conservation of mass–energy. Therefore, the *absolute* value of the thermodynamic functions are not observable, and its extensivity is merely a convenient choice. However, the extensivity is imposed on the differences of the thermodynamic observables.
(ii) In quantum mechanics, the individuality of identical particles is lost upon collision due to the particle–wave duality. But also in classical mechanics, the individuality is lost when we describe the chemical reactions or the processes of open systems in terms of the number of each molecular species.

Thus the factor of $n!$ in (5.22) appeared without evoking quantum mechanics.

---

[13] The duality asserts that the permutations among identical particles do not make new quantum states.

5.2.2 *The Criterion of the Quasistatic Process Refers to the Force on the External System*

We define quasistatic process by the limiting process that realizes the convergence of (5.12) or (5.15). In such processes, the work W does not depend on the protocol defined by $\tilde{a}(s)$ $(0 < s < 1)$, whatever is the number of components of a.[14]

A quasistatic process is reversible or retractable in the sense that the process that retraces the same pathway in the parameter space costs no work. The work ΔF to go is exactly compensated by the work $(-\Delta F)$ to return.

If the control process makes a closed loop and returns to the initial point, i.e., those $\tilde{a}(s)$ satisfying $\tilde{a}(0) = \tilde{a}(1)$, the quasistatic work is 0.[15] Whether or not such a process leaves any change after closure of the trajectory is a subtle question. We will discuss it later (5.2.3.4). That the Langevin equation derived by using the Markov approximation can realize the reversible process emphasizes the importance of being conscious about the scale of description.

In macroscopic thermodynamics, the quasistatic process is characterized such that "at each instant of time the system realizes the equilibrium state under a given constraints." However, the equilibrium state is defined as "the state which is realized in the system after infinitely long time under a given constraints." These two statements are incompatible unless we define the limiting procedures unambiguously. But macroscopic thermodynamics does not describe the temporal changes.

On the level of Langevin equation, one could consider the closeness to the quasistatic process by comparing the probability density $\mathcal{P}(X, a(t), t)$ obtained through the Fokker–Planck equation with the canonical equilibrium distribution, $\mathcal{P}^{\mathrm{eq}}(X, a; T)$, by using a suitable measure such as the Kullback–Leibler distance, $D(\mathcal{P}||\mathcal{P}^{\mathrm{eq}})$ (see (1.81)). However, the general theory in the previous section gives natural and operational criterion of the quasistatic process: We note that the difference $W - \Delta F$ in the continuous process is

$$W - \Delta F = \int_{a(0)=a_{\mathrm{i}}}^{a(\tau_{\mathrm{op}})=a_{\mathrm{f}}} \left[\left.\frac{\partial U(x(t), a)}{\partial a}\right|_{a=a(t)} - \left\langle \frac{\partial U(x, a)}{\partial a} \right\rangle_{\mathrm{eq}} \right] da(t). \qquad (5.23)$$

Thus we measure the approach to the quasistatic process by the effect of the replacement of force by its instantaneous equilibrium expectation value *in the integral* (5.23). That is

$$\left.\frac{\partial U(x(t), a)}{\partial a}\right|_{a=a(t)} \simeq \left.\left\langle \frac{\partial U(x, a)}{\partial a} \right\rangle_{\mathrm{eq}}\right|_{a=a(t)} \qquad (5.24)$$

---

[14] By "protocol" we distinguish, for example, $\tilde{a}(s)$ from $\tilde{a}(s^2)$.

[15] Precisely speaking, F and the parameter space should be such that the closed loop can be continuously shrunken to a point passing only the quasistatic processes.

or

$$\left\langle \left| \frac{d\mathsf{E}(a)}{da} \right| \psi_t \right\rangle \Bigg|_{a=a(t)} \simeq \left\langle \frac{d\mathsf{E}(a)}{da} \right\rangle_{\text{eq}} \Bigg|_{a=a(t)}, \tag{5.25}$$

in the integral by $a(t)$. This is a much more specific criterion than the comparison of the probability densities. The quantity $\frac{\partial U(x(t),a)}{\partial a}\Big|_{a=a(t)} - \left\langle \frac{\partial U(x,a)}{\partial a} \right\rangle_{\text{eq}}\Big|_{a=a(t)}$ will appear again in the context of asymptotic estimation of the "error," $W - \Delta F$ for the nonquasistatic process (Sects. 5.3.1).

5.2.3 Quasistatic Work Reflects Some Aspects of the System's State, but Not All

5.2.3.1 Simple Case 1: Deformation of an Ideal Chain

In the aforementioned example (Sects. 4.1.2.3 esp. Fig. 4.3), we can calculate the quasistatic work to stretch the chain. The Helmholtz free energy of the ideal chain is $F(a, \beta) = -k_B T \log Z(a)$, where $Z(a)$ is proportional to the number of configurations of the chains having the end-to-end distance a. Therefore, the work to displace the end point a from a_i to a_f is

$$W = \Delta F = k_B T \log \frac{Z(a_i)}{Z(a_f)} \qquad \text{(quasistatic)}.$$

As $Z(a)$ is decreasing function of $|a|$ (i.e., the chain is less flexible for large $|a|$), the external system does a positive work W to stretch the chain. According to the law of energy balance, $dE = d'W + d'Q$, this work is immediately released to the thermal environment ($-d'Q = d'W$), because the energy of the ideal chain is constant.

5.2.3.2 Simple Case 2: Van der Waals Forces

When two molecules are placed at the distance r, the induced and/or permanent dipoles of these molecules undergo thermal and quantum fluctuations. Because of long-range electrostatic interactions the fluctuations of the two molecules are correlated and, therefore, depend on the distance r. Interaction is attractive and its free energy $F_{VdW}(r, \beta)$ ($\beta = 1/k_B T$) writes $F(r, \beta) \sim -c(T) r^{-6}$ at (moderately) large distance, where $c(T)$ is a function of temperature and other molecular parameters.

On the level of description of fluctuations, the interaction force between the molecules fluctuates in time. The free energy $F(r, \beta)$ is measured by the time-averaged force on the external system. If we change the distance r from r_i to r_f quasistatically in a particular realization, the work needed W is

$$W = c(T)\left[\frac{1}{{r_i}^6} - \frac{1}{{r_f}^6}\right] \qquad \text{(quasistatic)}.$$

5.2.3.3 Single Molecule Ideal Gas

We will consider a single (Newtonian) particle confined in a 1D cylinder and piston at temperature T. A naive question is whether the equation of state of the ideal gas, $\mathsf{P}V = k_BT$, with $N = 1$ molecule between the pressure P and the 1D volume V holds. But the primary question is how we can set the problem up and how we can define the pressure and volume.

System: Let us consider a model schematized in Fig. 5.3. The molecule (filled disc) moves ballistically and collides elastically with the walls of the cylinder and piston (the T-shaped tip) and otherwise moves ballistically. The position x and the momentum p of the molecule obeys the following Newton equation:

$$\frac{dx}{dt} = \frac{p}{m}, \quad \frac{dp}{dt} = -\frac{\partial U_{\text{pis}}}{\partial x} - \frac{\partial U_{\text{T}}}{\partial x}, \tag{5.26}$$

where m is the mass of the molecule, and $U_{\text{pis}}(x, x_{\text{pis}})$ stands for the interaction energy between the molecule and the piston tip. x_{pis} is the position of the piston tip. We will define $U_{\text{T}}(x, x_{\text{T}})$ below.

System–thermal environment interface: We introduce a thermal wall (left vertical wall of the chamber). This is mechanically coupled to the thermal environment. We assume the overdamped Langevin equation for the position x_{T} of the thermal wall:

$$0 = -\gamma\frac{dx_{\text{T}}}{dt} + \xi_{\text{T}}(t) - \frac{\partial U_{\text{T}}}{\partial x_{\text{T}}}, \tag{5.27}$$

where $\xi_{\text{T}}(t)$ is the white Gaussian random noise with zero mean and $\langle \xi_{\text{T}}(t)\xi_{\text{T}}(t')\rangle = 2\gamma k_BT\delta(t - t')$. $U_{\text{T}} = U_{\text{T}}(x, x_{\text{T}})$ represents the interaction energy between the

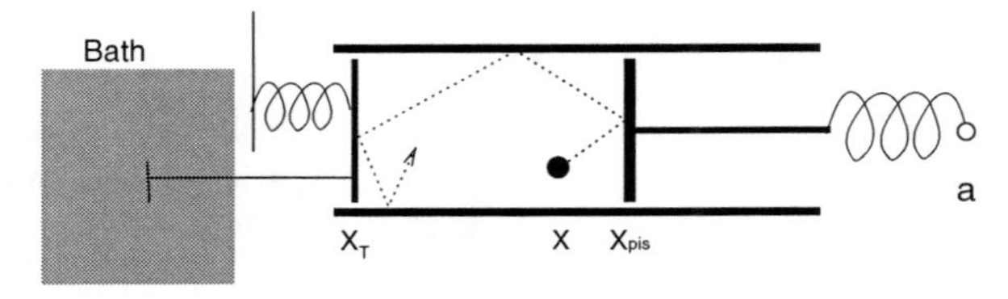

Fig. 5.3 Schematic setup of piston and cylinder system for a single particle. The particle (*thick dot*: position x) is confined within a volume *(central rectangle)* enclosed by (1) a thermal wall (*left vertical* wall: position x_{T}) which is linked to the thermal environment (*shaded rectangle* "bath") and is supported by a fixed point (spring to the *left* of the thermal wall), (2) cylinder walls (*upper* and *lower* horizontal walls), and (3) "piston tip" (*right vertical* wall: position x_{pis}), which is connected to the controlled point (*open circle*: position a) through a coupling potential (*spring* between x_{pis} and a)

molecule and the thermal wall as well as the supporting potential energy of the thermal wall.

System–external system interface: We assume that the piston is just a microscopic tip. This tip is connected through a spring to the macroscopic apparatus (the right-most open circle at a). The position $x_{\rm pis}$ and momentum $p_{\rm pis}$ of the piston tip obey the Newton equation:

$$\frac{dx_{\rm pis}}{dt} = \frac{p_{\rm pis}}{m_{\rm pis}}, \quad \frac{dp_{\rm pis}}{dt} = -\frac{\partial U_{\rm pis}}{\partial x_{\rm pis}} - \frac{\partial U_{\rm el}}{\partial x_{\rm pis}}, \tag{5.28}$$

where $m_{\rm pis}$ is the mass of the piston tip. $U_{\rm el} = U_{\rm el}(x_{\rm pis}, a)$ is the internal energy of the piston and depends only on $x_{\rm pis} - a$.The energies, $U_{\rm T}(x, x_{\rm T})$ and $U_{\rm pis}(x, x_{\rm pis})$, are expected to behave like sharp repulsive walls, blowing up as $x - x_{\rm T}$ or $x_{\rm pis} - x$ decrease to 0, respectively.

Balance of energy and quasistatic work: The balance of energy is found to be

$$d\left(\frac{p^2}{2m} + \frac{p_{\rm pis}^2}{2m_{\rm pis}} + U_{\rm T} + U_{\rm pis} + U_{\rm el}\right) = d'Q + d'W, \tag{5.29}$$

where

$$d'Q \equiv \left(-\gamma\frac{dx_{\rm T}}{dt} + \xi_{\rm T}(t)\right) \circ dx_{\rm T}\,, \qquad d'W \equiv \frac{\partial U_{\rm el}}{\partial a}da. \tag{5.30}$$

According to the general theory of Sects. 5.2.1.2, the quasistatic work for the displacement of the macroscopic apparatus a is the change of the Helmholtz free energy, $F(a, \beta)$, where

$$e^{-\beta F(a,\beta)} = C(\beta)\int e^{-\beta[U_{\rm T}(x,x_{\rm T})+U_{\rm pis}(x,x_{\rm pis})+U_{\rm el}(x_{\rm pis},a)]}dx\,dx_{\rm T}dx_{\rm pis}. \tag{5.31}$$

Here $C(\beta)$ is a factor independent of the external parameter, a.

Thermodynamic pressure: On the other hand, we can define pressure by an analogy to macroscopic thermodynamics:

$$dW \equiv -\mathsf{P}(a, \beta)\,da \qquad \text{(quasistatic process).} \tag{5.32}$$

The pressure P thus defined is the time-averaged force that the macroscopic apparatus receives at a. Using the law $W = \Delta F$ for the quasistatic process, we can identify this pressure P with the thermodynamic pressure:

$$\mathsf{P} = -\frac{\partial F(a,\beta)}{\partial a}. \tag{5.33}$$

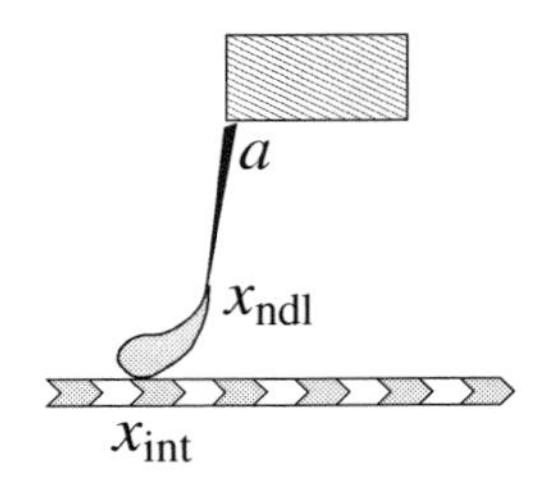

Fig. 5.4 Force apparatus (position a) binds the protein motor head through a needle (position $x_{\rm ndl}$). The motor interacts with filament (position $x_{\rm int}$)

Volume: It is not evident how a can be related to the "volume" of the chamber. Unless the repulsive walls in $U_{\rm T}(x, x_{\rm T})$ and $U_{\rm pis}(x, x_{\rm pis})$ are infinitely steep (i.e., the rigid wall), the volume cannot be unambiguously defined. The ambiguity due to the finite gradient of the potential energies, $\partial U_{\rm T}/\partial x$ and $\partial U_{\rm pis}/\partial x$, is important when the system's spatial extent, $x_{\rm pis} - x_{\rm T}$, is small. $a - \langle x_{\rm pis}\rangle$ depends on a as well as temperature. Therefore, we cannot identify da with $d\langle x_{\rm pis}\rangle$. That is, we cannot replace $dW = \mathsf{P}da$ by "$dW = \mathsf{P}d\langle x_{\rm pis}\rangle$."

In summary, controlling displacement and controlling force for small systems are not equivalent. In statistical mechanics, the saddle-point method or Darwin–Fowler method assured that these two are equivalent. (See a related discussion in [5].)

Figure 5.3 was a toy model. But a somehow similar setup is used to measure the interaction between a single head of myosin (protein motor) and an actin filament, see Fig. 5.4. The motor–filament interaction occurs at $x_{\rm int}$ while the AFM apparatus controls the position a. The motor head is bound to the latter by a needle at $x_{\rm ndl}$. If the position of $x_{\rm ndl}$ is optically measured, the result reflects both the motor–filament interaction and the thermal fluctuations of the motor and of the needle. An optical technique has been developed to suppress the thermal fluctuation of measuring devices (down to 5K!) [6]. The energetics of the feed-back-controlled system is discussed in [7].

5.2.3.4 Work-Free Transport of Heat and Particles

Suppose that the control parameter a has more than one component and that it is changed along a closed loop, $\hat{a}(0) = \hat{a}(1)$.[16] If the process is quasistatic, the work is 0. But the system's state can undergo a nontrivial change [8].

We consider the cyclic change of potential energy profile, $U(x, \hat{a})$, as shown in Fig. 5.5. We will regard the state point $x(t)$ as the position of a Brownian particle under the potential energy U. We impose a periodic boundary condition, i.e., the rightmost end (R) is continued to the leftmost end (L). Or, we assume that the system is open toward the reservoirs of particles in (L) and (R). In any case this quasistatic cycle transports particles from the left (L) to the right one (R) on the average. The calculus is given in Appendix A.5.2.

This transport without costing work indicates subtlety of the quasistatic process:

[16] We use the scaled protocol $\tilde{a}(s)$ introduced in Sects. 5.2.2.

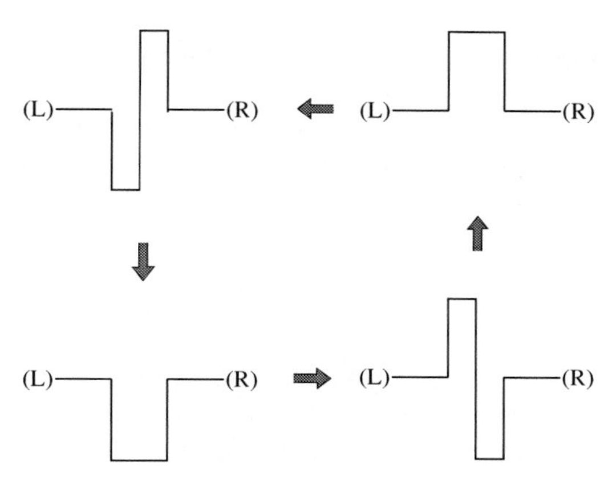

Fig. 5.5 Profile of potential $U(x, a)$ vs. x between the *left*-particle environment (L) to the *right* one (R). *Thick arrow* indicates the progress of the potential profile. (Figure adapted from Fig. 2 of [8])

(i) Even in the quasistatic process, we cannot always approximate $P(x, t)$ of the Fokker–Planck equation by the equilibrium density, $P_{\mathrm{eq}}(x, a; T)$. If we did it, the probability flux $J[P] \equiv -\frac{1}{\gamma}\left[\frac{\partial U}{\partial x}P + k_{\mathrm{B}}T\frac{\partial P}{\partial x}\right]$ is always 0, and we would not have transport. [17]

(ii) In the quasistatic limit, work has a potential function, i.e., the free energy $F(a, \beta)$. However, the probability flux J is not necessary the gradient of a potential function.

Remarks:

1. It is essential that the potential profile undergoes a cyclic change. If the profile change is a simple go-and-back along the same pathway, there is no net transport. The left–right asymmetry of the potential profile is a necessary but not sufficient condition.
2. Experimental demonstrations of such work-free transport must face the fluctuating part of the transport: in taking the quasistatic limit, the fluctuation in the number of transported particles per cycle will diverge as $\sqrt{t_{\mathrm{cyc}}}$ with the time spent for a cycle, t_{cyc}. The prefactor of $\sqrt{t_{\mathrm{cyc}}}$ can be decreased by raising the energy scales of the potential U. However, it then makes the condition for the quasistatic process more and more stringent, or it requires larger cycle time, t_{cyc}. More discussion will be given in Chap. 7.
3. The work-free quasistatic transport can be realized also in conventional thermodynamics by, for example, using the (macroscopic) Carnot cycle. In the formula of the reversible efficiency, $\eta_{\mathrm{rev}} = (T_{\mathrm{h}} - T_{\mathrm{l}})/T_{\mathrm{h}}$ (see Sect. 2.3.3), the case $T_{\mathrm{h}} = T_{\mathrm{l}}$

[17] This is general remark when we use linear nonequilibrium thermodynamics. The flux of energy or mass is caused by their small spatial gradients across the local regions in which the equilibrium is assumed.

assures the transport of heat between the two thermal environments of identical temperature without work $\eta_{\rm rev} = 0$.

4. The work-free transport discussed above does not contradict the second law of thermodynamics: the heat or particles transported between the environments of the same temperature or chemical potential cannot be the source of later work.

5.3 Work Under Very Slow Variation of Parameters

We will analyze the processes taking a finite time $\tau_{\rm op}$ from $a(0) = a_{\rm i}$ to $a(\tau_{\rm op}) = a_{\rm f}$. The quantity of interest is the difference between the work W and the increment of the Helmholtz free energy, ΔF. We call this difference the *irreversible work*,

$$W_{\rm irr} \equiv W - \Delta F. \tag{5.34}$$

See Fig. 5.6. For continuous process described by Langevin equation, it is written as (see (5.12))

$$W_{\rm irr} = \int_{a_{\rm i}}^{a_{\rm f}} \left[\frac{\partial U(x(t), a(t))}{\partial a} - \left\langle \frac{\partial U(x, a(t))}{\partial a} \right\rangle_{\rm eq} \right] da(t). \tag{5.35}$$

For finite $\tau_{\rm op}$, the irreversible work $W_{\rm irr}$ is a random variable, whose value varies from one realization to the other. In this section and next section we deal with the average of $W_{\rm irr}$ over the ensemble of paths, $\langle W_{\rm irr} \rangle$.

5.3.1 *The Average Irreversible Work and the Time Spent for the Work are Complementary*

When the interval $\tau_{\rm op}$ to change the parameter $a(t)$ is large, a general law of $\langle W_{\rm irr} \rangle$ is [9]

(1) The product, $\langle W_{\rm irr} \rangle \, \tau_{\rm op}$, is bounded below for $\tau_{\rm op} \to \infty$.
(2) This lower bound, which we denote by $S(a_{\rm i}, a_{\rm f})$, is positive for $a_{\rm i} \neq a_{\rm f}$.

That is,

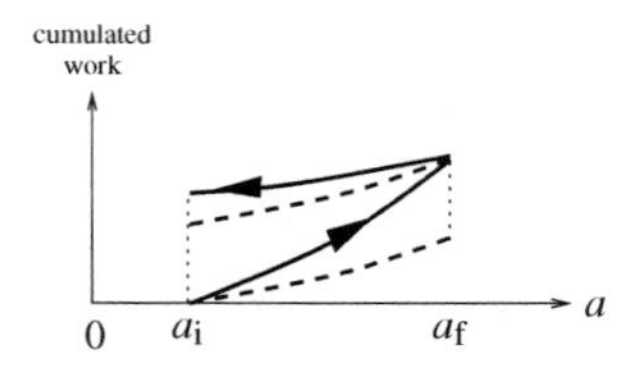

Fig. 5.6 Cumulated work *(solid curves)* along the change of parameter, $a_{\rm i} \to a_{\rm f} \to a_{\rm i}$. The difference from the quasistatic work *(dashed curves)* gives the irreversible work, $W_{\rm irr}$

$$\langle W_{\mathrm{irr}} \rangle \, \tau_{\mathrm{op}} \geq S(a_{\mathrm{i}}, a_{\mathrm{f}}) \qquad (\tau_{\mathrm{op}} \to \infty). \tag{5.36}$$

5.3.1.1 Origin of the Complementarity Relation

The above result comes out from the following expression of $\langle W_{\mathrm{irr}} \rangle$. The details of calculation is given in Appendix. A.5.3.

$$\langle W_{\mathrm{irr}} \rangle = \int_0^{\tau_{\mathrm{op}}} \frac{da}{dt} \mathbf{\Lambda}(a) \frac{da}{dt} \, dt + \mathcal{O}((\tau_{\mathrm{op}})^{-2}). \tag{5.37}$$

Here the symmetric matrix[18] $\mathbf{\Lambda}(a)$ is defined by

$$\mathbf{\Lambda}(a) \equiv \beta \left\langle\!\!\left\langle \left[\frac{\partial U}{\partial a} - \left\langle \frac{\partial U}{\partial a} \right\rangle_{\mathrm{eq}} \right] \bullet (-g) \bullet \left[\frac{\partial U}{\partial a} - \left\langle \frac{\partial U}{\partial a} \right\rangle_{\mathrm{eq}} \right] \right\rangle\!\!\right\rangle_{\mathrm{eq}}, \tag{5.38}$$

with $\langle\!\langle A \bullet (-g) \bullet A \rangle\!\rangle_{\mathrm{eq}}$, defined by [19]

$$\langle\!\langle A \bullet (-g) \bullet A \rangle\!\rangle_{\mathrm{eq}} \equiv \iint P^{(\mathrm{eq})}(x) A(x) (-g(x, x', a)) P^{(\mathrm{eq})}(x') A(x') \, dx \, dx'. \tag{5.39}$$

The Green function $g(x, x'; a)$ is defined by

$$\frac{1}{\gamma} \frac{\partial}{\partial x} P^{(\mathrm{eq})}(x) \frac{\partial}{\partial x} g(x, x'; a) = \delta(x - x'). \tag{5.40}$$

The integral on the right-hand side of (5.37) is always nonnegative.[20]

To derive the complementarity relation (5.36) we will extract τ_{op} from (5.37): we represent $a(t)$ in (5.37) by the rescaled protocol, $\tilde{a}(s) \equiv a(s\tau_{\mathrm{op}})$ with $0 \leq s \leq 1$ (see, Sects. 5.2.1.1). We then have the following asymptotic relation:

$$\langle W_{\mathrm{irr}} \rangle \, \tau_{\mathrm{op}} = \int_0^1 \frac{d\tilde{a}}{ds} \mathbf{\Lambda}(a) \frac{d\tilde{a}}{ds} \, ds + \mathcal{O}((\tau_{\mathrm{op}})^{-1}). \tag{5.41}$$

The first term on the right-hand side is positive and a functional of the rescaled protocol, $\tilde{a}(s)$. We can, therefore, define the lower bound of this integral as $S(a_{\mathrm{i}}, a_{\mathrm{f}})$:

---

[18] when a has more than one component.

[19] $P^{(\mathrm{eq})}(x) \equiv P^{(\mathrm{eq})}(x, a; T)$ is the canonical equilibrium density for a given parameter a.

[20] The operator on the left-hand side of (5.40) is self-adjoint and, therefore, has real spectra. This operator is of the form of diffusion operator with an inhomogeneous diffusion constant, $D(x) = \frac{1}{\gamma} P^{(\mathrm{eq})}(x, a)$. Since diffusion is a purely relaxing phenomena, the spectra of the above operator are all negative except of a single 0, corresponding to the constant eigenfunction. As the inverse operator of this diffusion operator, the Green function g, is a symmetric function with respect to x and x'. Moreover, the spectra of g are the inverse of the diffusion operator and, hence, all real and negative. From the last fact, $\mathbf{\Lambda}(a)$ is positive definite.

$$S(a_\mathrm{i}, a_\mathrm{f}) \equiv \min_{\tilde{a}} \left[\int_0^1 \frac{d\tilde{a}}{ds} \mathbf{\Lambda}(a) \frac{d\tilde{a}}{ds} ds \right], \quad (5.42)$$

where the minimum is sought for with all the continuous rescaled protocols $\tilde{a}(s)$ under the conditions, $\tilde{a}(0) = a_\mathrm{i}$ and $\tilde{a}(1) = a_\mathrm{f}$. Since $S(a_\mathrm{i}, a_\mathrm{f})$ does not depend on the protocol between these end points, this is what we should have in (5.36).

5.3.1.2 Implication of the Complementarity Relation

1. An interpretation of the complementarity relation (5.36) is that the loss in the work becomes large if the operation is done in haste (i.e., with small τ_op). The *total* loss is proportional to the average rate of the change of the parameter.
2. Another interpretation is to regard $\langle W_\mathrm{irr} \rangle \equiv \langle W \rangle - \Delta F$ as an error of the measurement of the thermodynamic information, ΔF.[21] Then (5.36) is reminiscent of the complementarity relations of quantum mechanics between the energy and the time, $\Delta E \Delta t \geq \hbar/2$. Unlike the quantum mechanical principle of uncertainty, the "Planck constant" $S(a_\mathrm{i}, a_\mathrm{f})$ depends on the system and the temperature. In particular, $S(a_\mathrm{i}, a_\mathrm{f})$ depends linearly on γ through the Green function, g.
3. The formula (5.38) expresses $\mathbf{\Lambda}(a)$ in terms of the correlation function of the "deviative force," $\frac{\partial U}{\partial a} - \left\langle \frac{\partial U}{\partial a} \right\rangle_\mathrm{eq}$. In the next section we will see that $\mathbf{\Lambda}(a)$ plays the role of (linear) friction coefficient relating the rate of change of the parameter, da/dt, to its conjugate frictional force.
4. What type of the protocol $a(t)$ realizes the lower bound of the complementarity relation (5.36)? Given a total time τ_op, the expression (5.41) implies that we should avoid the route of a along which the friction coefficient $\mathbf{\Lambda}(a)$ is large.[22] If we cannot avoid such region, e.g., when a has only one component, we should spend more time in that region than elsewhere.

5.3.2 * *For the External System the Weak Irreversible Work is Ascribed to a (Macro) Frictional Force*

From the standpoint of the external system ("Ext"), the quasistatic work W is apparently stored in the system as the increment of the potential energy, ΔF (Sect. 2.2). There is a parallelism: the external system "Ext" does not see the degrees of freedom of the system on the one hand, and the system ("Sys") does not see those degrees of freedom in the thermal environment on the other hand.[23]

---

[21] cf. The standard deviation of W_irr decreases with τ_op as $\mathcal{O}((\tau_\mathrm{op})^{-1/2})$. See Sects. 5.2.1.1.

[22] If $\mathbf{\Lambda}(a)$ has anisotropy, the orientation of the route as well as its location should be optimized.

[23] Below is an example of how the world looks differently from different viewpoints: study of the fluctuations of cell motility is an "activity measurement" for biologists but "passive measure-

The study of the irreversible work W_{irr} now provides a further parallelism: $\mathbf{\Lambda}(a)$ for the "Ext" corresponds to γ (friction coefficient) for "Sys." In fact, if we combine the results of Sects. 5.2 and 5.3, the average work $\langle W\rangle$ done through a slow change of the parameter a is

$$\begin{aligned}\langle W\rangle &= \Delta F + \int_0^{\tau_{\text{op}}} \frac{da}{dt}\mathbf{\Lambda}(a)\frac{da}{dt}\,dt \\ &= \int_{a_{\text{i}}}^{a_{\text{f}}} da(t)\left[\frac{\partial F}{\partial a} + \mathbf{\Lambda}(a)\frac{da}{dt}\right],\end{aligned} \tag{5.43}$$

with an error of $\mathcal{O}((\tau_{\text{op}})^{-1})$. The second line of (5.43) allows the interpretation that "Ext" applies the force, $\frac{\partial F}{\partial a} + \mathbf{\Lambda}(a)\frac{da}{dt}$, onto "Sys." See Fig. 5.7. By the law of action–reaction, the external system receives the potential force, $-\frac{\partial F}{\partial a}$, and the friction force, $-\mathbf{\Lambda}(a)\frac{da}{dt}$. Therefore, $\mathbf{\Lambda}(a)$ is the friction constant for the parameter $a(t)$.

To take into account the deviation of W from $\langle W\rangle$, we introduce a noise term $\Xi(t)$ such that

$$W = \int_{a_{\text{i}}}^{a_{\text{f}}} da(t)\left[\frac{\partial F}{\partial a} + \mathbf{\Lambda}(a)\frac{da}{dt} - \Xi(t)\right], \tag{5.44}$$

where the noise term behaves as[24] $\int_{a_{\text{i}}}^{a_{\text{f}}} \Xi(t)da(t) \sim \mathcal{O}((\tau_{\text{op}})^{-1/2})$. We can rewrite (5.44) as

$$\Delta F = \int_{a_{\text{i}}}^{a_{\text{f}}} \left[-\mathbf{\Lambda}(a)\frac{da}{dt} + \Xi(t)\right]da(t) + W, \tag{5.45}$$

The last expression is similar to the law of energy balance for the Langevin equation, $dE = d'Q + d'W$. Schematically, the parallelism is

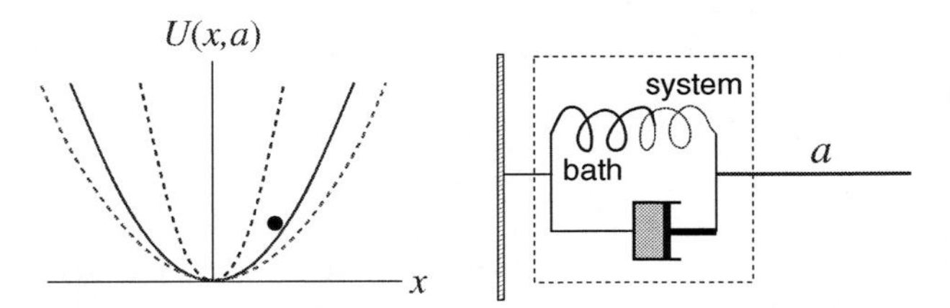

Fig. 5.7 (*Left*) A system and its control by an external parameter a. (*Right*) The system viewed from the external system as a black box

ment" for rheologists. Study of the response of cell against external perturbations is a "passive measurement" for biologists and "active measurement" for rheologists.

[24] cf. Footnote 21.

$$\begin{aligned}
\text{"Sys"} &\longleftrightarrow \text{"Ext"} \\
x &\longleftrightarrow a \\
U(x) &\longleftrightarrow F(a, \beta)+\text{"?"} \\
\gamma &\longleftrightarrow \mathbf{\Lambda}(a) \\
\text{random thermal force: } \xi &\longleftrightarrow \text{system noise: } \Xi \\
\text{time resolution} &\longleftrightarrow \tau_{\text{op}}
\end{aligned}$$

In the above schema, we have not considered the dynamical evolution of the parameter $a(t)$. Therefore, "?" in the above is not specified. On the last row, the time resolution of the Langevin equation should be larger than the bath's relaxation time on the one hand, and τ_{op} should be larger than the system's relaxation time on the other hand. If the "time resolution" τ_{op} is too large, there is no noise Ξ. A question is whether there is a smooth limit from stochastic energetics to macroscopic thermodynamics when the size of the system goes to infinity. The parallelism across different scales can also be found between the system and its subsystem. This issue is addressed in the next chapter.
Note: Throughout this section we have assumed that the temperature is constant. The process including the time-dependent temperature, $T(t)$, has been studied in Matsuo (1999, unpublished paper). The author showed the Clausius' inequality, $\oint d'Q/T \leq 0$, using stochastic energetics and statistical entropy.

5.4 Work Under the Change of Parameter at Arbitrary Rates

5.4.1 Jarzynski's Nonequilibrium Work Relation Leads to the Nonnegativity of the Average Irreversible Work

The Jarzynski's nonequilibrium work relation [10] is an important equation to assure the nonnegativity of the average irreversible work, $\langle W_{\text{irr}} \rangle$. If a Markov process with parameter(s) a has an equilibrium state for each value of a, and if the initial state obeys a canonical distribution, the irreversible work, W_{irr}, for the process between t_0 and t satisfies

$$1 = \langle e^{-\beta W_{\text{irr}}} \rangle_{\text{eq}}. \tag{5.46}$$

First it has been demonstrated for a thermally isolated system to which work is added mechanically. Later the relation turned out to be valid more generally. We show a brief demonstration using a form of the Feynman–Kac formula.[25] Those who are not interested in the mathematical details may skip to Sects. 5.4.1.2.

[25] The description of this section is based on [11] and the series of lectures by C. Jarzynski at Institut Henri Poincaré (Oct. 2007).

5.4.1.1 Jarzynski's Nonequilibrium Work Relation

Path Probability of Markov Process and Generating Operator

Suppose there is a Markov process, whose transition rate depends on a parameter $a_t \equiv a(t)$. We denote by $K(x,t|x_0,t_0)$ the conditional probability to find $\hat{x}_t$ at x at the time t, given that it started from x_0 at the initial time, t_0: $K(x,t_0|x_0,t_0) = \delta(x-x_0)$. Formally $K(x,t|x_0,t_0)$ can be written as

$$K(x,t|x_0,t_0) = \int_{(x_0,t_0)} P[X|x_0,t_0]\mathcal{D}X, \tag{5.47}$$

where $P[X|x_0,t_0]$ is the probability for the path X over the time interval $[t_0,t]$, and $\int_{(x_0,t_0)}\cdots\mathcal{D}X$ denotes the path integral with the initial condition, (x_0,t_0). We define the generating operator $\mathcal{L}(a_t)$ of $K(x,t|x_0,t_0)$ through

$$\frac{\partial K(x,t|x_0,t_0)}{\partial t} = \mathcal{L}(a_t)K(x,t|x_0,t_0). \tag{5.48}$$

Weighed Path Probability

Now we consider another Markov process whose transition probability, $G(x,t|x_0,t_0)$, is

$$G(x,t|x_0,t_0) = \int_{(x_0,t_0)} e^{\int_0^t w_{t'}dt'} P[X|x_0,t_0]\mathcal{D}X, \tag{5.49}$$

where $P[X|x_0,t_0]$ is always the path probability for the path X governed by $\mathcal{L}(a_t)$. w_t can depend on x and on a. $G(x,t|x_0,t_0)$ satisfies $G(x,t_0|x_0,t_0) = \delta(x-x_0)$.

From (5.49) $G(x,t+dt|x',t)$ can be written as

$$\begin{aligned} G(x,t+dt|x',t) &\simeq (1+w_t dt)K(x,t+dt|x',t) \\ &\simeq (1+w_t dt)[1+dt\mathcal{L}(a_t)]\delta(x-x'). \end{aligned} \tag{5.50}$$

To go to the second line (5.48) has been used. This formula will be used later.

Feynman–Kac Formula

A simple version of Feynman and Kac tells[26] that $G(x,t|x_0,t_0)$ obeys

$$\frac{\partial G}{\partial t} = (\mathcal{L}(a_t) + w_t)G. \tag{5.51}$$

That is, $G(x,t|x_0,t_0)$ is generated by $\mathcal{L}(a_t) + \omega_t$.

[26] In physicists' language, the general Feynman–Kac formula gives the (Feynman's) path integral representation of the solution of an SDE of Itô type. In the path integral, the "action" in the exponential is the sum of the kinetic part $\propto \gamma\dot{x}^2$ and the potential part $\propto U(x,a)$. To apply to the SDE of Stratonovich type, the action should be "corrected" by $\propto \partial^2 U/\partial x^2$.

The outline of the proof is as follows. As a Markov process $G(x, t + dt|x_0, 0)$ obeys the Chapman–Kolmogorov equation,[27] i.e.,

$$G(x, t + dt|x_0, 0) = \int G(x, t + dt|x', t)G(x', t|x_0, 0)dx'. \tag{5.52}$$

Substituting (5.50) into (5.52) and keeping up to the order of $\mathcal{O}(dt)$, we have $G(x, t + dt|x_0, t_0) \simeq (1 + w_t dt + dt\mathcal{L}(a_t))G(x, t|x_0, t_0)$. This means (5.51).[28] The $G(x, t|x_0, t_0)$ defined above is, therefore, the Green's function of (5.51).

Evolution of $e^{-\beta\mathcal{H}_t}$

We apply the Feynman–Kac formula to the case where a "Hamiltonian" $\mathcal{H}_t \equiv \mathcal{H}(x, a_t)$ satisfies $\mathcal{L}(a_t)e^{-\beta\mathcal{H}_t} = 0$ for each t. We then define w_t by

$$w_t \equiv -\beta \frac{\partial\mathcal{H}(x, a_t)}{\partial a_t}\frac{da_t}{dt}. \tag{5.53}$$

From w_t, the work done to the system by an external system during the interval $[t_0, t]$ is given by

$$\beta W_{t,t_0} = -\int_{t_0}^{t} w_{t'}\, dt'. \tag{5.54}$$

Then $e^{-\beta\mathcal{H}_t}$ satisfies $\frac{\partial}{\partial t}e^{-\beta\mathcal{H}_t} = (\mathcal{L}(a_t) + w_t)e^{-\beta\mathcal{H}_t}$. In fact $\frac{\partial}{\partial t}e^{-\beta\mathcal{H}_t} = w_t e^{-\beta\mathcal{H}_t}$ is an identity, and we can add $0 = \mathcal{L}(a_t)e^{-\beta\mathcal{H}_t}$ to each side of this equation.

$e^{-\beta\mathcal{H}_t}$ as a solution of $\frac{\partial}{\partial t}e^{-\beta\mathcal{H}_t} = (\mathcal{L}(a_t) + w_t)e^{-\beta\mathcal{H}_t}$ can be expressed using the Green's function $G(x, t|x_0, t_0)$:

$$e^{-\beta\mathcal{H}_t} = \int G(x, t|x_0, t_0)e^{-\beta\mathcal{H}_{t_0}}dx_0. \tag{5.55}$$

Jarzynski Nonequilibrium Work Relation

We divide both sides of (5.55) by $e^{-\beta F_{t_0}} \equiv \int e^{-\beta\mathcal{H}_{t_0}}dx_0$, and there substitute (5.49) and (5.54) for $G(x, t|x_0, t_0)$. The result is

---

[27] The Chapman–Kolmogorov equation means the following. The totality of the paths from (x_0, t_0) to (x, t) is given as the sum of those paths that pass through a "gate" at x_g at a fixed time t_g, then summed over all x_g. For the Markov process the probability weight for the paths from (x_0, t_0) to (x, t) via (x_g, t_g) can be factorized into those weights of each segments.

[28] If (5.51) is a Fokker–Planck equation, then (5.49) gives its formal explicit solution using $P[X|x_0, t_0]$. If (5.51) is a Schrödinger equation, (5.49) again gives its formal explicit solution in the same manner [12].

$$e^{\beta F_{t_0}} e^{-\beta \mathcal{H}_t} = \int \left[\int_{(x_0,t_0)} P[X|x_0,t_0] e^{-\beta W_{t,t_0}} \mathcal{D}X \right] e^{\beta(F_{t_0}-\mathcal{H}_{t_0})} dx_0. \tag{5.56}$$

Integration of (5.56) over x gives

$$e^{-\beta(F_t - F_{t_0})} = \left\langle e^{-\beta W_{t,t_0}} \right\rangle_{\text{eq}}, \tag{5.57}$$

where $e^{-\beta F_t} \equiv \int e^{-\beta \mathcal{H}_t} dx$, and $\langle \cdot \rangle_{\text{eq}}$ denotes the path average starting from the initial canonical probability density $e^{\beta(F_{t_0}-\mathcal{H}_{t_0})}$ at t_0. Equation (5.57) is called Jarzynski nonequilibrium work relation. Note that $\mathcal{L}(a_t)$ can be either the Liouville operator[29] of conserved dynamical process or the Fokker–Planck operator of stochastic process.

The result (5.59) due to Jarzynski [10] is very general since it holds for any protocol of $a(t)$ with any finite time τ_{op} of the process. The Jarzynski nonequilibrium work relation can be used to measure ΔF from the protocol of $a(t)$ at finite rate of change:

$$\Delta F = -k_{\text{B}} T \ln \langle e^{-\beta W} \rangle_{\text{eq}}. \tag{5.58}$$

Here the average is taken over the paths starting from canonical equilibrium. This relation works very well for small systems [13, 14]. With increasing number of degrees of freedom, this method requires a lot of data for a good statistics. The reason is that very rare events for W is dominantly important in the average because of its exponential dependence, $\langle e^{-\beta W} \rangle$ [15].

The precision of "canonical" initial condition in the above is important: a counterexample has been demonstrated for the "microcanonical" initial condition [16]. We come back to the implication of this example later (see, the end of Sect. 7.1.3).

5.4.1.2 Nonnegativity of $\langle W_{\text{irr}} \rangle$

We will show the nonnegativity, $\langle W_{\text{irr}} \rangle \geq 0$, in condition that the process starts with the canonical equilibrium state with a given initial parameter, $a = a(t_0)$.

In (5.57), $F_t - F_{t_0}$ is ΔF and W_{t,t_0} is a work of a particular realization, W. We can, therefore, identify $W_{t,t_0} - (F_t - F_{t_0}) = W - \Delta F$ as the irreversible work, W_{irr}. Then (5.57) is

$$1 = \langle e^{-\beta W_{\text{irr}}} \rangle_{\text{eq}}. \tag{5.59}$$

We now apply Jensen's inequality, $\langle e^{-z} \rangle \geq e^{-\langle z \rangle}$ to (5.59), where $\langle \rangle$ is average over any normalized probability density of z.[30] The result yields the inequality for $\langle W_{\text{irr}} \rangle_{\text{eq}}$:

---

[29] See the paragraph containing (A.10).

[30] Jensen's inequality is the relation for any concaved function, $f(z)$, the function with $f''(z) \geq 0$. On the graph of $f(z)$ vs. z, the center of mass of the points $(z_1, f(z_1)), \ldots, (z_n, f(z_n))$

$$\langle W_{\rm irr}\rangle_{\rm eq} \geq 0. \tag{5.60}$$

The inequality (5.60) includes the case of quasistatic process, where $\langle W_{\rm irr}\rangle_{\rm eq} = 0$. Recall that in Sects. 5.2.1.2 we obtained a stronger statement, $W_{\rm irr} = 0$, for an *individual* quasistatic process.

Remark. There is a different definition of the average irreversible work, which is not directly related to the work measurement [17]. Using the probability density $P(x,t)$, we define the statistical entropy, $S \equiv -\int P \ln P\, dx$, and then the quasi-free energy, $\tilde{F} \equiv \langle U\rangle - TS$, where $\langle U\rangle = \int UPdx$. Then the following inequality is proven:

$$\frac{\langle d'W\rangle}{dt} - \frac{d\tilde{F}}{dt} = \int \gamma \frac{J[P]^2}{P} dx \geq 0, \tag{5.61}$$

where $J[P]$ is the probability current of the Fokker–Planck equation. The formula (5.61) has essentially the same content as the "H-theorem" (4.38) in Chap. 4.

5.4.2 The Fluctuation Theorem Leads to Jarzynski's Nonequilibrium Work Relation for Discrete Process

We consider the stochastic processes characterized by the transition rates such as $w_{i\to j}(a)$ from a discrete state i to another state j, where a is an external control parameter (see Sects. 3.3.1.4 and 3.3.1.3). We assume that for each value of a, the transition probability admits the canonical equilibrium probability $P_i^{\rm eq}(a)$ satisfying the detailed balance condition:

$$P_i^{\rm eq}(a) w_{i\to j}(a) = P_j^{\rm eq}(a) w_{j\to i}(a). \tag{5.62}$$

The so-called (a version of) fluctuation theorem (FT) or Crook's relation for the irreversible work, $W_{\rm irr}$, is [18, 19].

$$\frac{P_R(-W_{\rm irr})}{P_R(W_{\rm irr})} = e^{-\beta W_{\rm irr}}, \tag{5.63}$$

where $P_R(W_{\rm irr})$ is the probability density for $W_{\rm irr}$. Integration of $e^{-W_{\rm irr}} P_R(W_{\rm irr})$ gives the average, $\langle e^{-W_{\rm irr}}\rangle$, while the integration of $P_R(-W_{\rm irr})$ gives unity. Therefore, the Jarzynski nonequilibrium work relation for the discrete process is derived:

$$\left\langle e^{-\beta W_{\rm irr}}\right\rangle_{\rm eq} = 1. \tag{5.64}$$

Finally the nonnegativity of the average irreversible work, $\langle W_{\rm irr}\rangle_{\rm eq} \geq 0$, is derived.

(some can be redundant) is always found above this graph. Especially this center of mass, $(\frac{1}{n}\sum_{i=1}^n z_i, \frac{1}{n}\sum_{i=1}^n f(z_i))$, is vertically above $(\frac{1}{n}\sum_{i=1}^n z_i, f(\frac{1}{n}\sum_{i=1}^n z_i))$.

In Appendix A.5.4 we sketch the derivation of the fluctuation theorem. Jarzynski nonequilibrium work relation and the fluctuation theorem have emerged in the same epoch as the emergence of stochastic energetics. In addition to those papers cited above, we refer to the other essential papers that initiated the approach described in this section [20–24, 18, 25]. There are preceding studies such as Bochkov and Kuzovlev [26–29] (see a commentary by [30]) and Kawasaki and Gunton [31][31] in the 1970's. Comprehensive textbooks on these subjects are to be written by the original contributors. In this book we mentioned only briefly the outline of some demonstrations.

5.5 Discussion

5.5.1 How Fast Can the External Parameter Be Changed?

In Sect. 5.4, no constraints has been put on the maximum rate of the parameter change, $da(t)/dt$. There are situations where a very large value of $da(t)/dt$ is considered.

One case is the optimal control problem. The control protocol with the least cost can have discontinuities in $a(t)$ at the initial and final times.[32] When the inertia is neglected, the effect of the discontinuity of $a(t)$ on the eliminated momentum should be carefully analyzed (cf. the Büttiker and Landauer ratchet Sects. 4.2.2.2).

The other case is the numerical discretization. When we simulate a Langevin equation with time-dependent parameter $a(t)$, we introduce discontinuity in $a(t)$ through the temporal discretization and the cutoff error of $a(t)$. The actual protocol $a(t)$ includes very small but very frequent jumps. If the result of stochastic energetics were to be sensitive to the limit of fine discretization, all the numerical calculations and the modeling of experiments for fluctuating phenomena would be dubious. Fortunately, in most cases the energetics is robust against this limit, though we cannot yet define rigorously the general condition of validity.

Let us take as example a Brownian particle (position: $x(t)$) moving in a harmonic potential. The center of the potential is $a(t)$ and the "spring" constant is K. The Langevin equation is

$$-\gamma\frac{dx}{dt}+\xi(t)-K[x-a(t)]=0. \tag{5.65}$$

The general solution for $\langle x(t)\rangle$ of (5.65) is $\langle x(t)\rangle = a(t)+e^{-Kt/\gamma}(a_{\mathrm{i}}-a(t))+(K/\gamma)\int_0^t e^{-Ks/\gamma}(a(t-s)-a(t))ds$. We compare the two protocols: (i) a smooth linear protocol $a(t)=a_{\mathrm{i}}+V_a t$ with V_a constant and (ii) stepwise protocol with the

[31] S.I. Sasa brought me this link.

[32] The discontinuity is related to the intrinsic nonlocal characteristic of the optimization problem. See [32, 33]. This discontinuity modifies the minimum of the average irreversibility $\langle W_{\mathrm{irr}}\rangle$. But the effect on the complementarity relation (Sects. 5.3.1) is a higher order correction in $(\tau_{\mathrm{op}})^{-1}$.

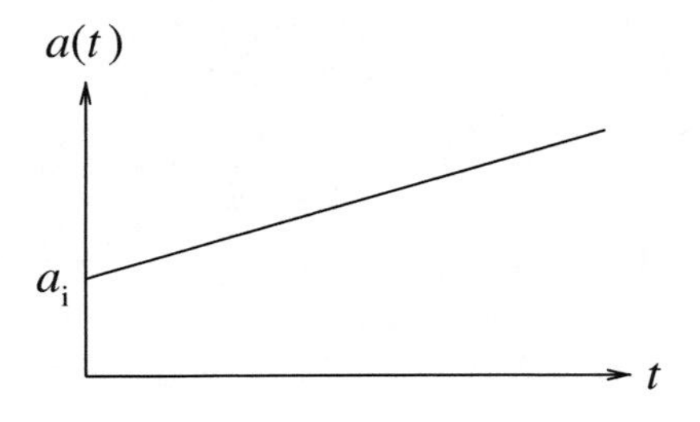

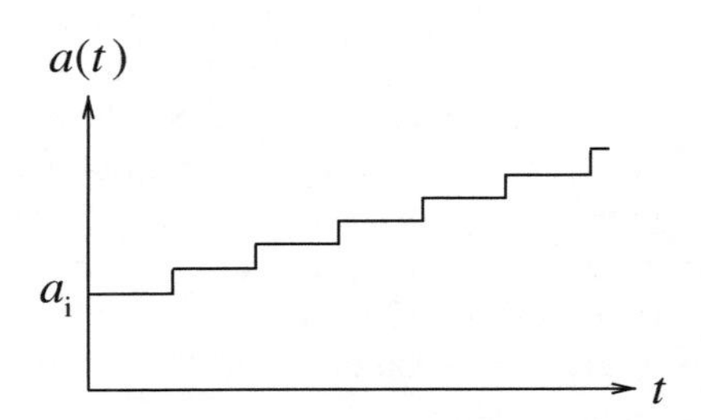

Fig. 5.8 (*Left*) Smooth linear protocol $a(t) = a_{\rm i} + V_a t$ and (*Right*) stepwise protocol $a(t) = a_{\rm i} + \delta a[V_a t/\delta a]$

same average rate, $a(t) = a_{\rm i} + \delta a[V_a t/\delta a]$, where $[z]$ denotes the integer part of $z(\leq 0)$. See Fig. 5.8. Using the above solution, the average work per unit time is (i) $\gamma {V_a}^2$ and (ii) $\bar{W} = {V_a}^2(K\delta t/2)\coth(\delta K t/(2\gamma))$ with $\delta t = \delta a/V_a$. The second $\bar{W}$ converges smoothly to the first in the limit of the fine steps $\delta t \to 0$.

5.5.2 Can We Change a Parameter Slowly Enough for the Quasistatic Process?

The answer is no. We will show in Chap. 7 that (i) there are cases where the quasistatic process is intrinsically impossible and (ii) but such nonquasistatic processes do not necessarily cause large irreversible work. The more important consequence of these processes is that the external system loses information and controllability of the system's state.

References

1. A.H. Nayfeh, *Perturbation Methods* (Wiley-VCH, New York, 2004)
2. C. Jarzynski, Phys. Rev. A **46**, 7498 (1992)
3. C. Jarzynski, Phys. Rev. Lett. **71**, 839 (1993)
4. T. Shibata, K. Sekimoto, J. Phys. Soc. Jpn. **69**, 2455 (2000)
5. T. Hondou, Europhys. Lett. **80**, 50001 (2007)
6. T. Aoki, M. Hiroshima, K. Kitamura, M. Tokunaga, T. Yanagida, Ultramicroscopy **70**, 45 (1997)
7. H. Qian, J. Math. Chem. **27**, 219 (2000)
8. J.M.R. Parrondo, Phys. Rev. E **57**, 7297 (1998)
9. K. Sekimoto, S.I. Sasa, J. Phys. Soc. Jpn. **66**, 3326 (1997)
10. C. Jarzynski, Phys. Rev. Lett. **78**, 2690 (1997)
11. G. Hummer, A. Szabo, PNAS **98**, 3658 (2001)
12. R.P. Feynman, A. Hibbs, *Quantum Mechanics and Path Integrals* (McGraw-Hill Inc. U.S., 1965)
13. F. Ritort, C. Bustamante, I. Tinoco Jr., PNAS **99**, 13544 (2002)
14. D. Collin et al., Nature **437**, 231 (2005)
15. G.E. Crooks, C. Jarzynski, Phys. Rev. E **75**, 021116 (2007)
16. K. Sato, J. Phys. Soc. Jpn. **71**, 1065 (2002)

17. H. Spohn, J.L. Lebowitz, Adv. Chem. Phy. **38**, 109 (1978)
18. J.L. Lebowitz, H. Spohn, J. Stat. Phys. **95**, 333 (1999)
19. G.E. Crooks, Phys. Rev. E **61**, 2361 (2000)
20. D.J. Evans, E.G.D. Cohen, G.P. Morriss, Phys. Rev. Lett. **71**, 2401 (1003)
21. D.J. Evans, D.J. Searles, Phys. Rev. E **50**, 1645 (1994)
22. G. Gallavotti, E.G.D. Cohen, Phys. Rev. Lett. **74**, 2694 (1995)
23. J. Kurchan, J. Phys. A **31**, 3719 (1998)
24. G.E. Crooks, Phys. Rev. E **60**, 2721 (1999)
25. C. Maes, J. Stat. Phys. **95**, 367 (1999)
26. G. Bochkov, Y. Kuzovlev, Sov. Phys. JETP **45**, 125 (1977)
27. G. Bochkov, Y. Kuzovlev, Sov. Phys. JETP **49**, 543 (1979)
28. G. Bochkov, Y. Kuzovlev, Physica A **106**, 443 (1981)
29. G. Bochkov, Y. Kuzovlev, Physica A **106**, 480 (1981)
30. C. Jarzynski, C. R. Physique **8**, 495 (2007)
31. K. Kawasaki, J.D. Gunton, Phys. Rev. A **8**, 2048 (1973)
32. Y.B. Band, O. Kafri, P. Salamon, J. Appl. Phys. **53**, 8 (1982)
33. T. Schmiedl, U. Seifert, Phys. Rev. Lett. **98**, 108301 (2007)

Chapter 6
Heat Viewed at Different Scales

This short chapter deals with the heat from different levels of description. We shall call the heat defined by the stochastic energetics *mesoscopic heat* and the heat defined by conventional statistical thermodynamics *calorimetric heat*. The latter should measured by conventional calorimetric methods. The main question of this chapter is to understand the connection between these two types. Simple thought experiments will help clarify the problem. "Work against a viscous medium" can be defined with the aid of the notion of transformation of the heat from one scale to another scale.

6.1 * Introduction – What Is Heat?

6.1.1 Each Description Scale Has its Own Definition of Heat

Roughly speaking, heat is the energy exchanged with or among the degrees of freedom that do not emerge in explicit observation and description. Once we fix the level of description, for example, of the Langevin equation or of the master equation, we retain certain degrees of freedom and eliminate other degrees of freedom from the evolution equation. Then heat is the work done by the retained degrees of freedom against the thermal environment that represents the eliminated degrees of freedom.

Mesoscopic heat has its proper significance in the context of irreversibility. Nevertheless, in experimental heat measurement we should be aware of the difference between mesoscopic heat and calorimetric heat. Depending on the system, mesoscopic heat can be only a small part of the total energy transfer involved in the process. Although the method of stochastic energetics has been applied to diverse fields (Sect. 4.3.2), little has been discussed about the relation between the mesoscopic heat and the total energy involved in the process. The meaning of the energy should also be carefully specified, see the next section.

Sekimoto, K.: *Heat Viewed at Different Scales*. Lect. Notes Phys. **799**, 203–220 (2010)
DOI 10.1007/978-3-642-05411-2_6

6.1.2 Examples of Nonuniqueness of Heat

Example 1. Macroscopically mimicked thermal environment and Langevin equation

In Sect. 1.2.1.2 we have seen that a thermal environment and a Langevin equation can be mechanically simulated with an effective temperature, which is unrelated to the (ambient) temperature of the constituting materials [1]. In fact the effective temperature is enormously higher than the ambient temperature. The mesoscopic heat defined from this virtual environment is, therefore, different from the heat that the materials exchange with their true environment.

Example 2. Coarse-grained Langevin equation for a diffusing particle in a mostly periodic potential

In Sect. 1.3.2.2 the Langevin equation (1.98), i.e.,

$$\gamma \dot{x} = -U_m'(x) + a_0 - V'(x) + [2\gamma k_{\mathrm{B}} T]^{\frac{1}{2}} \xi_m(t),$$

was coarse grained to eliminate the periodic potential, $U_m(x)$, which has a small period, ℓ. The resulting Langevin equation retaining only the slow modulation force, $a_0 - V'(x)$, could be written as (1.102), that is,

$$\Gamma(a_0)[\dot{\tilde{x}} - v_s(a_0)] = -V'(x) + [2\Gamma(a_0) k_{\mathrm{B}} \Theta(a_0)]^{\frac{1}{2}} \xi(t), \tag{6.1}$$

where $v_s(a_0)$ is the coarse-grained steady-state velocity at $V \equiv 0$. By this coarse graining a new friction coefficient, $\Gamma(a_0)$, and a new diffusion coefficient, $D_{\mathrm{eff}}(a_0)$, appeared.

In the case $a_0 = 0$ the system admits equilibrium state, and see how heat changes upon coarse graining. The Einstein relation holds for each level of description:

$$D_{\mathrm{eff}}(0)\Gamma(0) = k_{\mathrm{B}} T, \qquad D\gamma = k_{\mathrm{B}} T. \tag{6.2}$$

The heat for the coarse-grained Langevin equation is

$$d'Q_{\mathrm{eff}} \equiv \{-\Gamma(0)\dot{\tilde{x}} + [2\Gamma(0) k_{\mathrm{B}} T]^{\frac{1}{2}} \xi(t)\} \circ d\tilde{x}(t). \tag{6.3}$$

Since generally we have the inequalities, $\Gamma(0) \neq \gamma$, the heat Q_{eff} thus defined is different from the heat Q that we define for the original Langevin equation, $d'Q = \{-\gamma \dot{x} + [2\gamma k_{\mathrm{B}} T]^{\frac{1}{2}} \xi(t)\} \circ dx_m(t)$.

Example 3. Chain consisting of beads and springs

We shall consider the system consisting of N Brownian particles ("beads") with the positions, $\boldsymbol{x}_1, \ldots, \boldsymbol{x}_N$, and $(N+1)$ elastic springs (*not* rigid rods) of the spring constant K and zero natural length (Fig. 6.1). The springs join the beads linearly

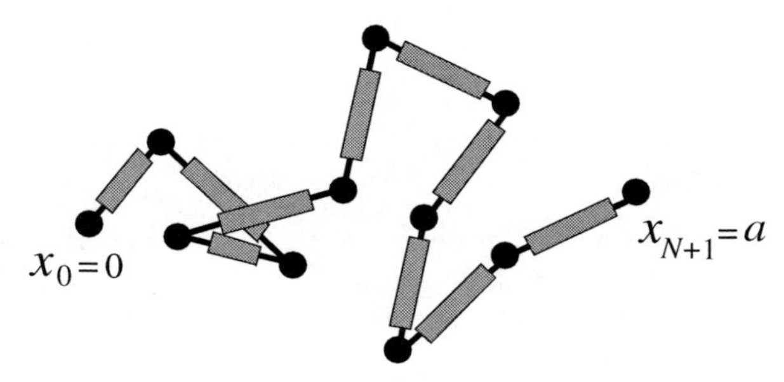

Fig. 6.1 A chain of beads (*thick dots* except at the ends) and springs (*shaded rectangles*). Each spring can be a purely mechanical spring, or it can be a purely entropic ideal chain. [The same figure as Fig. 4.3.]

except at the ends $\boldsymbol{x}_0$ and $\boldsymbol{x}_{N+1}$. Each bead is supposed to be a free joint for the two neighboring springs. We fix one of the end at the origin, $\boldsymbol{x}_0 \equiv \boldsymbol{0}$, while we control the other end, $\boldsymbol{x}_{N+1} \equiv \boldsymbol{a}$. The overdamped Langevin equations for the ith bead ($1 \leq i \leq N$) write as

$$0 = -\gamma \dot{\boldsymbol{x}}_i + \boldsymbol{\xi}_i(t) + K(\boldsymbol{x}_{i-1} + \boldsymbol{x}_{i+1} - 2\boldsymbol{x}_i), \tag{6.4}$$

where the random thermal forces, $\boldsymbol{\xi}_i(t)$, are i.i.d. Gaussian white noises with zero mean and $\langle \boldsymbol{\xi}_i(t)\boldsymbol{\xi}_i(t')\rangle = 2\gamma k_{\mathrm{B}}T\mathbf{1}\delta(t-t')$.[1]

The mesoscopic heat from the heat bath is

$$Q = \int \sum_{i=1}^{N} [-\gamma \dot{\boldsymbol{x}}_i + \boldsymbol{\xi}_i(t)] \circ d\boldsymbol{x}_i(t). \tag{6.5}$$

Using the Langevin equation (6.4), the balance of energy, $dU = d'W + d'Q$, holds with $U(\{\boldsymbol{x}_i\}, \boldsymbol{a}) = K/2 \sum_{i=0}^{N} (\boldsymbol{x}_{i+1} - \boldsymbol{x}_i)^2$ and[2] $d'W = K(\boldsymbol{x}_{N+1} - \boldsymbol{x}_N)d\boldsymbol{x}_{N+1} = (\partial U/\partial \boldsymbol{a})d\boldsymbol{a}$.

According to the result of Chap. 5, the quasistatic work W to pull the chain from $\boldsymbol{a} = \boldsymbol{a}_{\mathrm{i}}$ to $\boldsymbol{a}_{\mathrm{f}}$ is $\Delta F \equiv F(\boldsymbol{a}_{\mathrm{f}}) - F(\boldsymbol{a}_{\mathrm{i}})$, where the Helmholtz free energy $F(\boldsymbol{a})$ is *defined* by

$$e^{-\beta F(\boldsymbol{a},\beta)} = \int e^{-\frac{\beta K}{2} \sum_{i=0}^{N} (\boldsymbol{x}_{i+1} - \boldsymbol{x}_i)^2} \prod_{i=1}^{N} d\boldsymbol{x}_i = \left(\frac{2\pi}{\beta K}\right)^{3N/2} e^{-\frac{\beta K}{2(N+1)}\boldsymbol{a}^2}. \tag{6.6}$$

[1] The symbol $\mathbf{1}$ is the 3D unit tensor. In (6.4) we adopted the model of Rouse type [2], where the hydrodynamic effects ([3, 4], see also Sect. A.4.7.2), are ignored for simplicity.

[2] In order to avoid confusion with Itô-type product, we suppress the dot "·" for the 3D scalar product.

Therefore,

$$W = \Delta F = \left[\frac{K}{2(N+1)} \boldsymbol{a}^2 \right]_{\boldsymbol{a}_\mathrm{i}}^{\boldsymbol{a}_\mathrm{f}} \qquad \text{(quasistatic process).} \tag{6.7}$$

This result is valid independent of the origin of K. As for calorimetric heat, however, we need to consider the origin of the spring before comparing $d'Q$ defined above with the calorimetric measurement. It is enough to consider the quasistatic process to show the essential point:

Case 1: If the springs are made of thin stainless steel wire, the spring constant K can be regarded to be independent of the ambient temperature. Then the change of the average energy $\langle \Delta U \rangle$ can be identical to the prediction from the average energy $\langle U_\mathrm{m} \rangle$ of equilibrium statistical mechanics:

$$\langle U_\mathrm{m} \rangle = \frac{\partial(\beta F)}{\partial \beta} = \frac{3Nk_\mathrm{B}T}{2} + \frac{K}{2(N+1)} \boldsymbol{a}^2. \tag{6.8}$$

By comparing the $\Delta \langle U_\mathrm{m} \rangle$ from this result with (6.7), we find that $\Delta \langle U_\mathrm{m} \rangle = \langle \Delta U \rangle = \Delta F$, which then implies that Q is equal to the calorimetric heat, Q_m, and $\langle Q_\mathrm{m} \rangle = \langle Q \rangle = 0$ for the quasistatic process. This result could be confirmed by local calorimetric measurements or by numerical calculation.

Case 2: If each spring consists of an ideal polymer chain immersed in a fluid environment, then the calorimetric results will be different. Each spring is purely entropic with spring constant, $K = \frac{k_\mathrm{B}T}{nb^2}$, where n is the number of monomers in the chain, and b is the so-called Kuhn length. The second equality of (6.8) is no more valid since K depends on the temperature. From (6.6) with $\beta K = $ const., the product βF is found to be independent of temperature. Then (6.8) yields $\langle U \rangle = 0$ independent of $\boldsymbol{a}$. If we denote by Q_m the calorimetric heat expected from the calorimetric energy conservation, we have $\langle Q_\mathrm{m} \rangle = -W = -\Delta F$.

In summary, we have case-dependent relations:

$$\begin{aligned} &\langle Q_m \rangle = 0 = \langle Q \rangle, \Delta \langle U_\mathrm{m} \rangle = \Delta F = \langle \Delta U \rangle && \text{(steel wire spring),} \\ &\langle Q_\mathrm{m} \rangle = -\Delta F, \Delta \langle U_\mathrm{m} \rangle = 0 && \text{(ideal chain spring).} \end{aligned}$$

The relations between the mesoscopic heat and the calorimetric heat, therefore, depend on whether or not the spring constant K is of entropic origin. The argument, including the nonquasistatic case, will be developed in the next section.

Example 4. "Stochastic thermodynamics" of chemical reaction networks

For chemical reaction networks an alternative interpretation of work and heat has recently been proposed [5]. While it shares the basic idea of describing the energetics of individual stochastic process, their interpretation is different from our framework in Sect. 4.1.2.6.[3] For illustration purpose, we take a very simple situation as example.

A system is a vesicle which occupies the volume Ω, and Ω is subset of the entire volume $\Omega^{(0)}$. The vesicle is surrounded by the reservoir of particles of species A. The reservoir, occupying the volume $\Omega^{(0)} \setminus \Omega$, is big enough that it is characterized by a chemical potential, $\mu_{\mathrm{A}}^{(0)}$ [6]. The particle A can pass through the vesicle membrane. Using the notations of chemical reaction, $\mathrm{A} \to \mathrm{X}$ means that a particle A enters the vesicle from the reservoir, while $\mathrm{X} \to \mathrm{A}$ means the reverse process. The state of the system Ω is specified by the number of X particles, $n_{\mathrm{X}}(t)$. We denote by $n_{\mathrm{A}}(t)$ the number of A particles in the reservoir. Therefore, $\frac{n_{\mathrm{X}}(t)}{dt} + \frac{n_{\mathrm{A}}(t)}{dt} = 0$. The energy e_{X} is assigned to each particle inside the vesicle. The change of energy, ΔE, during $t = 0$ and $t = \Delta t$ is therefore

$$\Delta E = \int_0^{\Delta t} e_{\mathrm{X}} \frac{dn_{\mathrm{X}}(t)}{dt} dt = \int_{n_{\mathrm{X}}(0)}^{n_{\mathrm{X}}(\Delta t)} e_{\mathrm{X}} dn_{\mathrm{X}}(t). \tag{6.9}$$

The work, W_{chem}, is defined as follows [5]:

$$W_{\mathrm{chem}} \equiv - \int_{n_{\mathrm{A}}(0)}^{n_{\mathrm{A}}(\Delta t)} \mu_{\mathrm{A}}^{(0)} dn_{\mathrm{A}}(t). \tag{6.10}$$

The heat Q_{chem} is then defined so that the energy balance, $\Delta E = W_{\mathrm{chem}} + Q_{\mathrm{chem}}$, be satisfied[4] :

$$Q_{\mathrm{chem}} = \int_{n_{\mathrm{X}}(0)}^{n_{\mathrm{X}}(\Delta t)} e_{\mathrm{X}} dn_{\mathrm{X}}(t) + \int_{n_{\mathrm{A}}(0)}^{n_{\mathrm{A}}(\Delta t)} \mu_{\mathrm{A}}^{(0)} dn_{\mathrm{A}}(t). \tag{6.11}$$

This framework is utilized to derive several versions of the fluctuation theorems (FT) and Jarzynski nonequilibrium work relation, either around the equilibrium state or around a steady state. As for the heat Q_{chem}, however, it is not directly related to the (molecular) energy exchanged between the system and the thermal environment. For example, if the particle A has the same energy as that of X, i.e., $e_{\mathrm{A}} = e_{\mathrm{X}}$, the reaction $\mathrm{A} \rightleftharpoons \mathrm{X}$ accompanies no calorimetric heat. Even in such case Q_{chem} is generally nonzero since the chemical potential $\mu_{\mathrm{A}}^{(0)}$ depends on the density of A particles in the particle reservoir. That is, Q_{chem} is not the calorimetric heat.

Problems

Through the previous examples, we have seen different definitions of heat, depending on the scale of description, the assignment of energy, or identification of the state variables of the system. One might say that there are different semantic

[3] The authors of [5] use the word "stochastic thermodynamics" since they also introduce entropy for each point on each trajectory.

[4] [5] adopted the opposite-sign convention for the heat from the present text.

interpretations (models) of the syntax of thermodynamic structure (set of axioms). Conventional equilibrium thermodynamics is among those interpretations, the one in which heat is interpreted as calorimetric heat. Stochastic energetics also has its first law, the energy balance, and also the "second law". One of the basic questions is then how we can relate calorimetric heat with the mesoscopic heat of the stochastic energetics. Another, more ambiguous, question is whether the heat of a scale of description can evolve to become the heat of another description. We will address these two questions in the following two sections, respectively.

6.2 * Calorimetric Heat vs. the Heat of Stochastic Energetics

The energetics of the stochastic process has shown the balance of energy on the mesoscopic level. The heat and the energy defined are, however, generally different from their macroscopic counterparts. In this section we show that this discrepancy can be removed by adding to these quantities the reversible heat associated with the mesoscopic free energy.

6.2.1 System Can Have Different Levels of Random Variables

6.2.1.1 Background

Calorimetric heat is defined by its measurement. Usual measurement monitors the temperature of a probe material ("thermometer"). Calorimetric heat is then deduced from this temperature data, using various physical principles, such as heat diffusion, radiation, etc. It is the *microscopic* degrees of freedom of the probe material that equilibrate with the sample. Therefore, calorimetric heat is calculated from microscopic theories or simulations based on statistical thermodynamics.

Mesoscopic heat depends on the description level of the stochastic phenomena. Throughout this section, we use the notations with the tilde ($\tilde{F}$, $\tilde{f}$, $\tilde{Q}$, etc.) to mean those concepts of stochastic energetics.

If the mesoscopic energy comes from interactions with an external field (e.g., laser tweezers [7]) or with a nonentropic restoring force (e.g., a brass wire holding a pendulum [8]), then mesoscopic heat can be identified with the calorimetric (i.e., microscopic) heat. Contrastingly, they are different when mesoscopic heat contains the entropic contribution due to microscopic degrees of freedom which have been projected out to achieve the mesoscopic description. We have seen some examples in Sect. 6.1.2. Below we present another thought experiment analogous to the "jump-and-catch" mechanism in (Sect. 4.1.2.3).

6.2.1.2 A Thought Experiment

Let us suppose that a micron-sized bead in water is leashed at the point $\boldsymbol{x} = 0$ through a polymer chain and that an optical trap somehow constitutes a static potential well around $\boldsymbol{x} = \boldsymbol{a} \neq 0$ (see Fig. 6.2(a)). For simplicity we assume that the

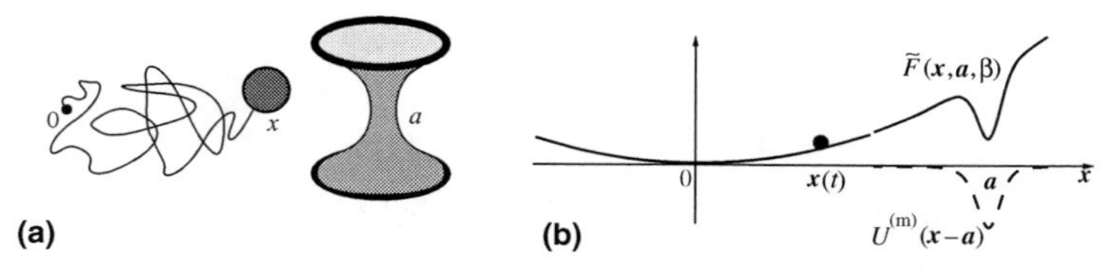

Fig. 6.2 (**a**) A bead (position: $\boldsymbol{x}$, a big *gray disk*) is leashed by an ideal polymer chain (*thin curve*) at the origin **0** (*thick dot*). A laser tweezer to trap the bead is focused around a distant point $\boldsymbol{a}$. (**b**) The mesoscopic potential energy $\tilde{F}(x, a, \beta)$ in (6.18) [*solid curve*] and the microscopic potential energy that accounts for the calorimetric measurement, (6.25) [*dashed curve*]

polymer is an ideal chain. We further assume that the arrangement is such that the bead undergoes bistable transitions, either being trapped around $\boldsymbol{x} = \boldsymbol{a}$ when the polymer chain is stretched to the distance $\simeq |\boldsymbol{a}|$, or wandering around $\boldsymbol{x} = 0$ when the chain is relaxed and fluctuating. We then assume that the stochastic behavior of the bead is well described by a Langevin equation for the bead position, $\boldsymbol{x}(t)$. The main question is how much heat is released to or absorbed from the thermal environment when the bead switches from one of the bistable states to the other.

The bistable states can be represented by a double-well mesoscopic potential energy $\tilde{F}(\boldsymbol{x}, \boldsymbol{a}, \beta)$ for the bead. See Fig. 6.2(b). However, it is only the optical trap that realizes a potential hole of microscopic energy; the ideal chain exerts purely entropic restoring forces. (Recall that the kinetic energy of ideal chain is independent of the conformation of the chain.) Microscopic theory should predict that calorimetric heat depends only on the microscopic potential energy of the optical trap. The framework of stochastic energetics predicts, however, that mesoscopic heat is absorbed from the environment whenever the bead climbs up the potential barrier of $\tilde{F}(\boldsymbol{x}, \boldsymbol{a}, \beta)$ and is released to the environment during the downhill motion.

6.2.2 *Equilibrium Statistical Mechanics of Mesoscopic Variables Can Have Different Level of Free-Energy Functions*

Conversion of mesoscopic heat into calorimetric heat is a straightforward generalization of what is known in equilibrium statistical mechanics. Therefore, we will first summarize the result of the latter discipline.

6.2.2.1 Hamiltonian, Landau Free Energy, and Helmholtz Free Energy

Suppose that the total system consists of the *system* whose Hamiltonian is $H(x, y, a)$ and a heat bath of the temperature T.[5] Here a stands for the external control parameter(s), and we have split, for the later use, the system's degrees of freedom into two groups, x and y. The Helmholtz free energy $F(a, \beta)$ ($\beta = (k_B T)^{-1}$) is defined through the canonical partition function, $Z(a, \beta)$:

[5] We could start from an entirely isolated whole system, except that the argument is more complicated.

$$e^{-\beta F(a,\beta)} = Z(a, \beta) = \mathrm{Tr}_{x,y} e^{-\beta H(x,y,a)}, \tag{6.12}$$

where the suffices x, y of $\mathrm{Tr}_{x,y}$ indicate the degrees of freedom over which the trace should be taken. We also introduce $\tilde{F}(x, a, \beta)$, sometimes called Landau free energy, by eliminating only the degree(s) of freedom, y:

$$e^{-\beta \tilde{F}(x,a,\beta)} = \mathrm{Tr}_y e^{-\beta H(x,y,a)}. \tag{6.13}$$

By definition we have the relation

$$e^{-\beta F(a,\beta)} = \mathrm{Tr}_x e^{-\beta \tilde{F}(x,a,\beta)}. \tag{6.14}$$

6.2.2.2 Fast and Slow Variables

We suppose that the variables x and y represent, respectively, the *slow* and *fast* variables of the system.[6] In such case, $\tilde{F}(x, a, \beta)$ is the mesoscopic potential energy which appears in the Langevin equation for x as the result of projection of fast degrees of freedom.

In the context of the example of Fig. 6.2, the slow variable, x, denotes the position of the bead, while the fast variables, y, describe the local movements of the monomers of the polymer chain and possibly the motion of the surrounding water molecules. Then $e^{-\beta \tilde{F}(x,a,\beta)}$ gives the relative probability density for the slow variable, x, under the given parameters a and β.

6.2.2.3 The Objectivity of Force and Energy

We will define *"objectivity"* as the character that satisfies the following two conditions:

(I) It can be defined on the three levels of descriptions, $\{x, y, a, \beta\}$, $\{x, a, \beta\}$, and $\{a, \beta\}$, corresponding to Eqs. (6.12), (6.13), and (6.14), respectively,
(II) The magnitudes of the quantity for these three descriptions are essentially the same, except for the fluctuations inherent to the description levels.

The first example is the force conjugate to the parameter a: We define

$$\begin{aligned}
\hat{f}(x, y, a) &\equiv \partial H(x, y, a)/\partial a, \\
\tilde{f}(x, a, \beta) &\equiv \partial \tilde{F}(x, a, \beta)/\partial a, \\
f(a, \beta) &\equiv \partial F(a, \beta)/\partial a.
\end{aligned}$$

Then, from (6.12), (6.13), and (6.14), we can verify the relations:

[6] [9] studied two-component (fast and slow) Brownian system, using the Fokker–Planck equation.

$$\tilde{f}(x,a,\beta) = \mathrm{Tr}_y[e^{\beta(F-H)}\hat{f}],$$
$$f(a,\beta) = \mathrm{Tr}_x[e^{\beta(F-\tilde{F})}\tilde{f}] = \mathrm{Tr}_{x,y}[e^{\beta(F-H)}\hat{f}]. \quad (6.15)$$

The second quantity with the objectivity is the energy (not the mesoscopic energy). In addition to $H(x, y, a)$, we define

$$\tilde{E}(x,a,\beta) \equiv \partial[\beta\tilde{F}(x,a,\beta)]/\partial\beta,$$
$$E(a,\beta) \equiv \partial[\beta F(a,\beta)]/\partial\beta.$$

Again, from (6.12), (6.13), and (6.14), we can verify

$$\tilde{E}(x,a,\beta) = \mathrm{Tr}_y[e^{\beta(F-H)}H],$$
$$E(a,\beta) = \mathrm{Tr}_x[e^{\beta(F-\tilde{F})}\tilde{E}] = \mathrm{Tr}_{x,y}[e^{\beta(F-H)}H]. \quad (6.16)$$

Symbolically we can write the above relations as $f = \langle\tilde{f}\rangle_x = \langle\hat{f}\rangle_{x,y}$ and $E = \langle\tilde{E}\rangle_x = \langle H\rangle_{x,y}$. The force and the energy of different level of descriptions, therefore, do not need the correction terms upon averaging.

6.2.2.4 Relation Between Mesoscopic Energy $\tilde{F}$ and (Calorimetric) Energy $\tilde{E}$

The above relationships indicate that (i) it is $\tilde{F}(x, a, \beta)$ that governs the probability weight of x on the mesoscopic level, while (ii) it is $\tilde{E}$ whose equilibrium average over x coincides with the thermodynamic energy E. The difference between these two quantities is nothing but the entropic term, which we obtain by rewriting slightly the definition of $\tilde{E}$ mentioned above:

$$\tilde{E} - \tilde{F} = -T\frac{\partial\tilde{F}}{\partial T}. \quad (6.17)$$

6.2.3 Calorimetric Heat Can Be Deduced from Stochastic Energetics

6.2.3.1 Case of Continuous Langevin Equation

Framework

If the timescale of the slow variable(s) x is well separated from that of fast variable(s) y, and if the temperature of the environment can be regarded as constant, we can use the Markovian description such as the Langevin equation to simulate the fluctuations of x. In the overdamped case, the equation is

$$0 = \gamma\frac{dx}{dt} + \xi(t) - \frac{\partial\tilde{F}(x,a,\beta)}{\partial x}, \quad (6.18)$$

where γ is the friction constant for x, and $\xi(t)$ is the white Gaussian random force with zero mean and the correlation, $\langle \xi(t)\xi(t')\rangle = 2\gamma k_B T \delta(t-t')$. From the standpoint of x, the fast fluctuations of the heat and energy due to the change in y are averaged over the time resolution of (6.18). As in the static case summarized above, it is the mesoscopic energy, $\tilde{F}(x, a, \beta)$, that gives the bias for the variable x. The mesoscopic energy balance along a particular realization of the stochastic process is

$$d\tilde{F} = d'\tilde{W} + d'\tilde{Q}, \tag{6.19}$$

where we use d (not d') to mean the total differential at constant temperature, i.e.,

$$d \equiv dx \frac{\partial}{\partial x} + da \frac{\partial}{\partial a}, \tag{6.20}$$

while the work $d'\tilde{W}$ and the "heat" $d'\tilde{Q}$ are defined by[7]

$$d'\tilde{W} \equiv \frac{\partial \tilde{F}}{\partial a} da, \tag{6.21}$$

$$d'\tilde{Q} \equiv \left[-\gamma \frac{dx}{dt} + \xi(t) \right] dx = \frac{\partial \tilde{F}}{\partial x} dx. \tag{6.22}$$

The core logic is the following: If the Langevin description (6.18) is a good model of a phenomenon, then the eliminated degree(s) of freedom y are supposed to follow x and a rapidly enough. From standpoint of y, the process of x is always quasistatic. It means that the heat released can be captured by the change in the pertinent entropy, $-\partial \tilde{F}/\partial T$ [10]. (A related argument is also found in [11].) In order to convert the mesoscopic heat $d'\tilde{Q}$ into the calorimetric heat, $d'Q_{\mathrm{m}}$, it is, therefore, sufficient to add to both $d'\tilde{Q}$ and $\mathrm{d}\tilde{F}$ the differential of the term found in (6.17), that is[8]

$$\begin{aligned} d'\tilde{Q} &\mapsto d'Q_{\mathrm{m}} \equiv d'\tilde{Q} - Td\frac{\partial \tilde{F}}{\partial T}, \\ d\tilde{F} &\mapsto d\tilde{E} \equiv d\tilde{F} - Td\frac{\partial \tilde{F}}{\partial T}. \end{aligned} \tag{6.23}$$

Now the mesoscopic energy balance equation (6.19) is converted to the new equation that includes only calorimetric heat and the quantities with objectivity:

$$d\tilde{E} = d'\tilde{W} + \mathrm{d}'Q_{\mathrm{m}}. \tag{6.24}$$

[7] N.B. all the products below should be interpreted as of Stratonovich type.

[8] We might call this "correction" term van't Hoff correction term. This is the reversible heat associated with the mesoscopic free energy.

The last expression holds for a particular realization of the Langevin equation (6.18). Equation (6.24) could be directly verified experimentally or calculated using the original Hamiltonian H. By the definition of the total differential (6.20), the term $-Td(\partial\tilde{F}/\partial T)$ in (6.23) has no cumulative effects for cyclic processes.

Application to the Thought Experiment

In the case of the thought experiment in Fig. 6.2, we assigned the variables y to the degrees of freedoms associated with the monomers of the ideal chain. For the mesoscopic potential energy, $\tilde{F}(\boldsymbol{x}, \boldsymbol{a}, \beta)$, we can write $\tilde{F}(\boldsymbol{x}, \boldsymbol{a}, \beta) = U^{(\mathrm{m})}(\boldsymbol{x} - \boldsymbol{a}) - TS^{(\mathrm{p})}(\boldsymbol{x})$, where $U^{(\mathrm{m})}(\boldsymbol{x} - \boldsymbol{a})$ represents the potential energy due to the optical trap, and $S^{(\mathrm{p})}(\boldsymbol{x})$ is the entropy of the ideal polymer chain. By substituting this form into (6.24), we find the concrete expression, term by term,

$$\begin{aligned} dU^{(\mathrm{m})} &= d'W + d'Q_{\mathrm{m}} \\ &= [-\nabla U^{(\mathrm{m})}(\boldsymbol{x} - \boldsymbol{a})d\boldsymbol{a}] + [\nabla U^{(\mathrm{m})}(\boldsymbol{x} - \boldsymbol{a})d\boldsymbol{x}], \end{aligned} \tag{6.25}$$

where ∇U denotes the gradient of U. The mathematical identity (6.25) is what we expected in the Introduction (*Example 3*). Experimentally, we should take account of the heat exchange with the bead as well as the effect of polymer–solvent interactions.

Further Considerations

The change of $\tilde{F}(x, a, \beta)$ through the change of x is a quasistatic work for the fast degrees of freedom, y. The ideal chain should, therefore, release heat even though the displacement of the bead $\boldsymbol{x}(t)$ is spontaneous. This statement looks somewhat paradoxical. But it does not contradict with the above analysis; it is the thermal environment that does the work to displace the bead, *gathering* the energy nearby.[9] The *released* heat $-TdS^{(\mathrm{p})}$ is, therefore, compensated.[10] If one can measure the heat at a very short distance, the local transfer of calorimetric heat around the chain and the bead should be observed (Fig. 6.3(a)). By contrast, if there is no bead at $\boldsymbol{x}(t)$, there is no such local transfer of heat (Fig. 6.3(b)). This should be checked experimentally.

6.2.3.2 Case of Discrete States

In Sect. 4.1.2.6 we have described stochastic energetics on discrete states (discrete Langevin equation). It is straightforward to generalize the above analysis to the

[9] To move a mesoscopic object, there should be the fluctuations, most probably, of the length scale of the object. See Sect. 6.3.1 below.

[10] Another heuristic argument could be to assume thermophoresis of the bead due to local warming up of solvent around the chain.

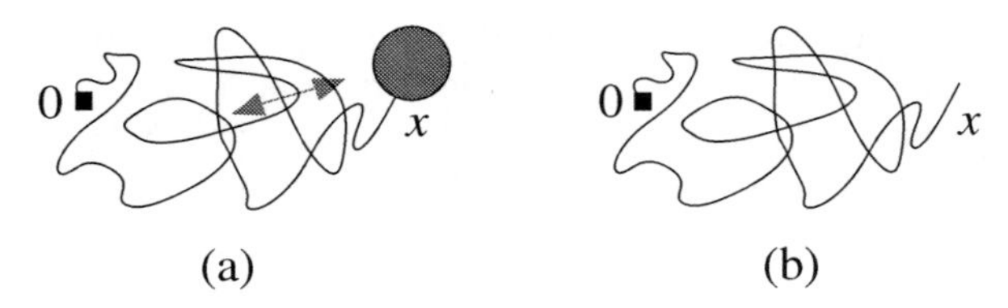

Fig. 6.3 Ideal chains with a bead and without bead (**b**). One end of the chains is fixed (**0**: *filled square*). The slow displacement of the bead (*gray disc*) in (**a**) accompanies the heat transfer between the chain and the neighborhood of the bead (*thick bidirectional arrow*), while the displacement of the free end (***x***) in (**b**) does not accompany the heat transfer

discrete stochastic process. We first adapt the notations of Eqs. (4.18), (4.19), (4.20), and (4.21) as follows to conform with the discrete case:

$$\Delta\tilde{F} = \Delta' W + \Delta'\tilde{Q}, \tag{6.26}$$

with

$$\Delta\tilde{F} = \tilde{F}_{i_{n+1}}(a(t), \beta) - \tilde{F}_{i_1}(a(0), \beta), \tag{6.27}$$

$$\Delta' W = \sum_{\alpha=1}^{n+1} \left[\tilde{F}_{i_\alpha}(a(t_\alpha), \beta) - \tilde{F}_{i_\alpha}(a(t_{\alpha-1}), \beta) \right], \tag{6.28}$$

$$\Delta'\tilde{Q} = \sum_{\alpha=1}^{n} \left[\tilde{F}_{i_{\alpha+1}}(a(t_\alpha), \beta) - \tilde{F}_{i_\alpha}(a(t_\alpha), \beta) \right]. \tag{6.29}$$

These relations correspond to (6.19) in the continuum case. In order to have the counterpart of (6.24) for the discrete process, we can again use the correspondence relations (6.23). As the result, the energy balance relation,

$$\Delta\tilde{E} = \Delta'\tilde{W} + \Delta' Q_{\mathrm{m}}, \tag{6.30}$$

holds with

$$\Delta\tilde{E} \equiv \Delta\tilde{F} - T\Delta\frac{\partial\tilde{F}}{\partial T}, \tag{6.31}$$

$$\Delta' Q_{\mathrm{m}} \equiv \Delta' Q - T\Delta\frac{\partial\tilde{F}}{\partial T}, \tag{6.32}$$

where the *total* difference in the correction term is defined by

$$T\Delta\frac{\partial\tilde{F}}{\partial T} \equiv T\left[\frac{\partial\tilde{F}_{i_{n+1}}(a(t), \beta)}{\partial T} - \frac{\partial\tilde{F}_{i_1}(a(0), \beta)}{\partial T} \right]. \tag{6.33}$$

To conclude this section, we have related mesoscopic heat of stochastic energetics with conventional heat along a single realization of the stochastic process. For the moment, mesoscopic potential energy and heat have only begun to be assessed experimentally [7, 8, 12, 13]. The direct measurement of the fluctuating observable heat, $d'Q_{\mathrm{m}}$, will be a future experimental challenge. The possibility to measure directly $d'\tilde{Q}$ is an open theoretical problem. We might need to extend our notion of the measurement so that the approach taken by the above references, i.e., the deduction of heat from the trajectory data *is* the mesoscopic measurement of the heat.

6.3 Change in the Scale of Heat

In the last section we established the relation between mesoscopic heat and calorimetric heat. The premise was clear separation of the timescales between the fast degrees of freedom (y) and the slow ones (x). Under this condition, the quasistatic treatment for the former degrees of freedom was justified. In the rest of this chapter, we will discuss the situations where this premise is not satisfied because the typical scale of thermal fluctuations changes in time. In such a case the most complete approach would be to go back to the microscopic mechanics and start everything from scratch. It is, however, not practical if the space-timescale of the slow degrees of freedom is far larger than the microscopic one. Our approach is to go back just one step: we remain in the mesoscopic description, but discuss the fate of the microscopic fluctuations whose space-time scale attain that of the slow variables.

6.3.1 *Fluid Fluctuations Causing Brownian Motion Have Memory*

The random force on a Brownian particle in a fluid reflects the temporarily *coherent* momentum transfer from the fluid molecules. Such spontaneous movements of fluid can be approximately modeled by the Langevin equation for the fluid, or, *fluctuating hydrodynamics* [14] (see also Appendix A.4.7.1). The evolution equation of fluctuating hydrodynamics is the Navier–Stokes equation complemented by the random dipolar source of momentum. The thermal motion of a Brownian particle is the passive reaction to the fluctuations in the fluid [15].[11] Although the fluctuating dipolar force (i.e., stress) in the fluid is assumed to be a white Gaussian process, the velocity of the Brownian particle has a long-term memory.

What causes memory in the velocity of Brownian particle? [12] Two mechanics coexist. The first factor is the inertia of the Brownian particle (and the fluid which is directly entrained by this particle). This effect is effectively included in the Langevin equation as the inertia term. The decay time[13] is characterized by $\tau_{\mathrm{p}} \equiv m/\gamma$, where m is the mass of the particle and $\gamma = 6\pi R\eta$ is the Stokes' friction constant with R and η being the radius of the particle and the viscosity of the fluid, respectively. If

[11] The result of [14] has been also found before [24] cf. [25].

[12] See also Sect. 1.1.3.1 *Remark about the inertia effect.*

[13] By solving $m\dot{v} = -\gamma v + \xi(t)$, the evolution of the particle velocity $v(t)$ contains the convolution of the random force $\xi(t)$ with the "memory kernel", $\theta(t)e^{-t/\tau_p}$.

the mass density of the particle is ρ_p, the time τ_p is $2R^2\rho_p/(9\eta)$. The second factor is the inertial motion of fluid around the Brownian particle. The momentum carried by a locally coherent motion of fluid can decay only by diffusion, since the momentum conservation law prohibits its individual decay. The diffusion coefficient of the fluid momentum is the kinetic viscosity, $\nu \equiv \eta/\rho_f$, where ρ_f is the mass density of the fluid. The decay of the average velocity in this region is then characterized by the time, $\tau_f = R^2/\nu = R^2\rho_f/\eta$.[14] If the mass density of the Brownian particle ρ_p is comparable to that of the fluid, ρ_f, we have $\tau_p \sim \tau_f$.

Moreover, the fluid's coherent momentum decays slowly (algebraic). This slow decay reflects the fact that the spatial range of this coherent motion changes in time: suppose that the fluid had an initial momentum P_0 around the Brownian particle. After the time t, this momentum spreads by diffusion over the range $\sim (\nu t)^{1/2}$. The mean local velocity $\bar{v}$ of the fluid is inversely proportional to the mass of the fluid within this range, i.e., $\bar{v} \sim P_0/(\rho_f(\nu t)^{d/2})$, where d is the spatial dimensionality. This algebraic decay $\sim t^{-d/2}$ is called the (hydrodynamic) *longtime tail* [16, 17].[15] This memory effect, which is neglected by the Langevin equation, is important over the timescales of $\tau_f \sim 10^2\tau_f$.[16] Recent experiments to trace the Brownian particle [18, 19] used a bead, for example, with $R \sim 0.5\,\mu$m. In that case $\tau_f = 0.25\,\mu$s. They have verified theoretical predictions [20] with the time resolution of $\sim \mu$s. However, if we used a protein of $R \sim 10$ nm, τ_f would be $\simeq 10^{-4}\,\mu$s. This is too small to resolve experimentally at present.[17] An evidence of the longtime tail is seen in the mean square displacement (MSD) $\langle[x(t)-x(0)]^2\rangle$ of the particle position $x(t)$. See Fig. 6.4 for the schematic behavior of the MSD. Due to the longtime tail effect, the ideal diffusion behavior $\langle[x(t)-x(0)]^2\rangle = 2Dt$ [18] has an algebraic correction [21–23, 15]:

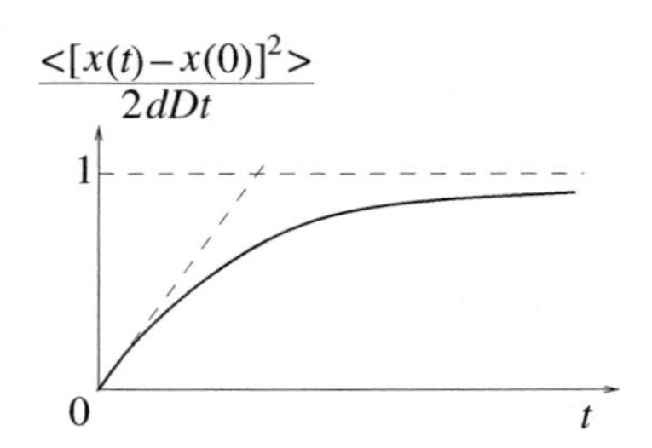

Fig. 6.4 Schematic representation of MSD, $\langle[x(t)-x(0)]^2\rangle$, normalized by its diffusion form, $2Dt$, is plotted against time t. Due to the (hydrodynamic) longtime tail, the approach of the MSD toward the latter value is algebraic, much slower than exponential

[14] If we replaced the fluid with a gas in the Knudsen limit, there is no such memory [26, 27].

[15] Rahman [16] found an anormaly in the fluctuation spectra in his molecular dynamic simulation. Then Alder and Wainwright [17] showed the power low decay of velocity correlation function.

[16] Even at $10^2\tau_f$ coherent momentum is spread over only $\sim 10R$.

[17] If a protein motor of several nanometer size wishes to make use of the fluctuations that last for $\sim \mu$s, one strategy would be to link itself with a bigger object of $\sim \mu$m size. cf. Sect. 1.3.1.3.

[18] We project on the x-axis.

$$\frac{\langle [x(t) - x(0)]^2 \rangle}{2Dt} = 1 - 2\sqrt{\frac{\tau_{\rm f}}{\pi\, t}} + \epsilon \quad (t > \tau_{\rm f}), \tag{6.34}$$

where $D = k_{\rm B}T/\gamma$ is the asymptotic diffusion coefficient and ϵ denotes the less dominant terms [15].

6.3.2 *Decay Cascade of Mesoscopic Fluctuations Resolves the Controversy on the Generalized Efficiency of Molecular Motor

A question has been raised about the efficiency of the molecular motor protein. When the motor protein hydrolyses an ATP molecule and swings a long filament, should we count the dissipated heat due to the filament motion as a part of work done by the motor? If yes, how can one estimate the performance of this work? This problem is relevant for scale-dependent description of fluctuations and heat.

F1ATPase is a protein complex that synthesizes ATP from ADP and Pi (inorganic phosphate) using the rotation of its central axial rod (γ-subunit) as the source of work [28]. F1ATPase can also work inversely, that is, as a rotary motor consuming the free energy of ATP hydrolysis [29]. To demonstrate this motor function, fluorescent actin filament was attached to the γ-subunit [30] (Fig. 6.5). In the presence of ATP (of concentration $<$ μM), the filament was observed to rotate stepwise against the viscous friction of the solvent. Each ATP hydrolysis amounts to the $2\pi/3$ rotation of the filament. An efficiency of (free) energy conversion, Θ, was defined as the ratio of the dissipation, $\zeta\langle\omega\rangle^2\tau_{\rm st}$, to the corresponding free-energy cost of ATP hydrolysis, $\Delta G_{\rm hyd} \simeq 20k_{\rm B}T$. Here ζ is a friction coefficient of the filament, $\tau_{\rm st}$ is the stepping time of the $2\pi/3$-rotation of γ-subunit, and $\langle\omega\rangle = 2\pi/(3\tau_{\rm st})$ is the *mean* angular velocity. The result [30] showed that the efficiency Θ is fairly close to unity, suggesting indirectly that the motor may well work inversely as the synthesizer.

Although apparently sound and useful, the above definition of efficiency requires justification on a physical basis. In general, the square of an average quantity, like

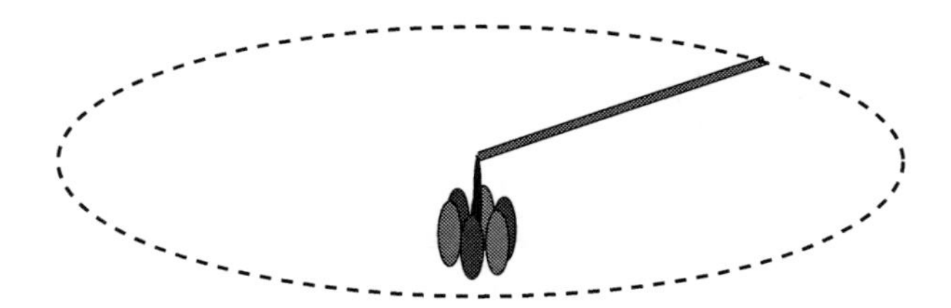

Fig. 6.5 Schematic setup of the F1ATPase (central cluster of oval modules) as rotary motor, which swings a long actin filament (*long bar*). The rotation of the actin filament is driven by the rotating axe ("γ-subunit": *short vertical bar*) of the F1ATPase. The spatial scale of the actin rotation (*dashed loop*) is of several microns and is intermediate between the mesoscopic scale of the rotary motor and the macroscopic scale

$\langle\omega\rangle^2$ above, cannot immediately correspond to the average of a physical observable. Moreover, the instantaneous angular velocity ω is a resolution-dependent quantity. A constructive argument was proposed from the engineering viewpoint [31]: In order to perform a $2\pi/3$-step within a *given* time τ_{st}, it incorporates inevitably a certain minimum dissipation. Such a dissipation is claimed to be taken as a part of necessary work in addition to the ordinary work against a conserved potential energy. This minimum dissipation is realized under the constant angular velocity ($\omega = \text{const.}$), i.e., when $\langle\omega^2\rangle = \langle\omega\rangle^2$ [31, 32]. Efficiency is thus calculated as mentioned above: $\Theta = \zeta\langle\omega\rangle^2\tau_{st}/\Delta G_{hyd}$. More generally, "generalized efficiency" Θ was defined as [31]

$$\Theta = \frac{\{\text{Minimum energy input required to accomplish a task as engine}\}}{\{\text{Actual energy input for the task}\}}, \tag{6.35}$$

where the task is required to finish within a given time. The above definition explains why the square of the average velocity $\langle\omega\rangle^2$ appears and can be useful for designing an optimum engine under a given task.

Still, a question remains: If we regard ω as fluctuating variable of mesoscopic scale, the least dissipating process, $\omega = \text{const.}$, is by no means realizable in the presence of thermal noise. Should we use such an idealized limit to define the efficiency? Is there an alternative, hopefully *scale conscious*, interpretation for the $\langle\omega\rangle^2$?

There is an insightful analysis on this issue [33]. The argument was done in the context of a thermal ratchet model (Sect. 1.3.4). This model is described by the Langevin equation with inertia,

$$\dot{p} = -\frac{\partial}{\partial x}(U_0(x,t) - Lx) - \gamma\frac{p}{m} + \xi(t), \qquad \dot{x} = \frac{p}{m}, \tag{6.36}$$

or its corresponding S.D.E.,

$$dp = -\frac{\partial}{\partial x}(U_0(x,t) - Lx)dt - \gamma\frac{p}{m}dt + dw_t, \qquad dx = \frac{p}{m}dt, \tag{6.37}$$

where $U_0(x,t)$ is periodic in both x and t, and nonsymmetric with respect to the inversion, $x \leftrightarrow (-x)$,[19] L is a constant load, and $dw_t \equiv \sqrt{2\gamma k_B T}dB_t$. The energy balance equation $dE = d'Q + d'W$ is written with the energy, $E = \frac{m}{2}\dot{x}^2 + U_0(x,t) - Lx$, and

$$d'W = \frac{\partial U_0(x,t)}{\partial t}dt, \qquad d'Q = -\frac{1}{\tau_S}\frac{p^2}{2m}dt + \frac{p}{m}\circ dw_t, \tag{6.38}$$

[19] After its sawtooth shape, this type of potential energy is named ratchet potential.

where $\tau_{\mathrm{S}} \equiv \frac{m}{2\gamma}$. When the potential $U_0(x,t)$ is externally changed in time, the model can transport the particle $x(t)$ against the load L. The crucial step is that the heat $d'Q$ is split into the "large-scale" part, $d'Q_{\mathrm{L}}$ and the "small-scale" part, $d'Q_{\mathrm{S}}$ [33], so that $d'Q = d'Q_{\mathrm{L}} + d'Q_{\mathrm{S}}$, with

$$
\begin{aligned}
d'Q_{\mathrm{L}} &= -\gamma v^2 dt, \\
d'Q_{\mathrm{S}} &= -\left\{\frac{m}{2}[\delta v(t)]^2 - \frac{k_{\mathrm{B}}T}{2}\right\}\frac{dt}{\tau_{\mathrm{S}}} + \frac{p}{m}\cdot dw_t,
\end{aligned}
\tag{6.39}
$$

where $v=$const. is the longtime average of $\dot{x}$, and $\delta v(t) \equiv \dot{x} - v$ is the deviation of the instantaneous velocity from v. The Itô-type product, $(p/m)\cdot dw_t$ vanishes if we take the average of $d'Q_{\mathrm{S}}$. The last decomposition (6.39) is of interest from the viewpoint of the scale of fluctuations: although a very crude approximation, the decomposition, $\dot{x} = v + \delta v(t)$, extracts the large-scale disturbances of thermal environment as separated from the other fluctuations. The overall displacement of the particle with the mean velocity v should cause *systematic* disturbances in the thermal environment. Contrastingly, the velocity fluctuations $\delta v(t)$ will be immediately dispersed in a short time $\sim \tau_{\mathrm{S}}$. The energy $(-d'Q_{\mathrm{L}})$ transmitted to the environment is in the form of large-scale disturbances and, therefore, is not immediately deteriorated as the heat of microscopic scale. Systematic disturbances can remain mesoscopic for some longtime $\tau_{\mathrm{L}}(\gg \tau_{\mathrm{S}})$ before they are thermalized by, for example, the cascade of vortices. Therefore, we can in principle devise a machinery to harness a part of this energy, $(-d'Q_{\mathrm{L}})$, as a systematic work. As long as the systematic disturbances break the detailed balance, the work extracted from the large-scale component of heat $d'Q_{\mathrm{L}}$ does not contradict with the second law of thermodynamics. The performance of such a machinery should depend crucially on the lifetime of those systematic disturbances.[20]

Coming back to the case of F1ATPase with a rotating actin filament, the idea of the generalized efficiency can be justified because this efficiency counts $(-d'Q_{\mathrm{L}}) = \zeta\langle\omega\rangle^2$ as a part of potentially systematic work done by the motor. The rotation of actin filament injects work $d'W$ as systematic disturbances in the environment over the length of the actin filament. If the lifetime of such disturbances in the

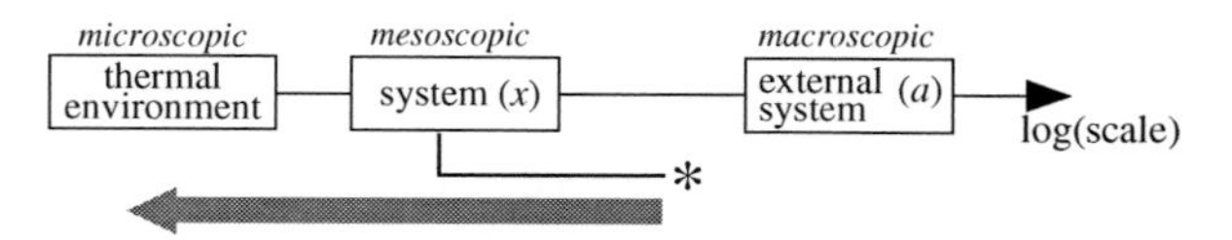

Fig. 6.6 Schematic representation of the spatio-temporal scales of fluctuations. The "L"-shaped symbol connecting the system (x) and "*" indicates the actin filament attached to F1ATPase motor. The disturbances made by the rotation of actin in the thermal environment decay from the scale * to microscopic scale along the *thick shaded arrow*

---

[20] We might recall that the eolian generator of electricity can operate under the winds even if the weeklong average velocity of wind is 0.

environment is mesoscopic, i.e., comparable or superior to the fluctuation timescale of motors motion, then the self-contained stochastic model should incorporate this large-scale disturbances among the slow variables of the model. If we put another actin filament alongside the rotating actin filament, the latter will be able to receive work through such slow variable. Figure 6.6 summarizes the notion of how the large-scale disturbances in the thermal environment decay in dimension, starting from the size of actin filament ($\gg$ motor size) down to the molecular scale.

References

1. G. Goupier, M. Saint Jean, C. Guthmann, Phys. Rev. E **73**, 031112 (2006)
2. I.P. Rouse, J. Chem. Phys. **21**, 1272 (1953)
3. B.H. Zimm, J. Chem. Phys. **24**, 269 (1956)
4. V. Lisy, J. Tothova, A.V. Zatovsky, J. Chem. Phys. **121**, 10699
5. T. Schmiedl, U. Seifert, J. Chem. Phys. **126**, 044101 (2007) (2004)
6. N.G. van Kampen, *Stochastic Processes in Physics and Chemistry*, Revised ed. (Elsevier Science, 2001)
7. V. Blickle, T. Speck, L. Helden, U. Seifert, C. Bechinger, Phys. Rev. Lett. **96**, 070603 (2006)
8. S. Joubaud, N.B. Garnier, S. Ciliberto, Journal of Statistical Mechanics **2007/09/,** P09018 (2007)
9. A.E. Allahverdyan, T.M. Nieuwenhuizen, Phys. Rev. E **62**, 845 (2000)
10. K. Sekimoto, Phys. Rev. E **76**, 060103(R) (2007)
11. R.A. Blythe, Phys. Rev. Lett. **100**, 010601 (2008)
12. S. Toyabe, T.N.H.-R. Jiang, Y. Murayama, M. Sano, Phys. Rev. E **75**, 011122 (2007)
13. S. Toyabe, M. Sano, Phys. Rev. E **77**, 041403 (2008)
14. L.D. Landau, E.M. Lifshitz, *Fluid Mechanics (Course of Theoretical Physics, Volume 6)*, 2nd edn. (Reed Educational and Professional Publishing Ltd, Oxford, 2000)
15. E.J. Hinch, J. Fluid Mech. **72**, 499 (1975)
16. A. Rahman, Phys. Rev. **136**, A405 (1964)
17. B.J. Alder, T.E. Wainwright, Phys. Rev. Lett. **18**, 988 (1967)
18. B. Lukić et al., Phys. Rev. Lett. **95**, 160601 (2005)
19. B. Lukić et al., Phys. Rev. E **76**, 011112 (2007)
20. H.J.H. Clercx, P.P.J.M. Schram, Phys. Rev. A **46**, 1942 (1992)
21. V. Vladimirsky, Y. Terletzky, Zh. Eksp. Teor. Fiz. **15**, 259 (1945)
22. A.V. Zatovsky, Izv. Vuzov. Fizika **10**, 13 (1969)
23. E.H. Hauge, A. Martin-Löf, J. Stat. Phys. **7**, 259 (1973)
24. V. Vladimirsky, Y. Terletzky, Zhur. Eksp Teor. Fiz. **15**, 259 (1945)
25. V. Lisy, J. Tothova, cond-mat/0410222
26. C. Van den Broeck, R. Kawai, P. Meurs, Phys. Rev. Lett. **93**, 090601 (2004)
27. C. Van den Broeck, R. Kawai, Phys. Rev. Lett. **96**, 210601 (2006)
28. H. Itoh et al., Nature **427**, 465 (2004)
29. H. Noji, R. Yasuda, M. Yoshida, K. Kinosita Jr., Nature **386**, 299 (1997)
30. R. Yasuda, H. Noji, K. Kinosita, Jr. M. Yoshida, Cell **93**, 1117 (1998)
31. I. Derényi, M. Bier, R.D. Astumian, Phys. Rev. Lett. **83**, 903 (1999)
32. H. Wang, G. Oster, Europhys. Lett. **57**, 134 (2002)
33. D. Suzuki, T. Munakata, Phys. Rev. E **68**, 021906 (2003)

Part III
Applications of Stochastic Energetics

Chapter 7
Control and Energetics

For a system external agents act as a main source of free energy, on one hand, and also as agents that control the processes, on the other hand. For example, when one drives a car, the car (system) needs two kinds of external agents; fuels (oil and oxygen) and a driver. From the viewpoint of energetics, the work for controlling the processes of a macroscopic system is usually negligible as compared with the main work, such as the combustion of fuel to keep a car going. It is, therefore, reasonable that standard textbooks of thermodynamics describe the Carnot heat engine without mentioning the work of attaching or detaching the engine with the heat baths.

In fluctuating mesoscopic systems, however, the situation is different. The work to control a fluctuating system can be an important part of the total work exchanged between the system and its external agents. Ignorance of this type of work would easily lead to paradoxes. If one were to invent a perpetual machine, and if one would check the consistency of this machine with the second law of thermodynamics, it would be better to make a mesoscopic model, because the energetics of control appears naturally in the mesoscopic description.

There have been many studies on the work related to control:

- *Paradox of Maxwell's demon [1]:* the "demon" which makes use of thermal fluctuations to realize a perpetual machine of the second kind (see Sect. 4.2.1.2).
- *Thermodynamics of computation [2, 3]:* theories revealing the minimal irreversible work to operate a binary digit memory (Sect. 7.1.2 below).
- *Feynman ratchet and pawl [4]:* a model of autonomous heat engine (see Sect. 1.3.4.2).
- *Motor proteins [5]:* autonomous chemical engines of molecular scale, such as linear or rotatory molecular motors or ion pumps. (See Chap. 8).
- *Signal transducing proteins [6]:* G-proteins, etc., which share a universal molecular architecture with the motor proteins [7].

One of the main questions about the energetics of control is "Can any type of operations to a system be done quasistatically?" (Sect. 7.1), because we know that quasistatic work is recoverable. Among the processes of control, there are certain important cases where the process can never be done quasistatically *by construction*. We will call such processes *essentially nonquasistatic processes*. There are two

Sekimoto, K.: *Control and Energetics*. Lect. Notes Phys. **799**, 223–253 (2010)
DOI 10.1007/978-3-642-05411-2_7

distinct mechanisms that disable a quasistatic process, and these two mechanisms come into play often together:

(Case 1) *The operation of external system imposes the crossover of timescales between the system's relaxation time and the time taken for the operation* (Sect. 7.1.1). Irreversible work directly associated with such operations can, however, be reduced as small as we want.

(Case 2) *A system loses the information about its past history when a system becomes equilibrated with a new environment* (Sect. 7.1.3). This type of nonquasistatic process costs certain irreversible work irrespective of the time of operation. Or, at least up to now we do not know how to reduce the irreversible work.

In 1960s Landauer [2], Bennett [3], and others have elucidated the minimal irreversible work required for a cycle of operations on a single-bit memory. While the operation of a single-bit memory includes essentially nonquasistatic processes of (Case 1), this minimal irreducible work and the above mentioned reducible irreversible work should be distinguished (Sect. 7.1.2).

The control of free-energy transducers will be discussed separately in Chap. 8.

Another important question of control concerns detection under fluctuations (Sect. 7.2). How can a mesoscopic system detect external signal particles with maximum certainty and minimum cost? While a "gate" correlates actively the objects of control with the subject system, the "sensor" or detector correlates passively the system with its surroundings. The two principles for avoiding the thermal noises from the detection use, respectively, (i) the steric repulsion or (ii) the compensation of interaction energies.

7.1 Limitations of Quasistatic Operations

7.1.1 * Essentially Nonquasistatic Process is Generally Caused by Crossover of Timescales τ_{op} and τ_{sys}

7.1.1.1 Two Timescales, τ_{op} and τ_{sys}

We will use the term *operations* to mean generically control and observation. The timescale of operations has an upper limit, as well as a lower limit (i.e., the time resolution). The upper limit is often the maximal time of tolerance, or the timescales beyond which the constituting elements of the system become unstable (see Sects. 1.3.3.1 and 3.1.1). We denote by τ_{op} this upper limit timescale.[1] The effect of operation on a system depends on the relaxation time of the system. We

[1] The "op" is for operation.

denote by τ_{sys} this time. The ratio between these two timescales is sometimes called the Deborah number, $De \equiv \tau_{sys}/\tau_{op}$.

7.1.1.2 Crossover of Timescales

The crossover of timescales is the phenomenon that the relative magnitudes of τ_{sys} and τ_{op} change between $\tau_{sys} \gg \tau_{op}$ and $\tau_{sys} \ll \tau_{op}$ (see Fig. 7.1) during an operation. More precisely,

(i) The relaxation time of the system, $\tau_{sys}(a)$, depends on the system parameter, a.
(ii) During a characteristic time of operation, τ_{op}, an external system changes the value of a.
(iii) The change in a is such that the relaxation time $\tau_{sys}(a)$ changes from the full relaxation regime, $\tau_{sys}(a) \ll \tau_{op}$, to the regime of "freezing", $\tau_{sys}(a) \gg \tau_{op}$, or the inverse. (Fig. 7.1 *right*)

This operation is, by definition, nonquasistatic. Especially, if $\tau_{sys}(a_f) = \infty$ at the end of the operation (Fig. 7.1 *left*), whatever large τ_{op} cannot satisfy the quasistatic condition $\tau_{sys}(a) \ll \tau_{op}$ throughout the operation.

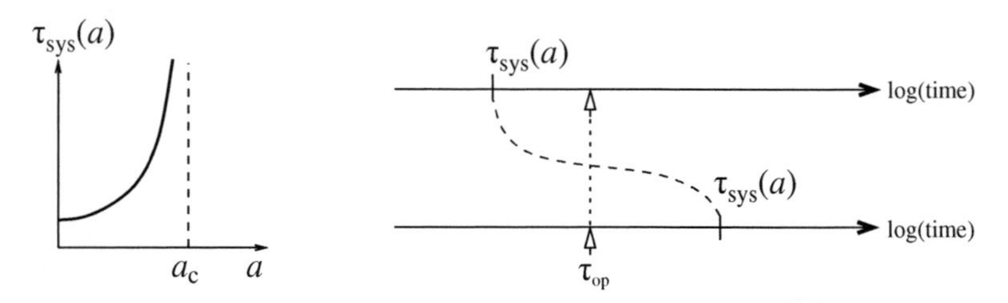

Fig. 7.1 *(Left)* $\tau_{sys}(a)$ vs. a. The figure shows the case where $\tau_{sys}(a)$ diverges for $a \uparrow a_c$. *(Right)* Change of $\tau_{sys}(a)$ across the operation time τ_{op} through the change of parameter a. The time axes should be regarded as logarithmic scale

7.1.1.3 Barrier Height and Relaxation Time

Let us look at a simple example of the crossover of timescales. The system is a single Brownian particle in the force potential $U(x, a)$ (see Fig. 7.2) and confined in a finite zone Ω_{tot} with $-L/2 \le x \le L/2$. Inside Ω_{tot} the potential energy $U(x, a)$ depends on the parameter a only in a small zone Ω of the width $|\Omega| = \delta(< L)$. Outside this zone, we assume that $U(x, a)$ =const. for any value of a. For simplicity we assume the symmetry, $U(-x, a) = U(x, a)$, with the maximum being located at $x = 0$.

The Brownian particle obeys the Langevin equation,

$$0 = -\frac{\partial U(x, a)}{\partial x} - \gamma \frac{dx}{dt} + \xi(t) \qquad (x \in \Omega_{tot}). \tag{7.1}$$

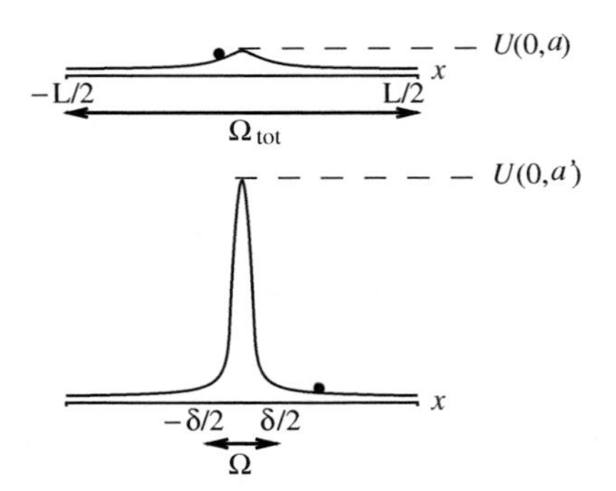

Fig. 7.2 A Brownian particle (*thick black dot*) in a zone $\Omega_{\mathrm{tot}} \equiv \{x| - L/2 \leq x \leq L/2\}$ along the x-axis. The curves show the profiles of the potential energy $U(x, a)$ with $a_{\mathrm{i}} \leq a < a^*$ (*top*) and $U(x, a')$ with $a^* < a' \leq a_{\mathrm{f}}$ (*bottom*) vs. x. The maximum value of the potential is $U(0, a) \sim k_{\mathrm{B}}T$ and $U(0, a') \gg k_{\mathrm{B}}T$, respectively. The profile of the potential $U(x, a)$ changes only in the zone, $\Omega \equiv \{x| - \delta/2 \leq x \leq \delta/2\}$

If the height of the potential barrier, $U(0, a)$, is small, i.e., $U(0, a)/k_{\mathrm{B}}T \sim 1$, the time τ_{sys} is approximately the diffusion time, $\tau_0 \equiv L^2/(2D)$ ($D = k_{\mathrm{B}}T/\gamma$), for a Brownian particle to visit almost entire zone Ω_{tot}. By contrast, if $U(0, a)/k_{\mathrm{B}}T \gg 1$ the relaxation time of the system, $\tau_{\mathrm{sys}}(a)$, is essentially the inverse rate of thermally activated transition across the potential barrier. According to (3.40) the rate is roughly given by

$$\frac{1}{\tau_{\mathrm{sys}}(a)} \equiv \frac{1}{\tau_0} e^{-\frac{U(0,a)}{k_{\mathrm{B}}T}}, \tag{7.2}$$

where we have ignored the corrections in the prefactor due, for example, to the details of the potential profile.

Now suppose that $U(0, a_{\mathrm{i}}) \sim k_{\mathrm{B}}T$ for the initial value a_{i} and $U(0, a_{\mathrm{f}}) \gg k_{\mathrm{B}}T$ for the final value a_{f}. Then $\tau_{\mathrm{sys}}(a)$ increases from $\tau_{\mathrm{sys}}(a_{\mathrm{i}})/\tau_0 \sim 1$ to $\tau_{\mathrm{sys}}(a_{\mathrm{f}})/\tau_0 \gg 1$ during the operation time τ_{op}. For example, if $\tau_0 = 10^{-9}$s and $\tau_{\mathrm{op}} = 1$ h, the crossover point $\tau_{\mathrm{sys}}(a) \sim \tau_{\mathrm{op}}$ is attained for $U(0, a) \sim 30k_{\mathrm{B}}T$. In the regime of $\tau_{\mathrm{sys}}(a) \gg \tau_{\mathrm{op}}$ the particle is practically localized within either side of the potential barrier. Through the crossover of the timescales, the accessible phase space of the system shrinks almost discontinuously. Moreover, we cannot precisely predict which subspace the particle will be confined to. The probabilities that the particle is confined to the left ($x < 0$) and that to the right ($x > 0$) of the barrier are 1/2:1/2.

7.1.1.4 Work to Raise the Potential Barrier

Essentially nonquasistatic process does not necessarily imply a large irreversible work. We will show that the *irreversible* part of the work associated with raising the potential barrier of Fig. 7.2 can be made very small if we carefully choose a protocol of raising the potential barrier.[2] We will denote by a^* ($a_i < a^* < a_f$) the value of the parameter a at the crossover of timescales, i.e., $\tau_{sys}(a^*) \simeq \tau_{op}$.[3] For $a_i \le a(t) < a^*$ the almost quasistatic operation is realizable while for $a^* < a(t) \le a_f$ it is impossible in any case because $\tau_{sys}(a(t)) \gg \tau_{op}$.

The optimal strategy to minimize the irreversible work is that we spend most of the operation time in raising the potential barrier up to $U(0, a^*)$ so that the associated work is almost recoverable when the barrier will be lowered. During this stage, the probability density $\mathcal{P}(x, t)$ for the particle position x is close to the canonical distribution over the entire zone Ω_{tot}, $\mathcal{P}^{eq}(x, a; T) = e^{[F(a,T)-U(x,a)]/k_B T}$, where $e^{-F(a,T)/k_B T} \equiv \int_{\Omega_{tot}} e^{-U(x,a)/k_B T} dx$. The associated (almost quasistatic) work $W_{[a_i,a^*]}$ is (see (4.6))

$$W_{[a_i,a_*]} \simeq F(a^*, T) - F(a_i, T). \tag{7.3}$$

The error in (7.3), i.e., the irreversible work associated with this slow process, is approximately proportional to the inverse of the time spent, i.e., $\sim \tau_{op}{}^{-1}$, as discussed in Sect. 5.3.1.

After this almost quasistatic work, we will raise the barrier height from $U(0, a^*)$ to $U(0, a_f)$ in a short time. The average work $\langle W \rangle_{[a^*,a_f]}$ associated with this step during $a^* < a(t) \le a_f$ is estimated to be

$$\frac{\langle W \rangle_{[a^*,a_f]}}{k_B T} \le \frac{(\text{const.})\delta}{L} e^{-U(0,a^*)/k_B T} \frac{U(x, a_f) - U(x, a^*)}{k_B T}, \tag{7.4}$$

Leaving the derivation to Appendix A.7.1, the result (7.4) can be qualitatively explained: Since the height of the potential barrier $U(0, a^*)$ is already very large with respect to $k_B T$, the particle almost surely escapes from the region Ω where the barrier is rising. Therefore, in most cases virtually no work is needed to raise the barrier. However, in rare cases, with the probability of $\sim (\delta/L) e^{-U(0,a^*)/k_B T}$, a particle happens to remain within the zone Ω and is "lifted up." The last process costs the work of the order of $U(0, a_f) - U(0, a^*)$. Therefore, we have (7.4). The last work is not recoverable because in the reverse operation from a_f to a^*, the probability to find the particle within Ω at $a = a_f$ is extremely less than $\mathcal{P}^{eq}(x, a^*; T)$.

Equation (7.4) tells that $\langle W \rangle_{[a^*,a_f]}/k_B T$ can be made very small due to the exponential factor, $e^{-U(0,a^*)/k_B T}$. If we take $\tau_0 = 10^{-13}$ s and $U(0, a_f)/k_B T \simeq 10^3$, then $\tau_{op} = 1$ ms is sufficient to satisfy $\langle W \rangle_{[a^*,a_f]}/k_B T < 10^{-7}$. That is, we need the *irreversible* work of less than 10^{-5}% of $k_B T$ and the time of 1 ms to establish the barrier which a particle will not pass before $\tau_0 e^{U(0,a^*)/k_B T} > 10^{400}$ days. Moreover,

[2] We discuss here about the irreversible work being *directly* related to a single action of raising the barrier: This work should be distinguished from the irreversible work associated with the cyclic operation to erase a single-bit memory. See Sect. 7.1.2.

[3] Practically, it is better to include a "safety factor" to define $\tau_{sys}(a^*)/\tau_{op} \simeq 10 - 10^2$ rather than $\tau_{sys}(a^*)/\tau_{op} \simeq 1$. We, however, ignore this point for the simplicity of the argument.

this $10^{-7}k_BT$ of work is not typical work: for the most realizations, W is practically 0 ($\ll 10^{-7}$), while for one realization out of $e^{10^3} \sim 10^{430}$ the work is $W \sim 10^3 k_B T$.

The access of particles to an open system from particle environments (reservoirs) can be controlled by the gates which consist of variable potential barriers between the system and the reservoirs.[4] The crossover of timescales of the type (1), therefore, occurs inevitably whenever the open system is attached or detached with particle environments.

7.1.1.5 Attachment/Detachment with a Thermal Environment

Attaching or detaching a (closed) system with a thermal environment also causes the crossover of timescales. When a system is in contact with a thermal environment of temperature T, the system's energy undergoes temporal fluctuations. From the viewpoint of energetics, the relaxation time of a system, τ_{sys}, is the time over which the temporal history of the system's energy approaches the canonical distribution. If the system is completely detached from its thermal environment, the system's energy becomes fixed. The relaxation time, τ_{sys}, is then infinite. Therefore, the operation of detaching a system from its thermal environment inevitably causes crossover of timescales.

Can we calculate the work associated with attachment/detachment with a thermal environment? In the Langevin equation (7.1) only the parameter γ characterizes the interaction between the system and its thermal environment.[5] One might regard γ as an externally controlled parameter. But the framework of stochastic energetics has excluded the control of the interface between a system and thermal environments (Sect. 4.1.1.1). A way to go around this constraint is schematically shown in Fig 7.3. We will call this device the "clutch mechanism." We introduce an auxiliary degree of freedom, say y, which always stays in contact with a thermal environment, and whose fluctuating motion is described by a Langevin equation. The variable of the main system, say x, does not interact any more with thermal environments, but it interacts with this auxiliary degree of freedom, y, through an interaction potential energy, say $\phi(x, y, \chi)$, where χ is the control parameter of operation.[6] One may interpret that the auxiliary variable y stands for those fluctuation modes of the thermal environment which is coupled to the system variable x.[7] The system of Langevin equations is

$$\frac{dx}{dt} = \frac{p}{m}, \qquad \frac{dp}{dt} = -\frac{\partial U(x,a)}{\partial x} - \frac{\partial \phi(x,y,\chi)}{\partial x}, \tag{7.5}$$

---

4 Sect. 7.2.2 below.

5 Remember that the random force $\xi(t)$ is also characterized by $\gamma k_B T$.

6 We reserve a for the control parameter of the main system.

7 See the discussion at the end of Sect. 6.3.2.

Fig. 7.3 A clutch mechanism to attach/detach a system with a thermal environment using an auxiliary variable y and an interaction potential $\phi(x, y, \chi)$

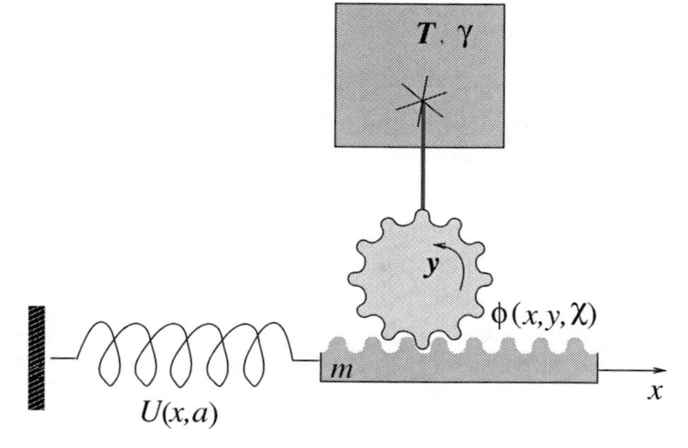

$$0 = -\gamma \frac{dy}{dt} + \xi(t) - \frac{\partial \phi(x, y, \chi)}{\partial y}. \tag{7.6}$$

In (7.6) and in Fig 7.3 we assumed that y is a rotational degree of freedom (i.e., angle) so that its Brownian motion is confined within a periodic domain of period 2π. We assume $\phi(x, y, \chi)$ such that (see Fig 7.4) for $\chi = 0$ there is no energetic coupling between x and y, i.e., $\phi(x, y, 0) = 0$, and for $\chi = 1$ the variation of y is highly correlated with that of x. For $0 < \chi < 1$, change of $\phi(x, y, \chi)$ is supposed to be monotonous with χ.

When the parameter χ is decreased down to the decoupling limit $\chi = 0$, the crossover of timescales is unavoidable with whatever large τ_{op}.[8] We will use the clutch mechanism to model the Carnot heat engine (Sect. 8.1.1) and the total work to operate χ parameters will be calculated in Sect. 8.1.1.2.

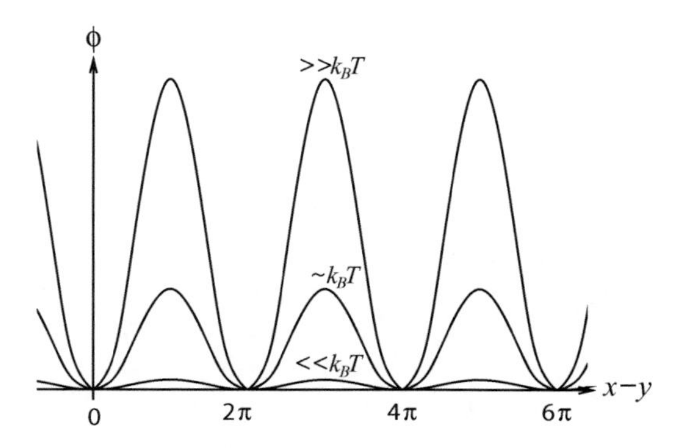

Fig. 7.4 Potential function used for the clutch mechanism, $\phi(x, y, \chi)$ vs. $x - y$, is shown for $\chi \simeq 0$ (*bottom curve*), $0 < \chi < 1$ (*middle curve*), and $\chi = 1$ (*top curve*)

[8] There is a secondary crossover of timescales in the opposite limit, $\chi \to 1$ if $\phi(x, y, 1) \gg k_{\text{B}} T$. For high barriers, the slippage between the variations of x and y is blocked.

7.1.1.6 Remarks

Generality of the Crossover of Timescales and Spatial Scales

Figure 7.2 represents a naïve oversimplified picture of the glass transition, where lowering temperature or changing other fields splits the equilibrium states into more than one (meta) stable states.

The crossover of timescales has long been known in quantum chemistry. The Born–Oppenheimer approximation assumes that the electrons' state follows quickly enough the movement of atoms or ions [8]. But when the two atoms are gradually separated, the rate of the electronic transition between the atoms becomes smaller and smaller with the increase of the interatomic distance. Therefore, the timescale of the electronic transition diverges during this process. Eventually the Born–Oppenheimer approximation becomes invalid.

The crossover of spatial scales has been studied in quantum mechanics. The Wentzel–Kramers–Brillouin (WKB) approximation [9] assumes that the de Broglie wavelength of the particle, $2\pi/k$, is much shorter than the length scale over which this wavelength varies. The de Broglie wavelength depends on the total energy E and the potential energy $U(x)$ through $\hbar k = \sqrt{2m(E - U(x))}$, where $\hbar$ is the Planck constant/2π and m is the mass of the particle. The *turning point* x^* is defined by $E = U(x^*)$. At this point the de Broglie wavelength diverges. The WKB approximation becomes, therefore, inevitably invalid as x approaches the turning point.

Auto-Adjusting Timescales

In the above examples the timescales are controlled by external operations. When the system's relaxation time τ_{sys} depends on the system's state, τ_{sys} can be adjusted by itself in the vicinity of the timescale of operation τ_{op}. A very simple model of aging and plastic flow show that τ_{sys} approaches to τ_{op} from below and from above, respectively. The models are described in Appendix A.7.2.

Consequences of the Discontinuous Change of Accessible Phase Space

As shown in Sect. 7.1.1.4, the irreversible work related to the crossover of timescales can be made small by suitably choosing the protocol of operation. Then what is the major outcome of this type of essentially nonquasistatic process?

The answer concerns the change of accessible phase spaces for the system. The final state of the system is confined to a subset of phase space which has been entirely available before. The reduction of accessible phase space contains the memory of either the external operation or the fluctuations in the environment when the crossover of the timescale took place. This memory can have an influence on the subsequent response of the system. In the following Sect. 7.1.2 we describe the irreversible work to operate a single-bit memory as a typical example.

7.1.2 Minimal Cost of the Operation of Single-Bit Memory is Related to the Second Law

Operation of a single-*bit* (binary digit) memory uses the crossover of timescales. The objects of the present section are:

I. To give a physical expression of the memory: A bit memory is a physical state of the system. This information can be changed as a physical process, either through the operations by the external system or through the fluctuations due to the interaction with the environment.
II. To relate the irreversible loss of information with the irreversible work: For a bit memory, the cyclic operation of *overwriting* is a basic physical process. Its energetics is described.
III. To describe "to know" in physics' language (Sect. 7.1.2.3): The process of copying the information is analyzed, including its minimal cost.

The basis of this subject was founded by Landauer [2] and Bennett [3] (see a survey [1]). They have shown that the work of overwriting a memory is no less than the work which "Maxwell's demon" can extract. Therefore, no perpetual machine of the second kind can be constructed on the basis of memory operations. Stochastic energetics provides an explicit formulation of the minimal irreversible work along a single realization of memory operation.

7.1.2.1 A Model of Bit Memory Operations

A single-bit memory is realized by a state point ("particle"), x, within a double-well potential, $U(x, a, b)$. See Fig. 7.5.

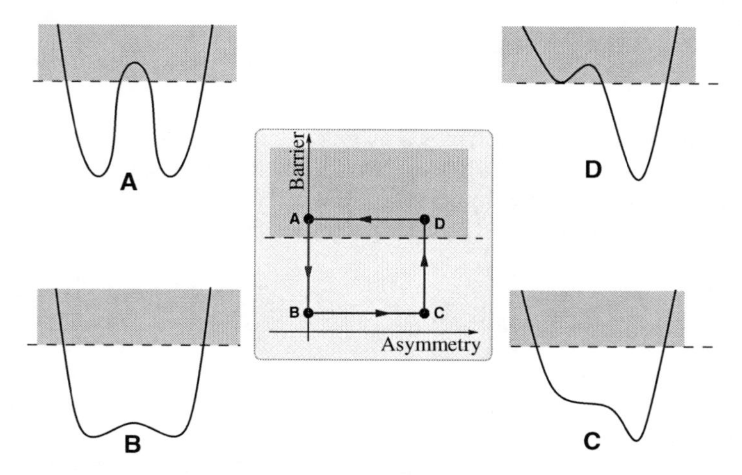

Fig. 7.5 A cycle of operation of overwriting, which consists of the erasure, $\mathcal{E}$: A $\rightarrow$ B, and the subsequent writing, $\mathcal{W}$: B $\rightarrow$ C $\rightarrow$ D $\rightarrow$ A

The particle is in contact with a thermal environment at temperature T. Neglecting the inertia effect, x obeys the following Langevin equation:

$$\gamma \frac{dx}{dt} = -\frac{\partial U(x, a, b)}{\partial x} + \xi(t), \tag{7.7}$$

where γ is the friction coefficient, $\xi(t)$ is the white Gaussian process of zero mean with $\langle \xi(t)\xi(t') \rangle = 2\gamma k_B T \delta(t - t')$. The operation requires two parameters, a and b [2]. The parameter a controls the asymmetry of the potential to bias the memory to take a desired state, while the parameter b controls the barrier height of the potential. $U(x, a, b) = -ax + b(x^2 - 1)^2$ is an example.

We split the x-axis into the left half $\Omega_0 = (-\infty, 0)$ and the right half $\Omega_1 = (0, \infty)$. Then the states $\sigma = 0$ [$\sigma = 1$] of the memory are assigned to $x \in \Omega_0$ [$x \in \Omega_1$], respectively.

By the dashed horizontal line and the shaded zone in Fig. 7.5 A–D, we denote the zone of (free) energy which the particle practically never attains within τ_{op} if it starts from the minimum of the potential. For example, in Fig 7.5A, the particle remains within one of the valleys, that is, the state of the memory σ is stable. Likewise, in Fig 7.5D the memory is in $\sigma = 1$ state. By contrast, in Fig 7.5B, C, the system's relaxation time τ_{sys} is less than τ_{op}. The memory σ can flip between 0 and 1.

The counterclockwise cycle in Fig. 7.5, A→B→C→D→A, is the operation of overwriting. This operation is decomposed into the process of erasing memory, $\mathcal{E}$, and the process of writing memory, $\mathcal{W}$:

$$\begin{aligned} \text{Erasing, } \mathcal{E}: &\quad \text{A} \to \text{B}, \\ \text{Writing, } \mathcal{W}: &\quad \text{B} \to \text{C} \to \text{D} \to \text{A}. \end{aligned} \tag{7.8}$$

Suppose that the system has retained a memory state, σ, in A. Through the process $\mathcal{E}$, this memory is lost. Then by the process $\mathcal{W}$, the memory is forcedly reset to the state $\sigma = 1$. In order to reset to the state $\sigma = 0$, the profiles of C and D should be replaced by their mirror images C* and D*, respectively.

The Timescales and Branching of Protocol

In Fig. 7.5 the operations A $\leftrightarrow$ B and A $\leftrightarrow$ D are the essentially nonquasistatic processes, as discussed in Sects. 7.1.1.2. Through these transitions, the accessible phase space for $x(t)$ changes discontinuously.

To look at the timescales more systematically, let us introduce two relaxation times: $\tau_{\text{sys}}^{(0\to 1)}$ associated with the transition of σ from 0 to 1, and $\tau_{\text{sys}}^{(1\to 0)}$ from 1 to 0. Therefore, four situations are possible with respect to the timescales:

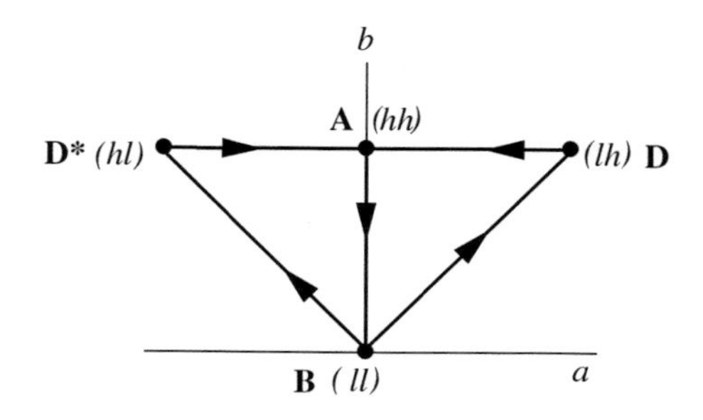

Fig. 7.6 The cycles of overwriting process viewed from the timescales. See the text

$$\begin{aligned}
(hh)&: \tau_{\mathrm{sys}}^{(0\to 1)}/\tau_{\mathrm{op}} \gg 1 \quad \text{and} \quad \tau_{\mathrm{sys}}^{(1\to 0)}/\tau_{\mathrm{op}} \gg 1\\
(h\ell)&: \tau_{\mathrm{sys}}^{(0\to 1)}/\tau_{\mathrm{op}} \gg 1 \quad \text{and} \quad \tau_{\mathrm{sys}}^{(1\to 0)}/\tau_{\mathrm{op}} \ll 1\\
(\ell h)&: \tau_{\mathrm{sys}}^{(0\to 1)}/\tau_{\mathrm{op}} \ll 1 \quad \text{and} \quad \tau_{\mathrm{sys}}^{(1\to 0)}/\tau_{\mathrm{op}} \gg 1\\
(\ell\ell)&: \tau_{\mathrm{sys}}^{(0\to 1)}/\tau_{\mathrm{op}} \ll 1 \quad \text{and} \quad \tau_{\mathrm{sys}}^{(1\to 0)}/\tau_{\mathrm{op}} \ll 1.
\end{aligned} \tag{7.9}$$

Now the two control parameters (a, b) are essentially reflected in $\ln[\tau_{\mathrm{sys}}^{(0\to 1)}/\tau_{\mathrm{sys}}^{(1\to 0)}]$ and $\ln[\tau_{\mathrm{sys}}^{(0\to 1)}\tau_{\mathrm{sys}}^{(1\to 0)}/\tau_{\mathrm{op}}^2]$, respectively. Therefore, the operations of overwriting and the timescales can be related as shown in Fig. 7.6. In this schema, the important part of the cycle is the vertical process, A→B. While the external system does the same operation along $a = 0$ line, the memory state at A depends on the previous operation, either from D or from D*. Therefore, it is often said that this erasure process $\mathcal{E}$ is the origin of irreversibility.

7.1.2.2 Energetics of the Writing and Erasure of a Bit Memory

Hereafter we consider only slow operations to change a and b.[9] Then we ignore the irreversible work of almost quasistatic process. Also, we ignore the irreversible work directly associated to the essentially nonquasistatic operations based on the argument in Sect. 7.1.1.4.

Irreversibility of the Operation of Overwriting

If we do the erasing process $\mathcal{E}$ and immediately retrace it, i.e., A→B→A, there is practically no irreversible work. The same is true if we do the writing process $\mathcal{W}$ and immediately retrace it, i.e., B→C→D→A→D→C→B, because the state is definitely $\sigma = 1$ at the turning stage A. Furthermore, if we do the sequential processes, $\mathcal{E} \to \mathcal{W}$, and immediately retrace them in the reverse order, i.e., A→B→C→D→A→D→C→B→A, there is no irreversible work. One might then expect that any process along this circle is reversible. In fact it is not the case. For example, along the path B→C→D→A→B→A→D→C→B, the external

[9] cf. Sect. 5.3.1.

system loses control of the state σ after the process, A$\rightarrow$B, as discussed in the above. If upon the operation of B$\rightarrow$A the memory comes back to $\sigma = 0$, then the subsequent operations, A$\rightarrow$D$\rightarrow$C$\rightarrow$B, will cost a work much more than $k_B T$. This is an example of the consequence of the discontinuous change of accessible phase space (Sect. 7.1.1.6).

Furthermore, we will see below that the total work for the overwriting, A$\rightarrow$B$\rightarrow$C$\rightarrow$D$\rightarrow$A, is nonzero, bounded below by $k_B T \ln 2$ per cycle.

Estimation of the Work of Writing: $\mathcal{W}$

The operation B$\rightarrow$C$\rightarrow$D can be done quasistatically. Also the operation D$\rightarrow$A costs no work since, after the process B$\rightarrow$C$\rightarrow$D, the state is surely $\sigma = 1$. In the last process, D$\rightarrow$A, the potential energy $U(x, a, b)$ can, therefore, be replaced by an effective potential $U^{\text{eff},1}(x, a, b)$, where $U^{\text{eff},\sigma}(x, a, b)$ ($\sigma = 0$ or 1) is defined by

$$U^{\text{eff},\sigma}(x, a, b) = \begin{cases} U(x, a, b) & (x \in \Omega_\sigma) \\ \infty & (x \in \Omega_{1-\sigma}) \end{cases}. \tag{7.10}$$

The work for the writing process, $W_{\mathbf{BCDA}}$, can therefore be given as the quasistatic work: Let us introduce the configurational free energies, $F(a, b; \Omega_0 \cup \Omega_1)$ and $F(a, b; \Omega_\sigma)$, corresponding to $U(x, a, b)$ and $U^{\text{eff},\sigma}(x, a, b)$, respectively, by

$$e^{-F(a,b;\Omega_0\cup\Omega_1)/k_B T} \equiv \int_{\Omega_0\cup\Omega_1} e^{-U(x,a,b)/k_B T} dx, \tag{7.11}$$

$$\begin{aligned} e^{-F(a,b;\Omega_\sigma)/k_B T} &\equiv \int_{\Omega_0\cup\Omega_1} e^{-U^{\text{eff},\sigma}(x,a,b)/k_B T} dx \\ &= \int_{\Omega_\sigma} e^{-U(x,a,b)/k_B T} dx. \end{aligned} \tag{7.12}$$

Then the work of writing is expressed as follows with probability 1:[10]

$$W_{\mathbf{BCDA}} = F(a_A, b_A; \Omega_1) - F(a_B, b_B; \Omega_0 \cup \Omega_1), \tag{7.13}$$

Estimation of the Work of Erasure: $\mathcal{E}$

Let us calculate the work for the reverse operation of erasure, $W_{\mathbf{BA}}$, because it is intuitively simpler than the erasure process $W_{\mathbf{AB}}$, and because we know that these two works cancel with each other: $W_{\mathbf{AB}} = -W_{\mathbf{BA}}$. $W_{\mathbf{BA}}$ writes

[10] Hereafter a_A, etc., denotes the value of parameter a at the stage A, etc.

$$W_{\mathbf{BA}} = \int_{b_{\mathrm{B}}}^{b_{\mathrm{A}}} \frac{\partial U(x(t), a_{\mathrm{A}}, b(t))}{\partial b} db(t), \tag{7.14}$$

where we used $a_{\mathrm{A}} = a_{\mathrm{B}}$. By an analogy to Sect. 7.1.1.4, we raise quasistatically the barrier from $b = b_{\mathrm{B}}$ up to b^* such that $\tau_{\mathrm{sys}} \simeq \tau_{\mathrm{op}}$ for $b = b^*$. The corresponding work is

$$W_{\mathbf{B}*} = \int_{b_{\mathrm{B}}}^{b^*} \left[\int_{\Omega_0 \cup \Omega_1} \frac{\partial U(X, a_{\mathrm{A}}, b)}{\partial b} \mathcal{P}^{\mathrm{eq}}(X, a_{\mathrm{A}}, b) dX \right] db. \tag{7.15}$$

Then for $b^* < b \leq b_{\mathrm{A}}$, we have $\tau_{\mathrm{sys}} \gg \tau_{\mathrm{op}}$. Suppose that x stays in a domain Ω_σ during this regime. The work of this part can be calculated as a quasistatic work with the effective potential energy, $U^{\mathrm{eff},\sigma}(x, a, b)$. The point is that *twice* the canonical distribution function ($\mathcal{P}^{\mathrm{eq}}(X, a, b) = e^{-U(X,a,b)/k_{\mathrm{B}}T}/Z$) should be used because the state is confined in Ω_σ:

$$\begin{aligned} W_{*\mathbf{A}} &= \int_{b^*}^{b_{\mathrm{A}}} \left[\int_{\Omega_0 \cup \Omega_1} \frac{\partial U^{\mathrm{eff},\sigma}(X, a_{\mathrm{A}}, b)}{\partial b} 2\mathcal{P}^{\mathrm{eq}}(X, a_{\mathrm{A}}, b) dX \right] db \\ &= \int_{b^*}^{b_{\mathrm{A}}} \left[\int_{\Omega_0 \cup \Omega_1} \frac{\partial U(X, a_{\mathrm{A}}, b)}{\partial b} \mathcal{P}^{\mathrm{eq}}(X, a_{\mathrm{A}}, b) dX \right] db. \end{aligned} \tag{7.16}$$

To go to the second line, we have used the symmetry of $U(X, a_{\mathrm{A}}, b)$ and of $\mathcal{P}^{\mathrm{eq}}(X, a_{\mathrm{A}}, b)$. Adding $W_{\mathbf{B}*}$ and $W_{*\mathbf{A}}$, we obtain the total work $W_{\mathbf{BA}}$

$$\begin{aligned} W_{\mathbf{BA}} &= F(a_{\mathrm{A}}, b_{\mathrm{A}}; \Omega_0 \cup \Omega_1) - F(a_{\mathrm{B}}, b_{\mathrm{B}}; \Omega_0 \cup \Omega_1) \\ &= -W_{\mathbf{AB}}. \end{aligned} \tag{7.17}$$

Total Work of the Overwriting Cycle, $\mathcal{E}+\mathcal{W}$

By adding $W_{\mathbf{AB}}$ and $W_{\mathbf{BCDA}}$, we have

$$\begin{aligned} W_{\mathbf{AB}} + W_{\mathbf{BCDA}} &= F(a_{\mathrm{A}}, b_{\mathrm{A}}; \Omega_1) - F(a_{\mathrm{A}}, b_{\mathrm{A}}; \Omega_0 \cup \Omega_1) \\ &= k_{\mathrm{B}}T \ln 2, \end{aligned} \tag{7.18}$$

because (7.11) and (7.12) give $e^{-F(a,b;\Omega_0 \cup \Omega_1)/k_{\mathrm{B}}T} = 2\, e^{-F(a,b;\Omega_\sigma)/k_{\mathrm{B}}T}$. This is the main result of [2, 3]: Every operation of overwriting costs at least $k_{\mathrm{B}}T \ln 2$ of irreversible work. By the stochastic energetics the result holds for an individual process with the probability of one in the limit of slow operation.

The heat associated with this cycle has been analyzed using the stochastic energetics [10].[11] In the language of information, the memory σ specified through the

[11] [10] used the Bennett's definition of "copying" of memory. Since this definition does not complete a cycle, their results differ apparently from those in the text. However, these formulations are mathematically equivalent.

preceding writing process is lost irreversibly during the erasure process. The statistical entropy associated with the variable σ has been increased by $\ln 2$. The ensemble averaged irreversible work has been given in [11]. More recently [12] incorporated mechanical approach of the type of Jarzynski nonequilibrium work relation, and confirmed both analytically and numerically $k_B T \ln 2$ as the lower bound of the average irreversible work. [12] also analyzed the case where $U(0, a)/k_B T$ is moderately large.

7.1.2.3 Copying a Memory

What Does "To Know" Mean?

The "memory" is the state of a system ("bit memory"). The subject who "knows" the memory is, therefore, this system, not the external agent that controls the parameters a and b. For such ignorant external agent, the cycle of overwriting operation in Fig. 7.5 is optimal with the least average dissipation. Now we define "to know" (to acquire a knowledge) as the process by which a fixed memory of a system, called data bit [3], is correlated to the state of another system, called movable bit. In the ideal process "to know" the memory of a movable bit is rendered equal to the memory of the data bit with probability 1. We will say that the (memory of) data bit is *copied* to the (memory of) movable bit.

Cost-Free Copying Would Constitute a Perpetual Machine

A simple thought experiment shows that the external agent must pay a certain irreversible work to copy the data bit to the movable bit. Otherwise, the second law of thermodynamics would be violated.

Suppose that an external agent can copy the memory of a given data bit onto n ($\gg 1$) movable bits without cost. At this point, the external agent does not know the value σ of the copied bit. As the next step the external agent applies the reverse cycle of Fig. 7.5, i.e., A $\to$ D$\to$ C$\to$ B$\to$ A, to one of those movable bits. If $\sigma = 1$, the external agent will gain the work $\simeq k_B T \ln 2$ from the reverse cycle. If $\sigma = 0$, the work by the external agent is positive and much larger than $k_B T \ln 2$. The external agent thus knows the value of σ.[12] Then as the third step, the external agent applies to the remaining $(n - 1)$ copies the "correct" reverse cycle, either A $\to$ D$\to$ C$\to$ B$\to$ A of Fig. 7.5 if $\sigma = 1$, or its mirror image ($a \to -a$) if $\sigma = 0$, so that the external system gains work of $\simeq k_B T \ln 2$ from the $(n - 1)$ copies. This is possible because the external agent knows what σ is. In this way, the external agent could obtain a positive net work which increases with n. The whole operation would, therefore, constitute a perpetual machine of the second kind, which extracts a work from an isothermal environment.

[12] This operation need not be very efficient as long as $n \gg 1$.

The Cost of Copying is No Less Than $k_B T \ln 2$ Per Movable Bit

The operation of copying is a cycle: Every time a new data bit of unknown state σ is given, the external agent must erase the previous memory of a movable bit σ' to replace it with the value of σ. See Fig. 7.7. Therefore, the work of copying is derived in the same way as that of overwriting. We will show this below.

We will denote by x_0 and x the degrees of freedom of the data bit and the movable bit, respectively. To simplify the analysis, we will ignore the thermal fluctuation of x_0 (the left column in Fig. 7.7), although thermal fluctuations of x_0 within the domain Ω_σ will not change the conclusion. The potential energy for the movable bit, $U(x, a, b; \sigma)$, has two control parameters, a and b. The parameter b controls the barrier height of the double-well potential for x, and the parameter a controls the *distance* between this bit and the data bit.

Initially the two-bit memories are well isolated from each other, so that there is no interaction between the memories x_0 (or σ) and x. $U(x, a, b; \sigma)$ then takes a symmetric form (A in Fig. 7.7). At this position, the potential barrier for x is lowered through the parameter b (A→B in Fig. 7.7). Then the movable bit is brought into interaction with the data bit. $U(x, a, b; \sigma)$ then becomes asymmetric in a σ-dependent manner (B→C in Fig. 7.7).[13] At this position, the potential barrier for x is raised (C→D in Fig. 7.7). Finally the movable bit is brought apart from the data bit (D→A in Fig. 7.7).

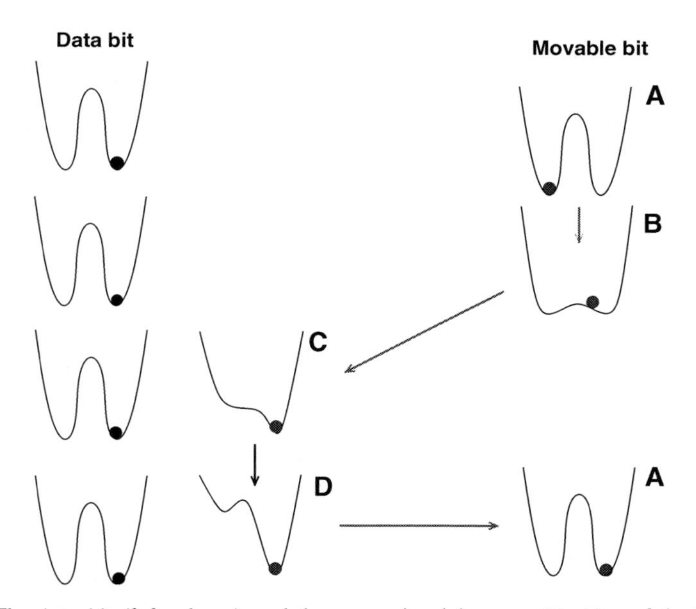

Fig. 7.7 The data bit (*left* column), and the energy involving movable bit *and* the interaction energy between the two bits, $U(x, a, b, \sigma)$ (*right*)

[13] We assume that the profile of $U(x, a, b; \sigma)$ and that of $U(x, a, b; 1-\sigma)$ are of the mirror images along the x-axis.

We realize that the complete cycle is essentially the same as the cycle of overwriting a bit memory shown in Fig. 7.5. The minimum cost of making copy of the data bit is, therefore, $k_B T \ln 2$. The only difference is that the asymmetry of the potential energy in the copying process is not determined by the control parameter a but is determined by the memory (σ) of the data bit. To conclude, we need the work to be no less than $k_B T \ln 2$ per movable bit. Therefore, we cannot make a perpetual machine.

7.1.3 * Essentially Nonquasistatic Process Can Take Place upon the Relaxation from Non-Gibbs Ensembles of Fluctuations

The second type of essentially nonquasistatic processes is related to the erasure process $\mathcal{E}$ in Sect. 7.1.2: The nonequilibrium state of the system is *irreversibly* equilibrated by an equilibrium environment.

7.1.3.1 Setup of the Problem and the Result

Let us consider the following cyclic operation of a system with two thermal environments. See Fig. 7.8(a) [13]. In (i) the system is thermalized with the thermal environment of temperature T_1. Then (ii) the system is slowly detached from the environment. (iii) Under complete isolation, the control parameter is slowly changed from a_1 to a_2. Then (iv) the system is slowly attached to the second thermal environment of temperature T_2. After the thermalization, the operations are retraced slowly back to the state (i). The value of a_2 is chosen so that, upon attachment to the environment of temperature T_2, no energy is exchanged *on the average* between the system and the new environment. The quantity of interest is the work needed to change the parameter a:

$$W = \int_{a_1:(ii)\to(iii)}^{a_2} \frac{\partial H(x(t), p(t); a(t))}{\partial a} \circ da(t) + \int_{a_2:(iii)\to(ii)}^{a_1} \frac{\partial H(x(t), p(t); a(t))}{\partial a} \circ da(t),$$

where $H(x, p; a)$ is the system's energy as a function of position x and the momentum p. We ignore the work of detachment and attachment with the thermal environments, since they were shown to be reducible as small as needed.

If the system is macroscopic, the choice of a_2 amounts to equalizing the system's temperature to T_2 before the contact with the second heat bath. Under this condition, the process is reversible in the limit of quasistatic adiabatic operation, and the work W is 0.[14]

For the mesoscopic system, several aspects are different from the macroscopic case:

1. The work W fluctuates from one realization to the other, however, slowly the operations are done.

[14] Unlike the Carnot cycle, the parameter a is not changed under isothermal condition.

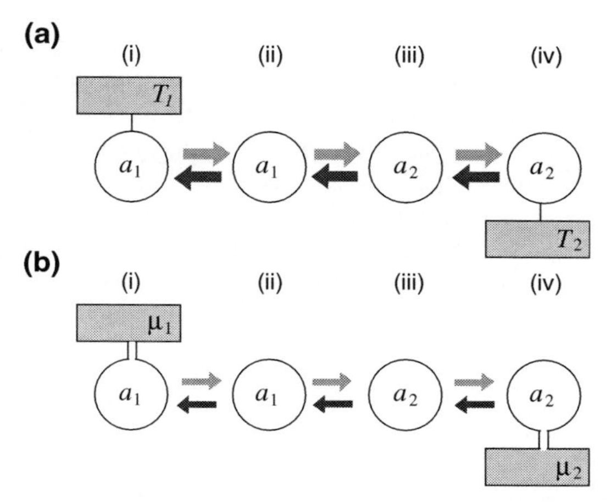

Fig. 7.8 A protocol including the detachment and attachment with heat baths (**a**) or with a particle reservoirs (**b**). The numbers, (*i*), etc., refer to those in the text. The system's parameter a is changed in the processes (*ii*)$\rightleftharpoons$(*iii*), where the system is isolated from these environments. (The figure is modified from Fig. 1 of [13])

2. The ensemble average of W is nonnegative even in the limit of slow operations.[15] Only for special class of systems (see below) the average work $\langle W \rangle$ can tend to 0 in the limit of slow operations.

A similar problem can be set up for the open system. See Fig. 7.8(b). Here the operations of detachment–attachment are done with the particle environments of different chemical potentials μ_1 and μ_2. And the operation of (ii)$\rightleftharpoons$(iii) is quasistatic isothermal process without particle exchange. The value of a_2 is chosen so that, upon the making contact with the environment of chemical potential μ_2, no particles is exchanged *on the average* between the system and the new environment. Again the average work $\langle W \rangle$ is positive, except for the special systems where $\langle W \rangle$ tends to 0.

7.1.3.2 Analysis of the Operations

Below we will analyze the elementary physical processes included in this intriguing cycle. As for the proof of the above statements, interested readers may consult [13, 14]. The paper uses the Kullback–Leibler distance defined for the ensemble of processes.

Slow Detachment from an Environment

This is essentially a nonquasistatic process. Although the associated work can be made negligibly small, the system's state becomes constrained after this operation:

---

[15] Roughly speaking, the system's energy at the end of return operation, (iii)→(ii), is on average higher than the initial thermalized state. Therefore, during the operation (ii)→(i) the heat $\langle (-Q) \rangle = \langle W \rangle$ is dissipated to the thermal environment.

in Fig. 7.8(a) the system's energy E is constrained to obey $dE = dW$ because the heat exchange is blocked, and in Fig. 7.8(b) the number of particles n is fixed. The work to change the parameter a depends on these fixed values. Therefore, W fluctuates from one realization to the other, however, slowly the operations are done.

Just after the detachment the distributions of E in Fig. 7.8(a) or n in Fig. 7.8(b) obey, respectively, the canonical or grand canonical distributions if the detachment is slow enough. In fact, the end of the detaching process (either (i)→(ii) or (iv)→(iii)) is very unpredictable for the external system. Therefore, the values of E or n at this last moment will be found with the probability proportional to their residence–time distribution. If the detaching process from an environment is quasistatic except at the last moment, then this distribution is the Gibbs statistical weight at the temperature or chemical potential of the (detaching) environment.

Adiabatic Process or Process Without Particle Exchange

The operations between (ii) and (iii) are deterministic in the energy E for Fig. 7.8(a) and in the particle number n for Fig. 7.8(b). For (a) the so-called adiabatic invariant remains constant during this process, while for (b) it is n that is invariant. In the course of this process, the probability distributions of these quantities are transformed accordingly from the initial Gibbs statistical weight to another distributions which depend on a.

At the end of these operations, distributions of E or n are generally different from the Gibbs statistical weight for the system at any T or μ, respectively. The exceptions are those systems in which (i) E or n take only two quantized values or (ii) the potential energy $U(x, a)$ takes the form of $k(a)|x|^{\alpha}$ with α a constant.[16] Generally, the system's E or n are brought to non-Gibbs distribution.

Attachment with a New Equilibrium Environment

This operation renders the above mentioned non-Gibbs distributions of energy E or number of particles n to the Gibbs statistical weight appropriate for the new environment. This process is irreversible.

We could avoid this irreversibility if there is a macroscopic and reversible operation that transforms the non-Gibbs distribution to a Gibbs distribution *before* the operation of attachment. Except for the special systems [15] there is no such protocol. The key issue is that it seems impossible to select reversibly those systems which have a prescribed energy.[17] Let us call such hypothetical reversible cycle *selecting operation*.

---

[16] For quantum system, this condition leads to the relations among the energy levels, $E_{\nu}(a') - E_0(a') = \phi(a, a')[E_{\nu}(a) - E_0(a)]$, where $\phi(a, a')$ is a given function. For the open system the list of exceptional cases is not exhaustive.

[17] A similar argument should be made for the open system.

The argument against this operation is as follows [16]. There is a concrete reversible operation to extract positive work from an initially microcanonical ensemble (i.e., the systems having the same energy).[18] Let us call this operation the *work-extracting operation*. Now this operation is incompatible with the existence of the selecting operation. In fact, if the both operations could be done reversibly, we could (i) detach a system from a thermal environment, (ii) use the selecting operation to check if the system has a prescribed energy, then (iii) if yes, apply the work-extracting operation to extract a work, or if otherwise, do nothing, and finally (iv) reattach the system to the original thermal environment. In this way we could constitute a perpetual machine of the second kind.

In conclusion, at least to our present understanding, the cyclic process in Fig. 7.8 is generally unavoidably irreversible because of an essentially nonquasistatic process of attachment to the environment.

7.2 Detection and Control Under Fluctuations

*7.2.1 * Two Types of Error-Free and Unbiased Detection Under Fluctuation are Possible*

Detection is a process that correlates the exterior of a system to the system's state. We focus on the detection of a particle coming randomly from outside to the system. Signal particles undergo random motion in the environment. When one of them happens to arrive at the detection site of a sensor, and if the sensor establishes a correlation with this event, the sensor acquires a signal as information.

Examples of signal detection are found everywhere. In biology, some motor proteins work at very good efficiency of free-energy conversion.[19] Those motor proteins must not spend a lot of energy source in the detection of the fuel particle (e.g., ATP) or of the object of work (e.g., actin filament). Some gene regulations are very precise, while they use highly irreversible reactions with kinase or phosphatase, etc. In optics, a CCD camera uses photoelectric transitions to detect light signals, and its signal-to-noise ratio depends on the temperature.

Now, questions are

(1) Is it possible to realize a mesoscopic single-particle sensor which works correctly without bias or errors under thermal fluctuations?
(2) If it is possible, what aspects of the usual sensors in the macroscopic world should be abandoned?

These are the main themes of this section.

---

[18] For the canonical ensemble, the similar operation immediately contradicts the second law of thermodynamics. But it is possible for microcanonical ensemble.

[19] Some claims that the motor proteins like F1ATPase are working at about 100% efficiency.

7.2.1.1 Half-Sensors

Let us distinguish the two notions about the detection, which we call IN/OUT and ON/OFF. A signal particle is said to be IN if it is present at the "detection site" of the sensor, and otherwise it is OUT. The sensor is assumed to take two internal states ON and OFF. The ON [OFF] states of the sensor are designed to be positively correlated to the signal particle's state, IN [OUT], respectively. IN and OUT are the objective fact about the presence or absence of the particle, while ON and OFF are the perception by the sensor about the particle.

Figure 7.9(a) shows those correlations in the ordinary sensor. The two shaded zones are prohibited to occur and these constraints allow one-to-one correspondence between the sensor's states and the arrival of signal particles.

In the mesoscopic scale, the unbiased and error-free detection is still possible if we sacrifice one of these two prohibition zones. We define two types of half-sensors by the tables Fig. 7.9b, c. In response to the particle's state IN/OUT, these sensors are multivalued functions in the range ON/OFF. The "half-absence sensor" can have the state OFF only when the particle is OUT. Thus if the sensor thinks that the particle is OFF, the signal particle is surely OUT without bias or errors. But this "absence" sensor can miss the particle's absence OUT by thinking it be IN. The "half-presence sensor" works completely complementary manner to the "half-absence sensor."

The half-sensors can be designed to work reversibly (see below).[20] The possibility of the usual sensor (Fig. 7.9a) on the mesoscopic scale is not known. A usual sensor can be made as the composite of Fig. 7.9b, c. This corresponds to the product of two Boolean variables. Since the product of two Boolean variables is an irreversible operation, the function of such composite sensor might need some free-energy resource (see [17]).

The utilities of the two types of half-sensors are different. For example, if the signal particle is "toxic" for a system, the system can use the OFF sign of the half-absence sensor. This OFF signal is analogous to the regulation of gene transcription by a repressor protein [18], since the transcription is surely prohibited

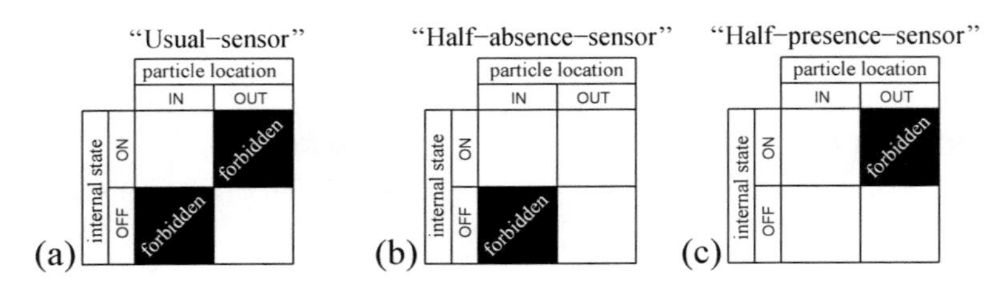

Fig. 7.9 The functions of the usual sensor (**a**) and the two kinds of half-sensors (**b**) and (**c**). The *black squares* indicate the forbidden situations

[20] Therefore, the half-sensors are different from "to know," which is irreversible and essentially nonquasistatic process.

by the repressor protein bound to the operator site on the DNA. By contrast, if the particle is the "food" for a system, it can use the ON sign of the half-presence sensor. This is analogous to the activation of the repressor protein, since the activation of the repressor surely requires the signal particle. Also the uptake of ATP molecule by a molecular motor will be analogous to the half-presence sensor.

7.2.1.2 Construction Principles of the Half-Sensors

On the scale of thermal fluctuations, the only way to avoid the errors of detection due to thermal noises is to use the high (free) energies of interaction with respect to $k_B T$. At the same time, the total free energy of the sensor *plus* the signal particle should be unbiased upon detection. In order to meet these two conditions at the same time, there are two principal ways. The half-absence sensor uses the short-range steric (repulsive) interaction, while the half-presence sensor uses the compensation between strong attraction and strong restoring force. Below we use a 1D representation for the particle's position x and use a sensor's state variable a as the second coordinate.

Interaction Potential of the Half-Absence Sensor

Steric interaction between two bodies excludes the coexistence of these two at the same position. For example, if a coffee cup and a coffee pot cannot be put on a saucer at the same time, the coffee cup on the tray implies the absence of the coffee pot. The half-absence sensor uses this principle.

We assume that the position of a signal particle, x, can move in a half space $0 \leq x < \infty$ and that $x = 0$ corresponds to the detection site (Fig. 7.10 *Left*). The state variable of the sensor, a, represents the leftmost point of a movable object (thick bars in Fig. 7.10 *Left*). a is allowed to move in the region of $-1 \leq a \leq 0$. A steric interaction is assumed between this object and the particle (thick dot). Then the signal particle is surely in the OUT state whenever the movable object is in the OFF state.[21] By construction, the total energy of the sensor plus the signal particle is constant over the allowed region on the reaction plane (shaded in Fig. 7.10 *Right*).

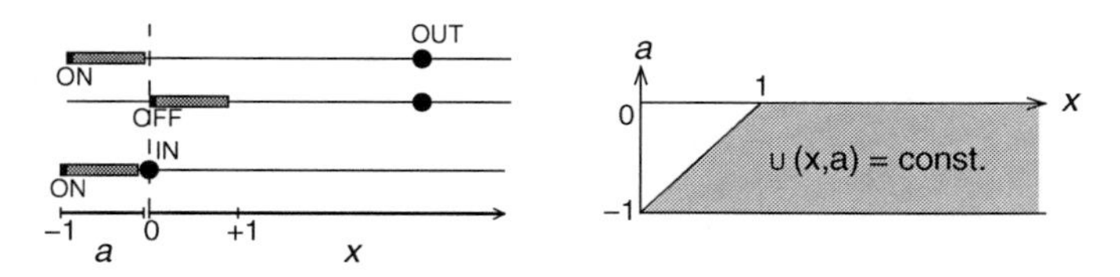

Fig. 7.10 *Left:* Schematic representation of the half-absence sensor. The *thick bar* (a: the *leftmost* position) and the *thick dot* (x) repel with each other by steric repulsive interaction. *Right:* On the (x, a)-plane the shadowed region is accessible without bias

[21] The total energy of the ligand–sensor system, $U(x, a)$, is written as $U(x, a) = U_0$ for $x \geq a+1$ and $U(x, a) = \infty$ otherwise. The sensor's states are assigned as OFF for $a \geq -1+\delta$ and as ON for $-1 \leq a < -1+\delta$, respectively, with a small $\delta > 0$.

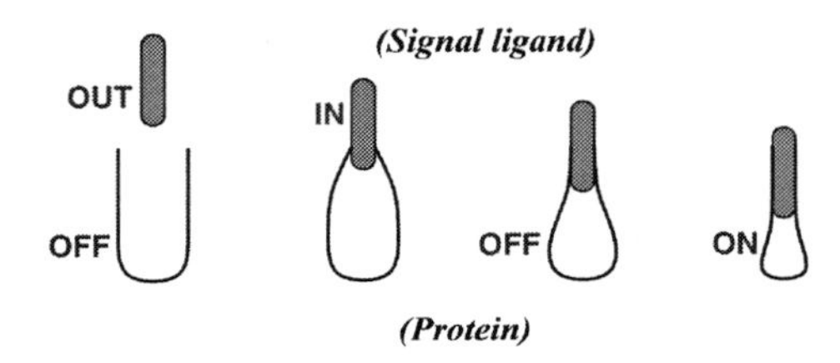

Fig. 7.11 The notion of *induced fit*. As the signal ligand is bound progressively by the protein, the protein deforms accordingly

Interaction Potential of the Half-Presence Sensor

The notion of the compensation of energies has long been discussed in biology [19]. Suppose that a signal ligand (particle) can gradually make strong attractive interactions with the sensor protein. Suppose also that this interaction causes the sensor to deform with a large free-energy cost. [22] See Fig. 7.11. Koshland called this deformation *induced fit* [19]. Now if the gain of the attractive interaction energy is almost compensated by the cost of sensor's deformation up to $\sim k_{\mathrm{B}}T$, the deformation of the sensor surely indicates the presence of the ligand particle without bias. In this case the induced fit realizes the half-presence sensor with a being the sensor's deformation.

Figure 7.12 *Left* shows schematically the relation between the signal particle x and the sensor's state a. If the signal particle is present ($x \sim 0$), the short-range attraction between x and a comes into action while the displacement of a from the resting position $a = -1$ costs deformation energy. As a result of the energy compensation, there appears an unbiased corridor of total free energy (shadowed region in Fig. 7.12 *Right*).[23] More precisely, the sensor's state is ON for $a \geq -1+\delta$

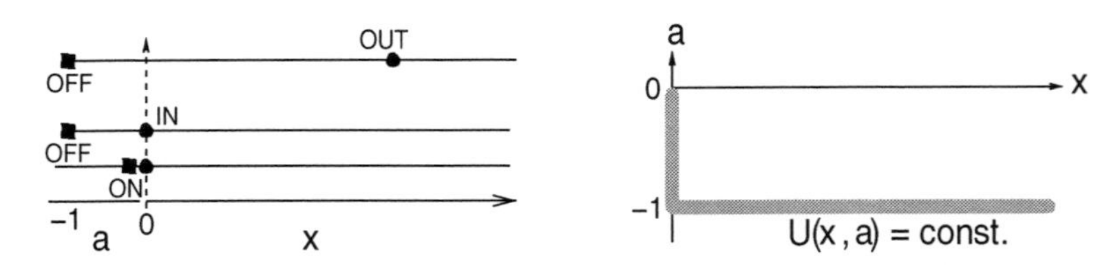

Fig. 7.12 *Left:* Schematic representation of the half-presence sensor. The *filled square* (a: sensor "tip") and the *thick dot* (x) attract each other for $|x-a| \leq 1$, while the displacement in the $a > -1$ region provokes a restoring force. *Right:* On the (x, a)-plane the shadowed region is accessible without bias

[22] The cost may be either energetic such as of mechanical deformations or entropic such as of folding. The detail of the process may contain the aforementioned jump-and-catch process.

[23] The total energy of the ligand–sensor system, $U(x, a)$, is the sum of short-range strong attractive interaction energy, $-M\phi(a-x)$ $(M \gg k_{\mathrm{B}}T)$, between the ligand particle and the sensor "tip" and also strong restoring potential $M\phi(a)$. Here $\phi(z)$ $(-\infty < z \leq 0)$ is $\phi(z) = 0$ for $z \leq -1$ and monotonically increasing from $\phi(-1) = 0$ to $\phi(0) = 1$.

and OFF for $-1 \leq a < -1 + \delta$, respectively, where δ is determined as the position where the deformation energy exceeds some $k_{\mathrm{B}}T$.[24]

Remarks

Discretization of state and the half-presence sensor: In looking at Fig. 7.12, one might wonder if the sensor is actually bijective, like Fig. 7.9a. However, the discretization of the sensor's state and the particle's position requires the *gray* zone of (IN, OFF), because the judgement of OFF→ON requires the presence of particle (IN) so that an energy more than $k_{\mathrm{B}}T$ is exchanged between the attracting potential and the restoring potential.

Timescale of sensing: The diffusion of small signal particles can be too rapid for the sensor to follow adiabatically.[25] This is also the reason to distinguish the particle's position and the sensor's state.

Relation between the half-absence sensor and half-presence sensor: The half-presence sensor is, in some sense, on the basis of the half-absence sensor. In fact, for the latter sensor, the object that interacts sterically with the signal particle (the thick bars in Fig. 7.10 *Left*) must be sensed elsewhere.

Signal sensing must have a consequence: An isolated sensor serves nothing. The state of the sensor must be coupled to its "downstream" mechanisms, e.g., by accelerating an enzyme reaction, modifying the access of another molecule to the system, etc. Through this coupling, the sensing process can become biased and irreversible.[26]

Technical Remarks

Reaction coordinate of the presence sensor: In Fig. 7.12 *right*, the detection of the particle occurs through an unbiased corridor corresponding to the unbiased valley of the total potential energy. Therefore, a single reaction coordinate

$$\tilde{x} \equiv x - a \qquad (0 \leq \tilde{x} < \infty) \tag{7.19}$$

can parametrize this pathway. The inversion to the values of x and a is given by

$$x(\tilde{x}) = \max(\tilde{x} - 1, 0), \quad a(\tilde{x}) = -\min(\tilde{x}, 1). \tag{7.20}$$

By definition $U(x(\tilde{x}), a(\tilde{x}))$ is constant.

---

[24] i.e., $M\phi(-1 + \delta) \sim k_{\mathrm{B}}T$. The consistency of these assignments with the above definition can be verified by scrutinizing the graphs of $M\{\phi(a) - \phi(a - x)\}$ vs. a for various values of x.

[25] Biological sensor may be designed to avoid this.

[26] Sect. 7.2.1.4 is an example.

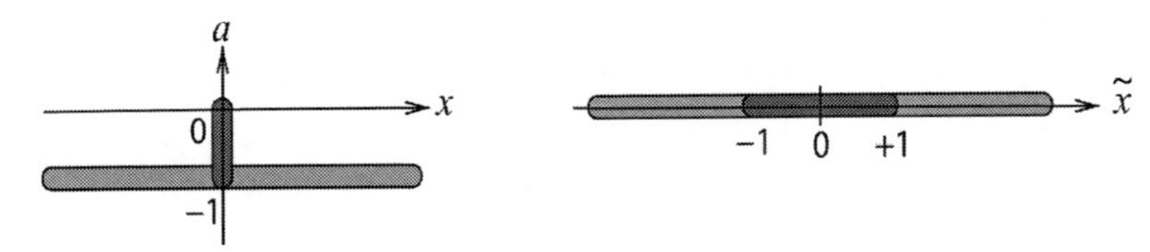

Fig. 7.13 *Left*: the (x, a) pathway between the right environment $(x > 0)$ and the *left* environment $(x < 0)$ through the detection site $x = 0$. *Right*: extended reaction coordinate $\tilde{x}$, where the a coordinate is duplicated to $-1 \leq \tilde{x} \leq 1$

Extension of the reaction coordinate: If a half-presence sensor is accessible from two-particle reservoirs, the two access routes can be distinguished as $x > 0$ and $x < 0$. See Fig. 7.13 *Left*. In that case the reaction coordinate $\tilde{x}$ $(-\infty < \tilde{x} < \infty)$ may represent x and a by

$$x = \frac{\tilde{x}}{|\tilde{x}|}\max(|\tilde{x}| - 1, 0), \quad a = -\min(|\tilde{x}|, 1). \tag{7.21}$$

See Fig. 7.13 *Right*. For $|\tilde{x}| \leq 1$, the pair of points $\{+\tilde{x}, -\tilde{x}\}$ should be identified as a single physical state, $(x(\tilde{x}), a(\tilde{x}))$. This technical trick is convenient for discussing the coupled transport in Chap. 8.

7.2.1.3 Balance of Forces in the Half-Presence Sensor

Apart from the energetic compensation, the mechanical force balance should hold when a half-presence sensor detects a particle. Figure 7.14 illustrates how the forces act among different elements while the sensor undergoes the transition, OFF→ON.

In this figure, the signal particle at the detection site $(x = 0)$ feels the attractive force from the sensor tip ("movable portion of the sensor"). The particle cannot enter into the $x < 0$ region because the "main body of the sensor" pushes back the particle through the steric repulsion. The sensor tip feels both the attractive force from the particle and the restoring force from the main body. These two forces should be

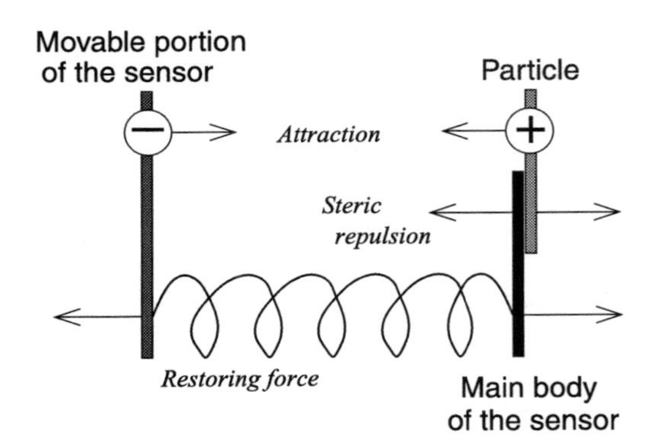

Fig. 7.14 The balance of forces among the three constituents of the half-presence sensor

balanced in an unbiased sensor. The main body of the sensor is pushed to the left by the particle, while it is pushed to the right by the restoring force of the sensor tip.

In terms of the flux of momentum in the x direction, Fig. 7.14 represents a permanent circulation of the momentum [20].[27] A similar mechanical analysis can be done for the induced fit in Fig. 7.11.

7.2.1.4 Functions of Coupled Sensors

If two half-sensors are energetically coupled, their individual functions as sensors are modified, on the one hand, but they can acquire various functions as a composite system. Two examples are discussed below.

Suppose that a composite system has two half-presence sensors, sensor 1 and sensor 2 with (x_1, a_1) and (x_2, a_2) being the pairs of signal particle position and state variable. As the coupling energy between these half-sensors, we take two examples of the quadratic form, $\Lambda(a_1 - a_2)^2$ or $\Lambda(a_1 + a_2 + 1)^2$, where $\Lambda > 0$ is a constant. In the biophysics of proteins, the coupling among different degrees of freedom in a protein is called allosteric coupling, and the effects due to this coupling is called *allosteric effects*. See, for example [21].

Case of $\Lambda(a_1 - a_2)^2$: allosteric transition [22, 23]

The coupling energy favors a synchronized response of the two sensors with $a_1 \simeq a_2$. See Fig. 7.15 *Top*. If the signal particles in the environment visit at random the detection sites, $x_1 = 0$ and $x_2 = 0$, the cooperative transition of the states (a_1, a_2) from $(\mathsf{OFF}_1,\mathsf{OFF}_2)$ to $(\mathsf{ON}_1,\mathsf{ON}_2)$ is likely to occur when the signal particles for both sensors are in the ON positions, $x_1 \simeq x_2 \simeq 0$. Especially when the two half-sensors detect the same species of particles, the probability of cooperative transition depends quadratically on the density of signal particles. Such nonlinear transitions are called *allosteric transition* Fig. 7.16 *Left* represents the process of allosteric transition on the plane of reaction coordinates, $(\tilde{x}_1, \tilde{x}_2)$. [22, 23]. [28]

Case of $\Lambda(a_1 + a_2 + 1)^2$: exchange of binding[5]

The coupling energy $\Lambda(a_1+a_2+1)^2$ takes the minimum value 0 when $a_1+a_2 = -1$. If $\Lambda \gg k_{\mathrm{B}}T$, this system is, therefore, no longer unbiased sensor, but it rather likes to fill particles in at least one of the half-sensors. As we will see below, this system functions as the *exchange of binding* between two allosteric sites.

Suppose that, at present, the system binds only one particle. The state of the composite system is, therefore, $(\mathsf{OFF}_1,\mathsf{ON}_2)$ or $(\mathsf{OFF}_1,\mathsf{ON}_2)$. This state will remain

---

[27] The $+x$-oriented momentum flows toward $+x$ along the attractive force and toward $-x$ along the repulsive (restoring) force.

[28] A similar phenomenon has been observed for the binding of ATP-activated kinesin molecules to a microtubule, called the *cooperative binding*, [24].

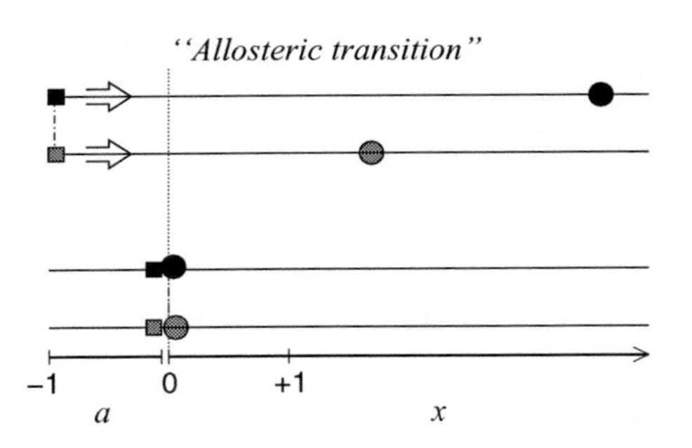

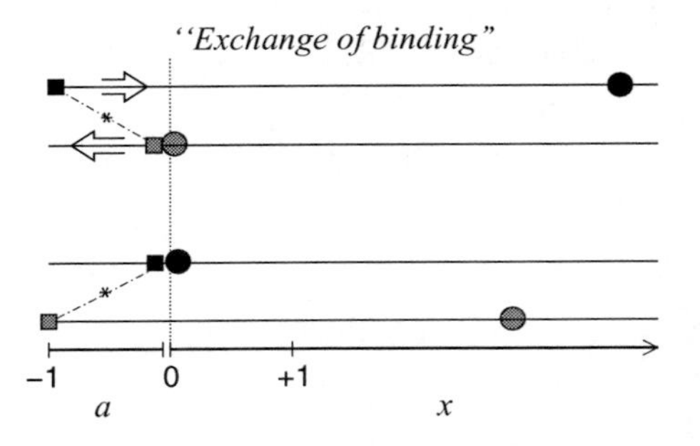

Fig. 7.15 *Top*: Allosteric transition by the coupling $\Lambda(a_1 - a_2)^2$. *Bottom*: Exchange of binding by the coupling $\Lambda(a_1 + a_2 + 1)^2$. For the symbols, see the caption of Fig. 7.12

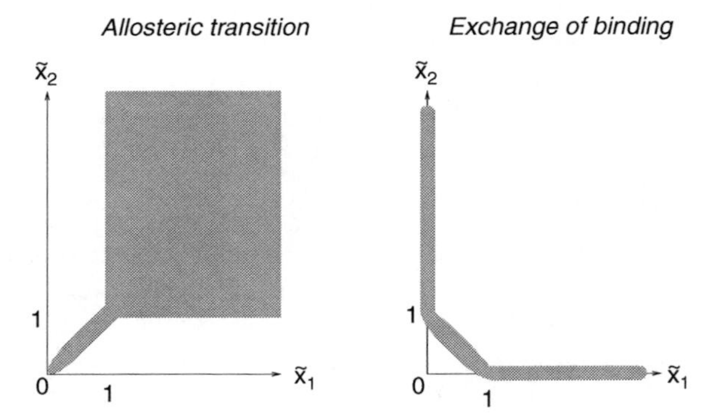

Fig. 7.16 *Left*: Allosteric transition by the coupling, $\Lambda(a_1 - a_2)^2$. *Right*: Exchange of binding by the coupling, $\Lambda(a_1 + a_2 + 1)^2$. $(\tilde{x}_1, \tilde{x}_2)$ are the reaction coordinates (7.19) for the two half-presence sensors. The shaded region is the region of a constant energy

stable until the second particle arrives at the half-sensor in the unoccupied (OFF) state. Once the two half-sensors bind their signal particles, the state variables (a_1, a_2) can diffuse along the line of $a_1 + a_2 = -1$. When (a_1, a_2) arrive either at $a_1 = -1$ or $a_2 = -1$, it is possible that one of the particles leaves the detection site. See Fig. 7.15 *Bottom*.[29] As a result, the switch between (OFF_1,ON_2) and (OFF_1,ON_2) can be realized without passing the (OFF_1,OFF_2) state. Figure 7.16 *Right* represents the process of exchange of binding on the plane of reaction coordinates, $(\tilde{x}_1, \tilde{x}_2)$. If there is a bias between $(a_1, a_2) = (-1, 0)$ and $(0, -1)$, the arrival of the particle with higher "affinity" can expel the previously bound particle by the allosteric effect.

7.2.2 * *The Gates to Control Particle's Access Can be Made Using Adjustable Potential Barriers*

Definition of Gate

Suppose that a detection site of a half-presence sensor is accessible both from the left-particle environment ($x < 0$) and from the right-particle environment ($x > 0$). The gate for these environments is a mechanism such that the signal particles can get access to the detection site exclusively from one of these environments. In the extended reaction coordinate, $\tilde{x}$ (see Fig. 7.13 and (7.21)), the gate implies a blockade either between $(-\infty, -1)$ and $[-1, 1]$ or between $[-1, 1]$ and $(1, \infty)$.

Gate Made by the Potential Energy Barriers

We introduce potential barriers localized around $x = \pm\epsilon$ with a small width $2\epsilon > 0$. In the extended reaction coordinate, the barriers are made around $\tilde{x} = \pm(1 + \epsilon)$, see Fig. 7.17. When only the left barrier is established with the height of $M \gg k_{\mathrm{B}}T$ (Fig. 7.17 *Top*), the access of the particles from the left environment is blocked. The situation is inverted when only the right barrier is established (Fig. 7.17 *Bottom*). The gate is realized by adjusting the heights of these two barriers so that at any time at least one of the barriers is high enough compared with $k_{\mathrm{B}}T$. The irreversible work to raise or lower the potential barriers is negligible for small ϵ (see Sect. 7.1.1.4). The reaction force from the particles against the operation of the barriers is negligible because of the steep gradient of the barrier profile.

Margins of Operation

A small positive number ϵ has been introduced to represent the fact that the potential barriers should be located outside but close to the detection sites. Although the irreversible work of operating the gate can be made negligible, this ϵ, as well as

[29] This is the first passage time (FPT) problem. See Sects. 1.3.3.3.

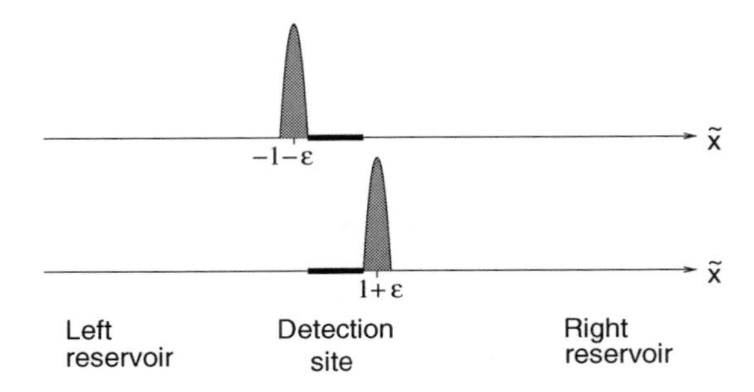

Fig. 7.17 Potential energy barriers of the gate for particles. $\tilde{x}$ is the extended reaction coordinate for the particle. *(Top)* Particles are accessible only from the *right*-particle environment. *(Bottom)* Particles are accessible only from the *left*-particle environment

the small δ introduced in the model of sensors (Sect. 7.2.1), constitute the source of small "leakage" when we construct free-energy transducers.

When we discretize the model in the form of the reaction networks, neglecting this marginal effect can sometimes lead to unphysical results. Especially caution must be maintained when one claims a strictly tight coupling or a 100% efficiency for a model of autonomous free-energy transducer.

7.3 Discussion

7.3.1 Control of Open System Has Both Common and Distinguished Features with Respect to the Control of Closed System

7.3.1.1 Parallelism with the Control of Closed System

The following list summarizes how the results of closed system can be generalized for the open system.

Quasistatic process: Quasistatic work under a fixed chemical potential μ of particle environment obeys the law, $W = \Delta J$, for each realization of process (Sect. 5.2.1.4).

Nonnegativity of the average irreversible work: When the external parameter is changed from a_{i} to a_{f} nonquasistatically, the average irreversible work $\langle W_{\mathrm{irr.}}\rangle \equiv \langle W\rangle - \Delta J$ is nonnegative. The nonequilibrium work relation applied to the entire system (see (5.59)[25]), $\langle e^{-W+\Delta F_{\mathrm{tot}}}\rangle = 1$, can be reduced to the equality for the open system,

$$\langle e^{-W+\Delta J}\rangle = 1, \tag{7.22}$$

where (5.18) is used.[30] By the Jensen inequality, $\langle e^{-x}\rangle \geq e^{-\langle x\rangle}$, we have

$$\langle W\rangle \geq \Delta J. \tag{7.23}$$

[30] The average $\langle\cdot\rangle$ is taken over all realizations starting from a pertinent equilibrium ensemble.

Complementarity relation: The average irreversible work $\langle W_{\mathrm{irr.}} \rangle$ and the time spent for the process, Δt, satisfy the relation parallel to (5.36) [26]:

$$\langle W_{\mathrm{irr.}} \rangle \, \Delta t \geq S(a_\mathrm{i}, a_\mathrm{f}) \qquad (\Delta t \to \infty), \tag{7.24}$$

where $S(a_\mathrm{i}, a_\mathrm{f})$ is independent of the (rescaled) protocol $\tilde{a}(s)$ between a_i and a_f.

Attachment/detachment with/from particle environment: The adiabatic process of the closed system corresponds to the closed isothermal process. The attachment with and detachment from the particle environment are realized by the raising and the lowering of the potential barrier (see Sect. 7.1.1.3). These processes are inevitably nonquasistatic (see Sect. 7.1.1.2).

Irreversible relaxation of probability distribution: See Fig. 7.8b.

"Carnot cycle": "Carnot cycle" for open systems extract work from the transport of particles between the particle environments of different chemical potentials μ_H and μ_L. See Fig. 7.18 below. In the next chapter we will present a concrete model of this cycle. The maximum available work is $(\mu_\mathrm{H} - \mu_\mathrm{L}) \times \langle n \rangle$, where $\langle n \rangle$ is the number of transfered particles.

7.3.1.2 Distinguishing Features of Open System

Basic facts are

(i) Unlike absolute temperature, chemical potential has a no absolute zero and admits indefinitely negative values.
(ii) Unlike a heat engine, the transferred particles of a particle engine do not necessarily carry energy.

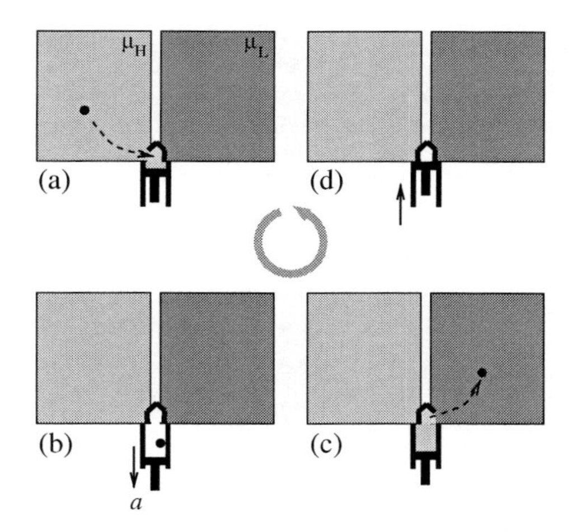

Fig. 7.18 A "Carnot cycle" to extract work. The particle (*thick dot*) enters from a dense particle environment of the chemical potential μ_H (**a**). The piston is pulled while the system is closed (**b**). The particle exits to the dilute particle environment of the chemical potential μ_H (**c**). The piston is pushed while the system is closed (**d**). If the dilute environment is vacuum ($\mu_\mathrm{L} = -\infty$), the infinite work can be extracted in principle

In regard to (i) a vacuum environment has the chemical potential of $\mu = -\infty$. Then thermodynamics implies that the maximum available work is positive infinity. By the fact (ii), the energy for this infinite work should be provided by the thermal environment, not by the transferred particles. See Fig. 7.18 for illustration. In this figure the open system consists of a cylinder furnished with a door and a movable piston. The starting point is a cylinder with the door being open to the particle environment of chemical potential $\mu_{\rm H}$. After a while the door is closed and the cylinder volume V is dilated indefinitely. At this point the cylinder contains particle(s) with a finite probability. The extracted work $(-W)$ can, therefore, be arbitrarily large by a large dilatation.[31] The flow of energy is from the thermal environment to the piston via the kinetic energy of the particles. After this dilation, the door is again opened, but now to the second particle environment of chemical potential $\mu_{\rm L}$, which has no particle ($\mu_{\rm L} = -\infty$). We wait patiently at this stage: Since the particles can only go out from the cylinder but cannot enter nor reenter, the cylinder becomes empty.[32] After closing the door, the next cycle restarts.

References

1. H.S. Leff, A.F. Rex, *Maxwell's Demon: Information, Entropy, Computing* (A Hilger (Europe) and Princeton U.P. (USA), 1990)
2. R. Landauer, IBM J. Res. Dev. **5**, 183 (1961) Reprint: **44**, 261–269 (2000)
3. C.H. Bennett, Int. J. Theor. Phys. **21**, 905 (1982)
4. R.P. Feynman, R.B. Leighton, M. Sands, *The Feynman Lectures on Physics – vol.1* (Addison Wesley, Reading, Massachusetts, 1963), §46.1–§46.9
5. E. Eisenberg, T.L. Hill, Science **227**, 999 (1985)
6. C. Branden, J. Tooze, *Introduction to Protein Structure*, 2nd edn. (Garland Science, New York, 1999), § 13
7. F.J. Kull, R.D. Vale, R.J. Fletterick, J. Muscle Res. Cell Motil. **97**, 877 (1998)
8. M. Born, R. Oppenheimer, Annalen der Physik **84**, 457 (1927)
9. A. Messiah, *Quantum Mechanics* (Dover, Mineola, 1999; reprint)
10. S. Ishioka, N. Fuchigami, Chaos **11**, 734 (2001)
11. K. Shizume, Phys. Rev. E **52**, 3495 (1995)
12. R. Kawai, J.M.R. Parrondo, C. Van den Broeck, Phys. Rev. Lett. **98**, 080602 (2007)
13. K. Sato, K. Sekimoto, T. Hondou, F. Takagi, Phys. Rev. E **66**, 06119 (2002)
14. H. Tasaki, cond-mat/0008420 v2 (2000)
15. I. Procaccia, D. Levine, J. Chem. Phys. **65**, 3357 (1976)
16. K. Sato, J. Phys. Soc. Jpn. **71**, 1065 (2002)
17. R. Landauer, Science **272**, 1914 (1996)
18. B. Lewin, *Genes IX* (Jones and Bartlett, U.S., 2007)
19. D.E. Koshland, Sci. Amer. **229**, 52 (1973)
20. K. Sekimoto, in *Chemomechanical Instabilities in Responsive Materials (NATO Science for Peace and Security Series -A: Chemistry and Biology)*, ed. by P. Borckmans, et al., (Springer, New York, 2009)

[31] For the ideal gas particle, the quasistatic isothermal work $(-W)$ is $(-W) \propto k_{\rm B}T \ln(V_{\rm f}/V_{\rm i})$, with $p \propto k_{\rm B}T/V$ the pressure and $V_{\rm i}$ [$V_{\rm f}$] are the initial and final values of volume, respectively.

[32] It may take a long time $\propto V_{\rm f}$.

21. B. Alberts et al., *Essential Cell Biology*, 3rd edn. (Garland Pub. Inc., New York & London, 2009)
22. J. Monod, J. Wyman, J.P. Changeux, J. Mol. Biol. **12**, 88 (1965)
23. D.E. Koshland, G. Nemethy, D. Filmer, Biochemistry **5**, 365 (1966)
24. E. Muto, H. Sakai, K. Kaseda, J. Chem. Biol. **168**, 691 (2005)
25. C. Jarzynski, Phys. Rev. Lett. **78,** 2690 (1997)
26. T. Shibata and S. Sasa, J. Phys. Soc Jpn. **67,** 2666 (1998)

Chapter 8
Free-Energy Transducers

Free-energy transduction on the mesoscopic scales is a theoretical and experimental challenge. Biological systems have machineries such as molecular motors, ion pumps, or polymerases which work through free-energy transduction. Small-scale technologies such as microfluidics, nanomachines, or biomimetic sciences have developed various artificial free-energy transducers.

There are two main categories of free-energy transduction: *externally controlled transduction* and *autonomous transduction*.

Miniaturized Carnot cycles [1] and their analog in the open system belong to the first category in the mesoscopic scale. Certain models of this type can attain the maximum efficiency of free-energy transduction (i.e., ideal Carnot efficiency) although finite (positive or negative) work of control intervenes along the cycle of process.

Feynman's pawl and ratchet wheel [2] and the Büttiker–Landauer ratchet [3, 4] (Sects. 1.3.4 and 4.2.2) belong to the second category in the mesoscopic scale. All biological machineries are basically autonomous transducers. One can design a mesoscopic pump which converts purely entropic free energy into another type of purely entropic free energy. Autonomous free-energy transducers do not attain the ideal efficiency limited by the second law of thermodynamics because of the intrinsic "leakage" associated with autonomous control. Understanding the structure–function relationship of the protein motors is among the future goals of mesoscopic sciences.

8.1 Externally Controlled Free-Energy Transducers

8.1.1 Mesoscopic Carnot Cycle Can Be Ideal Despite Finite Works of Control

Macroscopic thermodynamics is based on fundamental experimental facts. Carnot's experiment[1] and his analysis is among them. The study of Carnot cycle as a

[1] Sadi Carnot (1796–1832).

Sekimoto, K.: *Free-Energy Transducers*. Lect. Notes Phys. **799**, 255–279 (2010)
DOI 10.1007/978-3-642-05411-2_8

macroscopic thermodynamic system would be, therefore, a kind of tautology or self-reference. As for stochastic energetics, the framework is based on the dynamical model (the Langevin equation, etc.) and the detailed balance condition (Einstein relation). The study of Carnot cycle on the mesoscopic scale is, therefore, not a trivial job. The objects of this subsection are:

1. to understand that the works of control are finite but in principle recoverable,
2. to see that Carnot's maximal efficiency (Sect. 2.3.3), i.e., $\eta_{\rm rev} = (T_h - T_\ell)/T_h$, is attainable for an externally controlled mesoscopic heat engine, and
3. to notice that the efficiency of individual cycle fluctuates even above, $\eta_{\rm rev}$.[2]

8.1.1.1 * Composition and Control of the System

System Setup

A model of mesoscopic Carnot engine is schematically given in Fig. 8.1 [1]. The main system is a harmonic oscillator of mass m and spring constant k. It is an analog of the ideal gas in the sense that the average energy is $k_{\rm B}T$ independent of the "volume," $k^{-1/2}$. The energy of the main system is written as

$$H(x, p; k) = p^2/(2m) + kx^2/2,$$

where x and p are, respectively, the position and momentum of the oscillator. In this thought experiment we will not talk about how to isolate the main system. We will verify below that the usage of the harmonic potential avoids irreversible work upon relaxation from non-Gibbs ensembles of fluctuations (Sect. 7.1.3).

We use the two "clutches" which have been introduced in Sect. 7.1.1.5 (cf. Fig.7.4). The clutch $\phi_{\rm H}(x - y_{\rm H}, \chi_{\rm H})$ controls the contact with the high-temperature heat bath $T_{\rm H}$, and the other clutch $\phi_{\rm L}(x - y_{\rm L}, \chi_{\rm L})$ controls the contact with the low-temperature heat bath $T_{\rm L}$. We denote by $y_{\rm H}$ and $y_{\rm L}$ the auxiliary variables

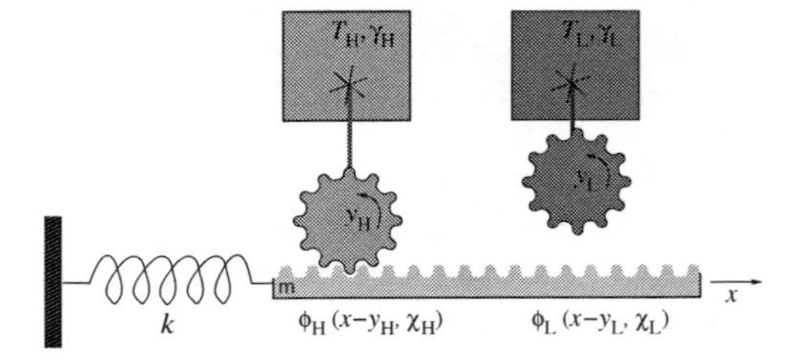

Fig. 8.1 Schematic composition of a mesoscopic Carnot engine. (Adapted from Fig. 1 of [1])

[2] About this fact, people sometimes say that the second law of thermodynamics is locally broken. If is so, it applies also to the free Brownian motion, whose local unidirectional motion takes place against the viscous force.

corresponding to the two clutches.[3] The system of stochastic equation of motion is written as follows:

$$\frac{dx}{dt} = \frac{p}{m}, \qquad \frac{dp}{dt} = -kx - \frac{\partial \phi_H}{\partial x} - \frac{\partial \phi_L}{\partial x}, \tag{8.1}$$

$$\gamma_H \frac{dy_H}{dt} = -\frac{\partial \phi_H}{\partial y_H} + \xi_H(t), \qquad \gamma_L \frac{dy_L}{dt} = -\frac{\partial \phi_L}{\partial y_L} + \xi_L(t), \tag{8.2}$$

where the thermal random forces $\xi_H(t)$ and $\xi_L(t)$ are white Gaussian processes with zero mean and correlation ($\alpha, \beta = \mathrm{H}$ or L):

$$\langle \xi_\alpha(t)\xi_\beta(t')\rangle = 2\gamma_\alpha T_\alpha \delta_{\alpha\beta}\delta(t-t').$$

The parameters χ_H and χ_L are introduced as in Sect. 7.1.1.5, so that $\chi_\alpha = 0$ ($\alpha = \mathrm{H}$ or L) implies that the system is isolated from the heat bath of the temperature T_α. The Carnot cycle is controlled by the three external parameters (k, χ_H, χ_L).

Protocol of Control

A cycle of the operation for the parameters is depicted in Fig. 8.2. As the macroscopic Carnot cycle, two isothermal processes and two adiabatic processes (along the k-axis) alternate. However, at every instance, there intervenes an explicit control

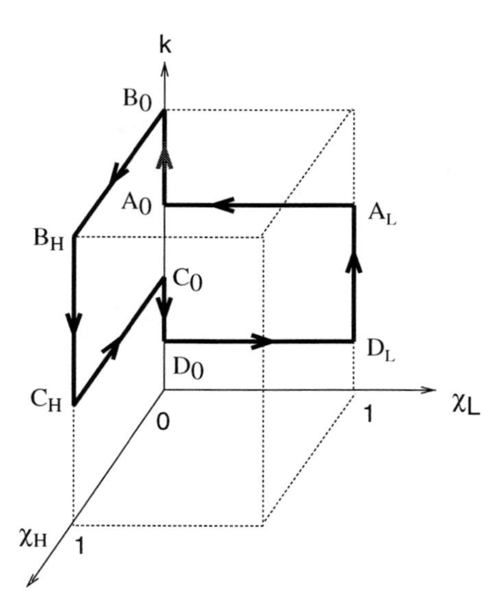

Fig. 8.2 Protocol of control for the mesoscopic Carnot cycle (*arrowed loop*). Horizontal paths (k =const.) correspond to the attachment and detachment with thermal environments. (Adapted from Fig. 2 of [1])

[3] Unlike Sect. 7.1.1.5, we explicitly assumed that the clutches depend on the differences of the variables, $x - y_H$ or $x - y_L$.

process of attachment/detachment with the thermal environments (the processes with k =const.).

8.1.1.2 * Energetics and Efficiency per Cycle

Attachment/Detachment with Thermal Environments

Hereafter we ignore the irreversible work associated with the essentially nonquasistatic character of the attachment and detachment with the thermal environments (Sect. 7.1.1.5). The reversible works for the quasistatic change of the clutches are summarized as follows:

$$
\begin{aligned}
W(A_{\rm L} \to A_0) &= F_{\rm L}(T_{\rm L}, k_{\rm A}, 0) - F_{\rm L}(T_{\rm L}, k_{\rm A}, 1),\\
W(D_0 \to D_{\rm L}) &= F_{\rm L}(T_{\rm L}, k_{\rm D}, 1) - F_{\rm L}(T_{\rm L}, k_{\rm D}, 0),\\
W(B_0 \to B_{\rm H}) &= F_{\rm H}(T_{\rm H}, k_{\rm D}, 1) - F_{\rm H}(T_{\rm H}, k_{\rm D}, 0),\\
W(C_{\rm H} \to C_0) &= F_{\rm H}(T_{\rm H}, k_{\rm D}, 0) - F_{\rm H}(T_{\rm H}, k_{\rm D}, 1),
\end{aligned} \tag{8.3}
$$

where the Helmholtz free energy $F_\alpha(T_\alpha, k, \chi_\alpha)$ is defined by ($\beta \equiv (k_{\rm B} T_\alpha)^{-1}$):

$$
\begin{aligned}
e^{-F_\alpha/k_{\rm B}T} &= \int dx \int dp \int dy_\alpha e^{-\beta_\alpha(H(x,p;k)+\phi_\alpha(x-y_\alpha,\chi_\alpha))}\\
&= \int dx \int dp e^{-\beta_\alpha H(x,p;k)} \int dz e^{-\beta_\alpha \phi_\alpha(z,\chi_\alpha)}.
\end{aligned} \tag{8.4}
$$

In order to go to the second line, we have replaced y_α by $x - z$. The results (8.3) hold with probability 1 for each realization. Substituting (8.4) for $F_\alpha(T_\alpha, k, \chi_\alpha)$ in (8.3), we find that *the reversible works to control the clutches cancel after a cycle*:

$$
\begin{aligned}
W(A_{\rm L} \to A_0) + W(D_0 \to D_{\rm L}) &= 0,\\
W(B_0 \to B_{\rm H}) + W(C_{\rm H} \to C_0) &= 0.
\end{aligned} \tag{8.5}
$$

Adiabatic Processes

In the limit of slow control of the external parameters, the only source of stochasticity is the detachment from the thermal environment. Upon detachment from the heat bath of temperature T, the energy of the main system is chosen randomly according to the canonical equilibrium distribution of this temperature. The probability density of the energy, $\mathcal{P}_{\rm eq,e}(E;T)$, is obtained as

$$
\begin{aligned}
\mathcal{P}_{\rm eq,e}(E;T) &= \text{const.} \int dx \int dp\, e^{-H(x,p;k)/k_{\rm B}T}\, \delta(E - H(x,p;k))\\
&= \frac{1}{k_{\rm B}T} e^{-E/k_{\rm B}T}.
\end{aligned} \tag{8.6}
$$

After the detachment, the oscillator's energy E changes with the quasistatic change of parameter k. Classical mechanics says [5] that the so-called the action integral, $I(E,k)$ remains constant (*adiabatic invariant*) during this process. Here $I(E,k)$ is defined by

$$I(E,k) \equiv \int_{H(x,p;k)\leq E} dqdp = \frac{E}{2\pi}\sqrt{\frac{m}{k}}. \tag{8.7}$$

Therefore, the energy of the main system E should change as

$$E \to E' = \sqrt{k'/k}\,E \tag{8.8}$$

when the spring constant k is changed quasistatically, $k \to k'$. Then the work done to the oscillator during an adiabatic process $k \to k'$ is

$$W_{\mathrm{ad}} \equiv E' - E = \left[\sqrt{k'/k} - 1\right] E. \tag{8.9}$$

The average of the adiabatic works, $W_{\mathrm{ad}}(\mathrm{A_0} \to \mathrm{B_0})$ and $W_{\mathrm{ad}}(\mathrm{C_0} \to \mathrm{D_0})$, is, therefore, given by (8.9) and (8.6):

$$\langle W_{\mathrm{ad}}(\mathrm{A_0} \to \mathrm{B_0})\rangle = T_{\mathrm{L}}\left[\sqrt{\frac{k_{\mathrm{B}}}{k_{\mathrm{A}}}} - 1\right], \quad \langle W_{\mathrm{ad}}(\mathrm{C_0} \to \mathrm{D_0})\rangle = T_{\mathrm{H}}\left[\sqrt{\frac{k_{\mathrm{D}}}{k_{\mathrm{C}}}} - 1\right]. \tag{8.10}$$

As for the energy distribution (8.6), the quasistatic adiabatic process (8.8) preserves the initial form of Gibbs equilibrium distribution with only a shift in temperature[4]:

$$\mathcal{P}_{\mathrm{eq,e}}(E;T) \to \mathcal{P}_{\mathrm{eq,e}}(E';\sqrt{\frac{k'}{k}}T). \tag{8.11}$$

This result shows that the irreversible work of attachment to the new thermal environment (Sect. 7.1.3) can be completely avoided by a proper choice of the ratio k'/k. Such a choice corresponds, naturally, to the adjustment of the temperature parameters:

$$\sqrt{\frac{k_{\mathrm{B}}}{k_{\mathrm{A}}}}T_{\mathrm{L}} = T_{\mathrm{H}} \quad (\mathrm{A_0} \to \mathrm{B_0}), \qquad \sqrt{\frac{k_{\mathrm{D}}}{k_{\mathrm{C}}}}T_{\mathrm{H}} = T_{\mathrm{L}} \quad (\mathrm{C_0} \to \mathrm{D_0}). \tag{8.12}$$

[4] The energy distribution after the adiabatic process, $\mathcal{P}'(E')$, should satisfy, $\mathcal{P}_{\mathrm{eq,e}}(E;T)dE = P'(E')dE'$. Substituting $E' = \sqrt{k/k'}\,E$ into this equation, we obtain (8.11).

By these choices, there is no net energy transfer upon reattachment with heat baths. Also, (8.12) leads to the cancellation of the average of the total adiabatic works (see (8.10):

$$\langle W_{\rm ad}({\rm A_O} \to {\rm B_O})\rangle + \langle W_{\rm ad}({\rm C_O} \to {\rm D_O})\rangle = 0. \tag{8.13}$$

Isothermal Process

The quasistatic isothermal work to change k is given as the change in the Helmholtz free energy, $F_\alpha(T_\alpha, k, \chi_\alpha)$ (Sect. 5.2.1.1). From the form of (8.4), we have

$$W_{\rm H}({\rm B_H} \to {\rm C_H}) = \frac{T_{\rm H}}{2}\log\left(\frac{k_{\rm C}}{k_{\rm B}}\right), \quad W_{\rm L}({\rm D_L} \to {\rm A_L}) = \frac{T_{\rm L}}{2}\log\left(\frac{k_{\rm A}}{k_{\rm D}}\right), \tag{8.14}$$

with probability 1. Using the relations (8.12) we have

$$W_{\rm H}({\rm B_H} \to {\rm C_H}) + W_{\rm L}({\rm D_L} \to {\rm A_L}) = \frac{T_{\rm H} - T_{\rm L}}{2}\log\left(\frac{k_{\rm C}}{k_{\rm B}}\right) \ (< 0). \tag{8.15}$$

The result is negative since the engine does work to the external system.

* Efficiency per Cycle

The average Carnot efficiency $\eta_{\rm av}$ is defined by the ratio of the extracted work per cycle (i.e., the work with minus sign) to the heat transmitted to the system from the high-temperature heat bath, $Q_{\rm H}({\rm B_H} \to {\rm C_H})$. This heat is also a random variable because of the essentially nonquasistatic character of the detachment process. The law of energy balance (Sect. 4.1.2) for the isothermal process ${\rm B_H} \to {\rm C_H}$ is

$$\Delta E({\rm B_H} \to {\rm C_H}) = Q_{\rm H}({\rm B_H} \to {\rm C_H}) + W_{\rm H}({\rm B_H} \to {\rm C_H}).$$

The average energy $\langle E({\rm B_H} \to {\rm C_H})\rangle = \frac{\partial(\beta F_{\rm H})}{\partial\beta}$ does not depend on k (see (8.4)), and therefore, $\langle \Delta E({\rm B_H} \to {\rm C_H})\rangle = 0$. Then

$$\langle Q_{\rm H}({\rm B_H} \to {\rm C_H})\rangle = -W_{\rm H}({\rm B_H} \to {\rm C_H}) = \frac{T_{\rm H}}{2}\log\left(\frac{k_{\rm B}}{k_{\rm C}}\right) \ (> 0). \tag{8.16}$$

We, therefore, have the efficiency:

$$\eta_{\rm av} = \frac{|W_{\rm H}({\rm B_H} \to {\rm C_H}) + W_{\rm L}({\rm D_L} \to {\rm A_L})|}{\langle Q_{\rm H}({\rm B_H} \to {\rm C_H})\rangle} = \frac{T_{\rm H} - T_{\rm L}}{T_{\rm H}} \tag{8.17}$$

in the limit of slow operation of mesoscopic Carnot cycle. In conclusion, the mesoscopic Carnot cycle can attain Carnot's maximal efficiency. The works of the control are nonzero but they cancel after an entire cycle.

8.1.1.3 Fluctuation of the Efficiency of Individual Cycle

On mesoscopic scales the Carnot efficiency of an individual cycle can exceed the ideal limit, $(T_{\mathrm{H}} - T_{\mathrm{L}})/T_{\mathrm{H}}$. Let us introduce the "gain" of work, $\hat{\Delta}$, of an individual cycle:

$$\begin{aligned}\hat{\Delta} \equiv &-[W_{\mathrm{ad}}(\mathrm{A_O} \to \mathrm{B_O}) + W_{\mathrm{H}}(\mathrm{B_H} \to \mathrm{C_H}) + W_{\mathrm{ad}}(\mathrm{C_O} \to \mathrm{D_O}) + W_{\mathrm{L}}(\mathrm{D_L} \to \mathrm{A_L})] \\ &-Q_{\mathrm{H}}(\mathrm{B_H} \to \mathrm{C_H})\frac{T_{\mathrm{H}} - T_{\mathrm{L}}}{T_{\mathrm{H}}}.\end{aligned} \tag{8.18}$$

The first line on the right-hand side is the net work extracted from the cycle, and the second line is the average extractable work in the ideal case. Therefore, $\langle\hat{\Delta}\rangle = 0$, but in general $\langle(\hat{\Delta})^2\rangle \geq 0$. The question is whether the fluctuations in $\hat{\Delta}$ can break the second law of thermodynamics. To find the answer, we introduce the cumulative gain over n consecutive cycles, $\hat{R}_n \equiv \sum_{i=1}^{n} \hat{\Delta}_i$, where $\hat{\Delta}_i$ is the gain of the ith cycle. The theory of unbiased random walk[5] tells that for some n the value of $\hat{R}_n$ will become positive with probability 1. Can we stop at that moment ($\hat{R}_n > 0$) and repeat the same operation again and again? If we could have $\hat{R}_n > 0$ for an *extensive* number of times with respect to the total time of experiment, then the mesoscopic Carnot cycle constitutes a perpetual machine of the second kind, which is impossible. Again the theory of unbiased random walk (discrete step size $= \pm 1$ with probability 1/2 for each sign) tells that the probability of having $\hat{R}_n \geq 1$ for the first time at $n = (2m-1)$ is $f_{2m} \equiv \frac{(2m-2)!}{2m[2^{m-1}(m-1)!]^2} \sim m^{-3/2}$. This f_{2m} has a long tail such that the mean number of $(2m-1)$ is infinite, $\sum_{m=1}^{\infty}(2m-1)f_{2m} = \infty$. That is, there is no chance to accumulate the gains ($\hat{R}_n > 0$) for extensive times.

8.1.2 *Mesoscopic Open "Carnot Cycle" Can Transform Heat into Work Without Stockage of Energy*

Here we consider the isothermal open "Carnot cycle"[6] that works between two solute particle reservoirs. In Sect. 7.3.1.2 we have already presented a model (see Fig. 7.18). There the principle was directly analogous to the mesoscopic Carnot engine (Fig. 8.1). In both cases heat is absorbed while it does the work, being mediated by the kinetic energy of gas particle. This mechanism using the kinetic energy is, however, not likely to be a prototype of molecular machine, notably for biological motors. Another principle using the binding energy is presented below. In this mechanism the heat is supplied *after* the work is done.

The objects of this section are:

1. to understand that the machinery uses the "*rareness*" of the particles to extract heat from the thermal environment,

---

[5] For example, see [6].

[6] Hereafter we call it simply open Carnot cycle.

2. to see that the machinery can work with the maximal efficiency allowed by the second law of thermodynamics.
3. to verify that there is no need to stock the energy in the system.

8.1.2.1 Composition and Control of the System

System Setup

Figure 8.3 illustrates the setup of the open "Carnot cycle." The main system is similar to the half-presence sensor (Sect. 7.2.1.1) and binds at most one particle.[7] Work is extracted by the "T"-shaped thick rod, which is similar to the sensor "tip" (see Fig. 7.12) of the half-presence sensor. We assume a large binding energy when this tip is close to the particle-binding site, as in Fig. 8.3C. The (negative) binding energy is parameterized by a (see below).

The particles can access the main system from the dense particle reservoir (left box) of chemical potential μ_H[8] and the dilute particle reservoir (left box) of chemical potential μ_L. This access is controlled by two parameters, χ_H and χ_L, as in the mesoscopic Carnot cycle. For $\chi_\alpha = 1$ the particle can access the main system from the α-environment (α = H or L), while for $\chi_\alpha = 0$ access is blocked. For $\chi_H = \chi_L = 0$ the main system is, therefore, closed. The entire system is at the temperature T.

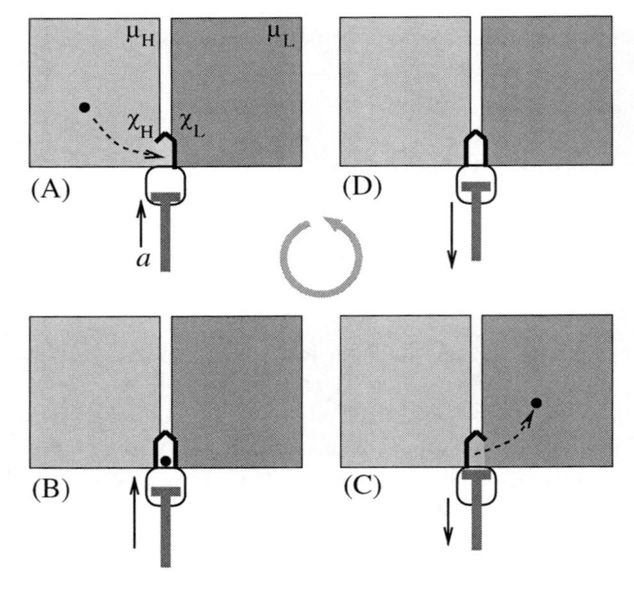

Fig. 8.3 Setup and typical process of an open Carnot cycle. A particle (*thick dot*) migrates from the dense reservoir (*left box*) to the dilute reservoir (*right box*). The system's binding strength (affinity) to the particle is controlled by a. The particle's access is controlled by χ_H and χ_L. (**A**) and (**C**) are open isothermal processes and (**B**) and (**D**) are closed isothermal processes

[7] Steric repulsion between particles is assumed.

[8] In this section we omit the upper suffix "^{res}," which we used in Sect. 3.3.3.

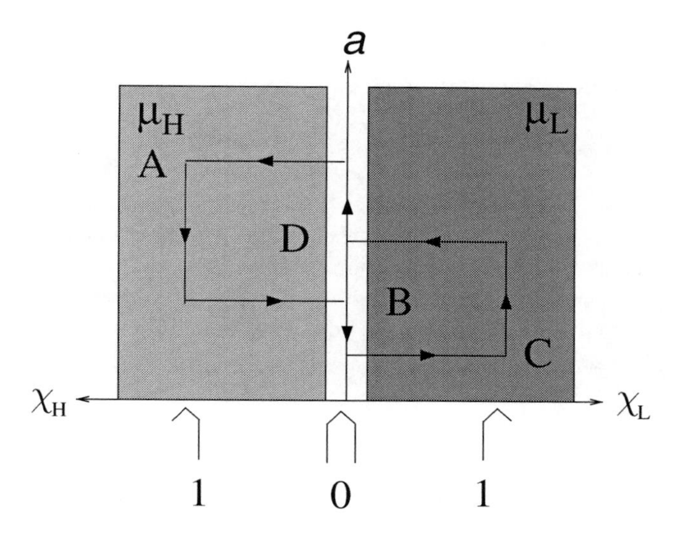

Fig. 8.4 The protocol of the control parameters (*arrowed loop*). A–D correspond to those in Fig. 8.3. Horizontal arrows (a =const.) correspond to the attachment and detachment with particle environments

Protocol of Control

Three parameters (a, χ_H, χ_L) are changed through the protocol of Fig. 8.4, which is essentially the same as Fig. 8.2 of (closed) Carnot cycle. We assume that all the changes are done slowly enough.

The main work is extracted during the stage of binding ((A) and (B) in Fig. 8.4), while heat is supplied later for the unbinding ((C) and (D) in Fig. 8.4). The dense reservoir "pushes" the particles into the main system despite the low binding energy (A), while the dilute reservoir favors the particles diffusing out despite the large binding energy (C).

8.1.2.2 Energetics and Efficiency per Cycle

Work for Detachment/Attachment with Particle Environments

We will not describe explicitly the work associated with the attachment/detachment with the particle environments. It is because we have seen (Sect. 8.1.1.2) that the reversible part of this work sums up to 0 after a cycle of operation, and because the irreversible part of this work, due to the essentially nonquasistatic process, can be made negligibly small by using the slow operations (Sect. 7.1.1.5).

With this understanding, we take the discrete description. The system's state is distinguished by the presence ($n = 1$) or absence ($n = 0$) of particle in the main system.

Definition of the Parameter a and Related Formulas

We define the control parameter a by the negative of the binding free energy of the particle,

$$a \equiv F^{(1)} - F^{(0)} \quad (< 0), \tag{8.19}$$

where $F^{(n)}$ is the free energy of the system (5.22) when it contains n particles. This definition should reflect the position of the sensor "tip" in Fig. 8.3.[9] For the later usage, we summarize several equilibrium properties of the system. When the system is open to the α-reservoir (α = H or L), the equilibrium probability to find a particle in the system, $P_\alpha^{(1)}$, is[10]

$$P_\alpha^{(1)}(a) = \frac{e^{(\mu_\alpha - a)/k_B T}}{1 + e^{(\mu_\alpha - a)/k_B T}} = 1 - P_\alpha^{(0)}(a). \tag{8.20}$$

Also the corresponding complete thermodynamic function of this open system, $J_\alpha(a, T, \mu_\alpha)$ (see (2.10)), is,

$$e^{-\beta J_\alpha} = e^{-F^{(0)}/k_B T} + e^{-(F^{(1)} - \mu_\alpha)/k_B T} \tag{8.21}$$

or[11]

$$\begin{aligned} J_\alpha &= F^{(0)}(a) + k_B T \log\left(1 + e^{\frac{\mu_\alpha - a}{k_B T}}\right) \\ &= F^{(0)}(a) - k_B T \log\left(1 - P_\alpha^{(1)}(a)\right). \end{aligned} \tag{8.22}$$

Isothermal Process as Closed System

By the general result, $W = \Delta F$, for the quasistatic work of closed isothermal process (see (5.12)), the work $W(\mathrm{B} \to \mathrm{C})$ is either $F^{(0)}(a_\mathrm{C}) - F^{(0)}(a_\mathrm{B})$ ($n = 0$ case) or $F^{(1)}(a_\mathrm{C}) - F^{(1)}(a_\mathrm{B})$ ($n = 1$ case). According to the previous analysis of (Sect. 7.1.3.2 *Slow detachment from an environment*), the probabilities for $n = 0$ and $n = 1$ is the equilibrium probabilities immediately before the isolation, i.e., $P_\mathrm{H}^{(0)}(a_\mathrm{B})$ and $P_\mathrm{H}^{(1)}(a_\mathrm{B})$, respectively. The same arguments apply for $W(\mathrm{D} \to \mathrm{A})$. The statistical average of the works is, therefore,

$$\begin{aligned} \langle W(\mathrm{B} \to \mathrm{C}) \rangle &= P_\mathrm{H}^{(0)}(a_\mathrm{B})(F^{(0)}(a_\mathrm{C}) - F^{(0)}(a_\mathrm{B})) + P_\mathrm{H}^{(1)}(a_\mathrm{B})(F^{(1)}(a_\mathrm{C}) - F^{(1)}(a_\mathrm{B})) \\ &= F^{(0)}(a_\mathrm{C}) - F^{(0)}(a_\mathrm{B}) - P_\mathrm{H}^{(1)}(a_\mathrm{B})(a_\mathrm{C} - a_\mathrm{B}), \\ \langle W(\mathrm{D} \to \mathrm{A}) \rangle &= F^{(0)}(a_\mathrm{A}) - F^{(0)}(a_\mathrm{D}) - P_\mathrm{L}^{(1)}(a_\mathrm{D})(a_\mathrm{A} - a_\mathrm{D}). \end{aligned} \tag{8.23}$$

---

[9] a_A and a_D in Fig. 8.3 correspond, respectively, to $a = -1$ and $a = 0$ of the sensor "tip" of Fig. 7.12.

[10] $P_\alpha^{(n)}$ is defined by (A.78).

[11] Once a is defined, $F^{(0)}$ and $F^{(1)}$ are also functions of a

Isothermal Processes as Open System

By the general result, $W = \Delta J$, for the quasistatic work of open isothermal process (5.20), the works are

$$W(\mathrm{A} \to \mathrm{B}) = F^{(0)}(a_{\mathrm{B}}) - F^{(0)}(a_{\mathrm{A}}) - T \log \left(\frac{1 - P_{\mathrm{H}}^{(1)}(a_{\mathrm{B}})}{1 - P_{\mathrm{H}}^{(1)}(a_{\mathrm{A}})} \right),$$

$$W(\mathrm{C} \to \mathrm{D}) = F^{(0)}(a_{\mathrm{D}}) - F^{(0)}(a_{\mathrm{C}}) - T \log \left(\frac{1 - P_{\mathrm{L}}^{(1)}(a_{\mathrm{D}})}{1 - P_{\mathrm{L}}^{(1)}(a_{\mathrm{C}})} \right), \qquad (8.24)$$

where we have used (8.22). These works are definite in the quasistatic limit, despite the fact that they are represented in terms of $P_{\alpha}^{(1)}(a)$.

Matching of Chemical Potentials and Efficiency per Cycle

The present system takes only two states, $n = 0$ and $n = 1$. This corresponds to the exceptional case that can avoid "relaxation from non-Gibbs ensembles of fluctuations" (Sect. 7.1.3). Therefore, we choose the values of a so that no irreversible work appears in the limit of slow operation. This is achieved by $P_{\mathrm{L}}^{(1)}(a_{\mathrm{D}}) = P_{\mathrm{H}}^{(1)}(a_{\mathrm{A}})$ and $P_{\mathrm{L}}^{(1)}(a_{\mathrm{C}}) = P_{\mathrm{H}}^{(1)}(a_{\mathrm{B}})$. Intuitively, these conditions assure that, upon attachment to a particle reservoir, the system behaves as if it were already in equilibrium with the environment. These equations are equivalent to the following:

$$a_{\mathrm{A}} - a_{\mathrm{D}} = -(a_{\mathrm{C}} - a_{\mathrm{B}}) = \mu_{\mathrm{H}} - \mu_{\mathrm{L}}. \qquad (8.25)$$

Equation (8.25) leaves $a_{\mathrm{A}} - a_{\mathrm{B}} (= a_{\mathrm{D}} - a_{\mathrm{C}})$ arbitrary.[12] That is, a need not to change while the system is opened to one of the particle reservoirs.[13]

Substituting (8.25) into (8.24) and (8.23), the average of the total extracted work per cycle, $\langle -W_{\mathrm{cyc}} \rangle$, reads

$$\langle -W_{\mathrm{cyc}} \rangle = [P_{\mathrm{H}}^{(1)}(a_{\mathrm{B}}) - P_{\mathrm{L}}^{(1)}(a_{\mathrm{D}})](\mu_{\mathrm{H}} - \mu_{\mathrm{L}}).$$

This result shows that this open Carnot cycle extracts the maximal work defined by (2.25), where the coefficient of $(\mu_{\mathrm{H}} - \mu_{\mathrm{L}})$ is the number of transported particles per cycle.[14] As the process is reversible, the inverse operation can work as a particle pump at maximal efficiency.

---

[12] Also the value of a_{A} is arbitrary. See below.

[13] As an afterthought it seems reasonable. But the analysis shows that $a_{\mathrm{A}} - a_{\mathrm{D}} (= a_{\mathrm{B}} - a_{\mathrm{C}})$ need not be 0.

[14] The system carries $P_{\mathrm{H}}^{(1)}(a_{\mathrm{B}})$ particle from H-reservoir to L-reservoir and carries back $P_{\mathrm{L}}^{(1)}(a_{\mathrm{D}})$ particle per cycle.

Conclusion

The open Carnot cycle could extract energy from the thermal environment of a unique temperature. This energy was used to unbind the particle. Since the work has already been extracted upon binding the particle, the thermal energy is effectively converted into work. With whatever large binding energy, the thermal energy can release a bound particle if only we wait for sufficiently long time. The rareness of particle in the dilute particle reservoir assures that the particle once unbound has little probability to return. One may say that open Carnot cycle synthesizes the first and second laws of thermodynamics.

Thanks to this mechanism, the main system need not stock high energies. In principle the constraint (8.25) does not exclude the possibility of $a > 0$ (energy stockage) [7, 8]. But molecular architectures to stock high energy seem less feasible than that to bind at low energies. The 3D X-ray structural data of protein motors show no magic structures to temporarily stock the energy.[15] Stochastic energetics will help decrypt the structure–function relationship of the protein motors.

8.2 Autonomous Free-Energy Transducers

Introduction

Autonomous systems are the systems which control themselves. In biology autonomous systems are found from proteins to societies. As a module the autonomous system works more or less independently of the others for a range of space–time. The collaboration among autonomous modules can create higher order functions and also protect themselves on the one hand, but it can impose constraints on the constituting modules on the other hand. The organisms consisting of modular structure are robust against growth and accidental injury.[16]

In the rest of this chapter we focus on mesoscopic autonomous systems that convert a form of free energy into another form in the presence of thermal fluctuations. The Feynman ratchet wheel and pawl [2] and the Büttiker–Landauer ratchet [3, 4] are classical examples (Sects. 1.3.4 and 4.2.2). Symmetry and the Curie principle capture a physical basis of autonomous free-energy transducers (Sect. 1.3.4.4). Protein motors, ion pumps, signal-processing proteins, etc., are biological autonomous free-energy transducers [9]. Many studies have been done to understand the structure–function relationship of specific biological free-energy transducers, such as myosin and kinesin. But, at present, only tentative studies have been made to explain the functions of these motors on the common structural basis [10].

---

[15] Inside the nucleotide pocket of a motor protein, the cleavage of the "high-energy bond" between β-Pi and γ-Pi of an ATP molecule occurs almost at equal free energy.

[16] The object-oriented architecture of information systems uses these properties.

The approach in this section is to construct an open autonomous system that transports a species of noninteracting particles against their density gradient using the diffusion of another species of noninteracting particles along their density gradient. Building blocks of the model (submodules) are the half-presence sensors and gates, whose structures were discussed in Sect. 7.2. The level of description distinguishes individual realizations, where the chemical potentials enter only through the probabilities.

8.2.1 Autonomous Free-Energy Transducer Functions Among Different Equilibrium Environments

In this section we discuss free-energy transducer as a black box from the functional point of view.

8.2.1.1 Definition of Autonomous Free-Energy Transducer

We define the autonomous free-energy transducer as a system working in contact with several *equilibrium* environments. Therefore, any external operating systems should not help the operation of this system, and Maxwell's demon (Sect. 4.2.1.2) is excluded.

As an example, ATPase-driven Ca^{2+} ion pump (e.g., [11]) works in contact with the particle environments of ATP, ADP and Pi (inorganic phosphate), water, Ca^{2+} ion inside a vesicle/cell, and Ca^{2+} ion outside the vesicle/cell. The first four substances, which participate in the hydrolysis reaction, can occupy the same 3D space because practically none of the spontaneous hydrolysis reactions (ATP+water $\rightleftharpoons$ADP+Pi) occur without catalysts.[17]

8.2.1.2 A Simple Scheme of Free-Energy Transduction

As a prototype of the free-energy transducer, we shall adopt again the scheme Fig. 8.5, which has been introduced for the thermodynamic description of chemical coupling (Sect. 2.3.2). The autonomous transducer is in contact with two equilibrium particle environments of fuel (F) and also with two equilibrium particle environments of load (L), all being under a uniform temperature, T. Within each pair of environments, (F_h,F_ℓ) or (L_h,L_ℓ), the suffixes h [ℓ] indicate, respectively, the higher [lower] chemical potential of the particle species.

This scheme is similar to the cotransporter ion channels [12], by either symport or antiport (e.g., lactose/proton symport [13]). Those ion channels carry two species of ions across a membrane along or against their density gradient. This schema also resembles the Ca^{2+} ion pump [11] mentioned above, or redox reaction (e.g., proton

[17] Rigorously speaking ATP-hydrolyzing protein motors are catalysts only when it does not work against macroscopic external forces.

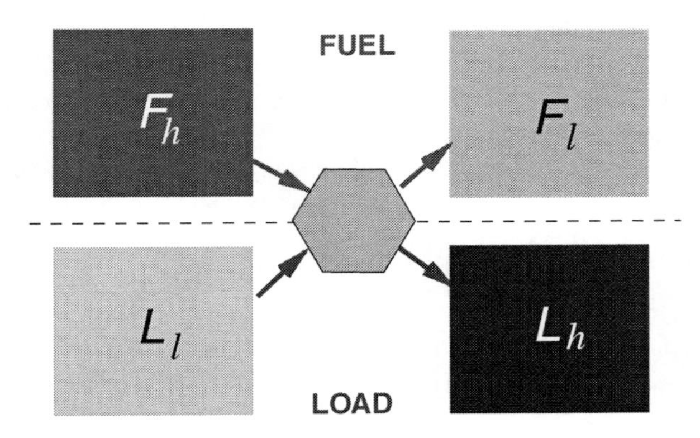

Fig. 8.5 The autonomous transducer (the hexagon at the *center*) is in contact with the particle environments of the fuel, (F_h,F_ℓ), and with particle environments of the load, (L_h,L_ℓ). The transducer couples the transport between the pair (F_h,F_ℓ) with that between the pair (L_h,L_ℓ). (The same figure as Fig. 2.7)

pumping [14]). In the latter case the substrates {ATP, water} and the products {ADP, Pi} are, respectively, identified as the h-side and on the ℓ-side of the fuel particles.[18] Mechanotrasduction by motor proteins, such as myosin or kinesin, is also similar to the above schema: The displacement dx along the filament can be regarded as the transport of the filament monomers (G-actin monomers for myosin, tubulin dimers for kinesin, etc.) between an "anterior monomer reservoir" and a "posterior monomer reservoir", see Fig. 8.6. If a mechanical load L is applied between the myosin molecule and the actin filament, the two reservoirs have different "chemical potentials" corresponding to the work done by a single-step displacement under the load.

The thermodynamic constraint on the function of this transducer is the decrease of the total Gibbs free energy,

$$\Delta G_{\text{tot}} \equiv \Delta G_{\text{F}} + \Delta G_{\text{L}} < 0,$$

with $\Delta G_{\text{F}} \equiv (\mu_{\mathsf{F}h} - \mu_{\mathsf{F}\ell})(-n_{\text{F}})(< 0)$ and $\Delta G_{\text{L}} \equiv (\mu_{\mathsf{L}h} - \mu_{\mathsf{L}\ell})n_{\text{L}}(> 0)$ (see Fig. 2.8). Here $\mu_{\mathsf{F}h}$, etc., are the chemical potentials of the particle environments, and n_{L} of the L particles are transported upward ($\mathsf{L}_\ell \to \mathsf{L}_h$) at the expense of the transport of n_{F} of the F particles downward ($\mathsf{F}_h \to \mathsf{F}_\ell$).

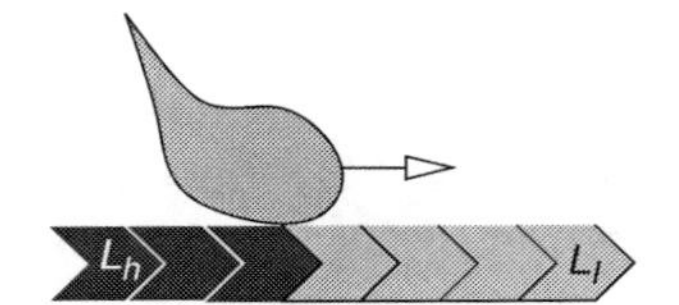

Fig. 8.6 Interpretation of the filament's monomers as particle reservoirs. A motor protein moves actively toward the right (*open arrow*), binding the monomer in the front (*bright unit*) and unbinding the monomer in the back (*dark unit*)

[18] The correspondence is not exact because of the four species of particles in place of two. However, the stoichiometry relation of the reaction allows us a simplified representation of Fig. 8.5.

8.2.1.3 Chemical Kinetics

The process of free-energy transduction can be described as a chemical kinetics. Figure 8.7 is an example. We assumed that at most one L particle and one F particle are admitted at a time and that the migration of these particles alternates between F and L.[19] The state of occupancy is denoted by ($0_\mathsf{F}\, 1_\mathsf{L}$), etc., where 0 and 1, respectively, denote the presence and absence of the particle on the transducer. The transducer works as a pump of L particles, if the chemical potentials of the environments satisfy $\mu_{\mathsf{F},h} - \mu_{\mathsf{F},\ell} \geq \mu_{\mathsf{L},h} - \mu_{\mathsf{L},\ell}$.

The rate constants of the state transitions are defined in the figure. We assume that the pair of rate constants associated to a migration event, e.g., $k_{\mathrm{F}h}$ and $k'_{\mathrm{F}h}$ between ($0_\mathsf{F}\, 1_\mathsf{L}$) and ($1_\mathsf{F}\, 1_\mathsf{L}$), is related through the detailed-balance (DB) relationships *when* this pair of states is hypothetically isolated from the other states.[15–17][20]:

$$k_\mathsf{L}^{\ell'} = k_\mathsf{L}^{\ell} \exp[\beta(\mu_\mathsf{L}^{\ell} + \epsilon_{0_\mathsf{L}}^{0_\mathsf{F}} - \epsilon_{1_\mathsf{L}}^{0_\mathsf{F}})], \qquad k_\mathsf{F}^{h'} = k_\mathsf{F}^{h} \exp[\beta(\mu_\mathsf{F}^{h} + \epsilon_{1_\mathsf{L}}^{0_\mathsf{F}} - \epsilon_{1_\mathsf{L}}^{1_\mathsf{F}})],$$
$$k_\mathsf{L}^{h'} = k_\mathsf{L}^{h} \exp[\beta(\mu_\mathsf{L}^{h} + \epsilon_{0_\mathsf{L}}^{1_\mathsf{F}} - \epsilon_{1_\mathsf{L}}^{1_\mathsf{F}})], \qquad k_\mathsf{F}^{\ell'} = k_\mathsf{F}^{\ell} \exp[\beta(\mu_\mathsf{F}^{\ell} + \epsilon_{0_\mathsf{L}}^{0_\mathsf{F}} - \epsilon_{0_\mathsf{L}}^{1_\mathsf{F}})],$$

where $\epsilon_{0_\mathsf{L}}^{1_\mathsf{F}}$, etc., are the free-energy levels of the transducer in each state.

The master equation for the schema Fig. 8.7 yields the steady-state rate of the counterclockwise reaction cycle, ν:

$$\nu = A(e^{(\mu_{\mathsf{F},h}-\mu_{\mathsf{F},\ell})/k_\mathrm{B}T} - e^{(\mu_{\mathsf{L},h}-\mu_{\mathsf{L},\ell})/k_\mathrm{B}T}) \qquad (A > 0).$$

The chemical kinetics approach has recently been put forward to describe single-molecule experiments, taking into account applied forces [18] and intermediate

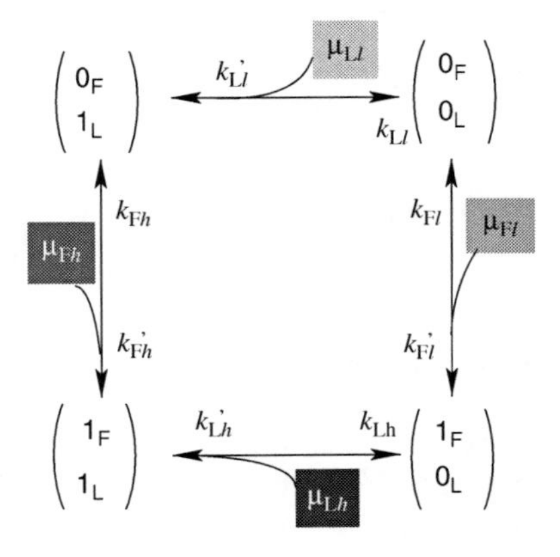

Fig. 8.7 Chemical kinetic diagram specifying a mesoscopic free-energy transducer. The rectangles branching from the *arrows* by solid curves indicate the export or import of particles between the transducer and the particle environment, which are indicated by their chemical potential, e.g., $\mu_{\mathsf{F}h}$ for the (F, h) environment, etc. k's denote the rate constant of the state transitions

[19] Other sequence of migration is also possible. In an ion pump [13], for example, the in-and-out of a proton occurs in between the in-and-out of a lactose molecule.

[20] Section 3.3.1.5 *Nonequilibrium settings II.*

steps [19, 20]. How well this phenomenology works depends on the choice of representative chemical states and of allowed transitions among them, and also depends on the identification of basic parameters from separate but consistent experiments. It is beyond the scope of the chemical kinetics approach to study what structural basis constitutes the representative chemical states and what mechanisms support state transitions.[21]

8.2.2 Pairs of Sensors and Gates are Enough to Constitute an Autonomous Free-Energy Transducer

8.2.2.1 From Function to Structure

There are several facts that motivate to construct autonomous transducers based on the elementary sensors and gates discussed in the previous chapter (Sect. 7.2).

1. In the nonautonomous free-energy transducers (mesoscopic Carnot cycle or open Carnot cycle), the detection of the point where a process is switched into another process is implicitly assumed. Attachment/detachment with the environments must also be done by the external system.
2. The chemical state diagram such as Fig. 8.7 implies that the release of an L particle to its dense reservoir (L_h) happens after the arrival of an F particle. Also the figure implies that the L particle is admitted only from its dilute reservoir (L_ℓ) when the F particle is absent.
3. The structure of the half-presence sensor (Fig. 7.12 *Left*) is very similar to the core module of the open Carnot cycle (Fig. 8.3).
4. The X-ray structural data of molecular motors are often described using the notions of sensors and/or of gates.

With these observations, we will discuss the possible structures that realize autonomous free-energy transducers.

8.2.2.2 Number of Internal Degrees of Freedom of Autonomous Transducers

There must be at least one internal degree of freedom (i.e., allosteric coupling) in order to correlate the fuel (F) side and the load side (L) of the autonomous free-energy transducer. There must also be a cyclic process that breaks the equilibrium detailed balance. A cycle needs at least two open coordinates to realize. The presence of cycle, however, does not necessitate the presence of a second <u>internal</u> degree of freedom. What is required is to realize cyclic process at the <u>interfaces</u> between the transducer and the particle reservoirs.[22] With only one internal degree of freedom

---

[21] One insightful paper has addressed this issue [11].

[22] cf. Fig. 8.3 for externally controlled case.

it is possible to realize such cycle.[23] Some people think that a class of biological transmembrane transporters function with a single principal degree of freedom. This hypothesis is called the alternative access model [21, 13].

However, the transducer with a single allosteric coupling may not be the simplest case, because a single degree of freedom must play double roles of detector and gate.[24]

Contrastingly, recent structural analyses of a single head of myosin suggests two independent internal degrees of freedom [22, 23]. Conceptually the transducer with two allosteric degrees of freedom is simpler than that with a single allosteric degree of freedom. The actions of sensor and detector at an interface of F particles can be directly communicated to those at the interface of L particles and vice versa if there are two allosteric degrees of freedom. We, therefore, focus on the free-energy transducer with two allosteric couplings.

8.2.2.3 * Bidirectional Control

We will consider a simple setup: (i) The particles are noninteracting, i.e., ideal gases, except for the steric repulsion. At most one F particle and one L particle can occupy the transducer. (ii) We use only the half-presence sensors and the gates. Hereafter, the "detection of a particle" implies the ON state of these half-sensors. (iii) No thermal activation process will intervene. It was the strategy of constructing the half-sensors and gates (Sect. 7.2).

It is then necessary to assign one sensor and one gate to the interface between the transducer and the F particle reservoir, and also assign one sensor and one gate to the interface between the transducer and the L particle reservoir.

As the allosteric coupling, the only choice is to connect the sensor of one interface with the gate the other interface. See Fig. 8.8. This symmetric rule is called bidirectional control [24, 10].

In bidirectional control, the detection of a particle (ON) can be associated with either one of the configurations of the gate, i.e., (*Top*) or (*Bottom*) of Fig. 7.17. Therefore, there are four possible combinations to design bidirectional control. Figure 8.9 shows one of these possibilities. In this option, the detection of a particle is associated with the accessibility of another species of particles from their dense reservoir.

Figure 8.10 graphically shows the consequence of these rules on the plane of the extended reaction coordinates[25] of L and F particles. On this plane it turns out that the process is constrained to a channel (gray zone), which couples the diffusion of F particles (downward) and the pumping up of L particles (rightward). During

---

[23] E. Muneyuki and K. Sekimoto, preprint.

[24] As a result the allosteric degree of freedom has a rate-limiting step.

[25] See Fig. 7.13.

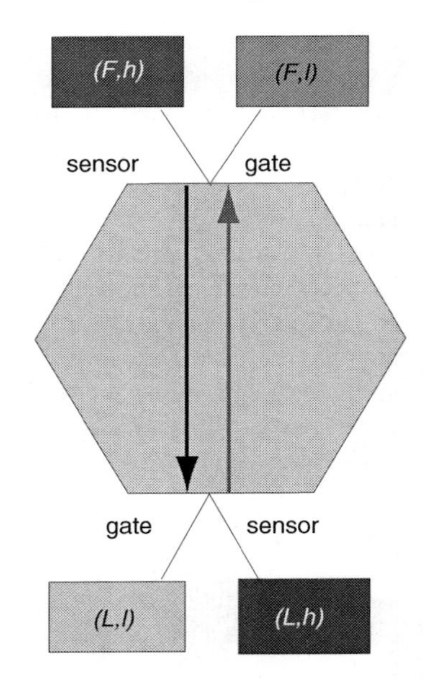

Fig. 8.8 Notion of the bidirectional control. The gate at the detection site of F particle is controlled by the half-sensor at the detection site of L particle (*upward arrow*), and vice versa (*downward arrow*)

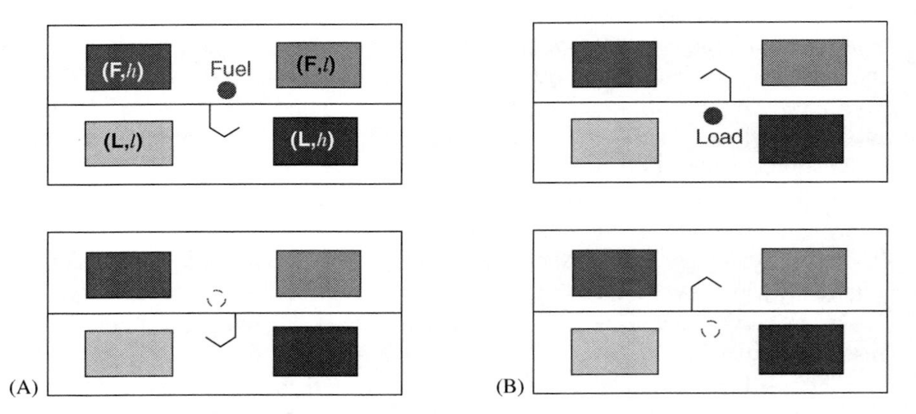

Fig. 8.9 A coupling rule between the sensor and the gate. The symbols for the gate are the same as Fig. 8.3. (**A**) The detection of F particle (*thick dot*) allows the access of L particles from their dense reservoir (L,h), and nondetection of F particle (*open dashed circle*) allows the access of L particles from their dilute reservoir (L,ℓ). (**B**) The completely symmetric rule applies for the socond allosteric coupling

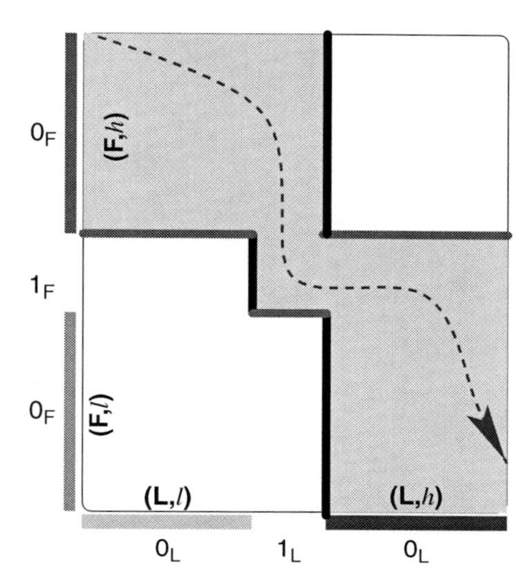

Fig. 8.10 Schematic representation of the function of bidirectional control by the rules of Fig. 8.9. The horizontal axis is the extended reaction coordinate for L particles, where the *left* [right] of the detection site correspond to the dilute [dense] environment, respectively. The vertical axis is the extended reaction coordinate for L particles, where the zone above [below] the detection site correspond to the dense [dilute] environment, respectively. Thick vertical [horizontal] borders of the gray region represent the blockade by the gate by the coupling rule Fig. 8.9(**A**) [(**B**)], respectively

a particular realization such as a dashed curve in Fig. 8.10, the occupancies of the particles complete a clockwise cycle along the schema of Fig. 8.7, that is,

$$(0_{\mathsf{F}}, 0_{\mathsf{L}}) \rightarrow (0_{\mathsf{F}}, 1_{\mathsf{L}}) \rightarrow (1_{\mathsf{F}}, 1_{\mathsf{L}}) \rightarrow (1_{\mathsf{F}}, 0_{\mathsf{L}}) \rightarrow (0_{\mathsf{F}}, 0_{\mathsf{L}}). \tag{8.26}$$

Throughout this cycle the potential energy is constant, and this free-energy transducer is, therefore, purely entropic.

8.2.2.4 Properties of Bidirectional Control

Invertibility

The rules of allosteric coupling realized in Fig. 8.10 are symmetric with respect to L and F particles. Therefore, if the relative magnitude between the chemical potential differences, $\mu_{\mathsf{F},h} - \mu_{\mathsf{F},\ell}$ and $\mu_{\mathsf{L},h} - \mu_{\mathsf{L},\ell}$, is inverted, the F particles can be pumped up by the passive diffusion of the L particles, so that the total Gibbs free energy decreases: $\Delta G_{\mathrm{F}} + \Delta G_{\mathrm{L}} < \mathrm{O}$.

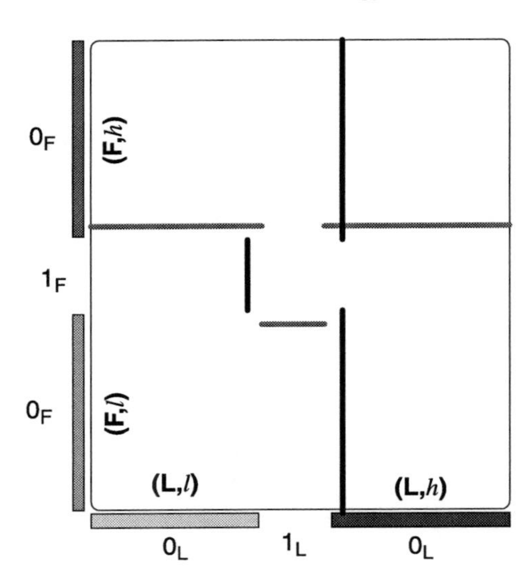

Fig. 8.11 Refinement of Fig. 8.10 incorporating the margins of operation

Viewpoint of a Load Particle

From the viewpoint of L particles, the process of Fig. 8.10 would look as if the transducer itself had an intelligence to pump up them: for the L particles in the dilute reservoir (L, ℓ), the transducer is often accessible. They, therefore, frequent the detection site. When an L particle is IN, it is from time to time bound (by the transition $0_L \rightarrow 1_L$). During such period it often happens that the gate from which the L particle has entered is closed and the exit opens toward the dense reservoir (by the transition, $0_L \rightarrow 1_L$ invisible by the L particle). After the L particle diffuses out, the gate for the dense reservoir (L, h) will close quickly, before another L particle enters from this reservoir. The transducer then becomes accessible for the dilute reservoir (L, ℓ) again.

Unattainability of Tight Coupling

Up to this point we have ignored the small margins of operation, δ and ϵ in the sensors and gates, respectively.[26] With these margins being taken into account, the blockade by the gate shown in Fig. 8.10 is modified as Fig. 8.11, where δ and ϵ are made large for visibility. The widths of detection are a little narrowed while the gaps between the gates are widened a little.

In this figure, the three gaps along a border of the channel constitute an intrinsic loss mechanism of this transducer. This leakage does not allow the free-energy transducer to attain the tight coupling of transport between L and F particles. Especially

[26] Section 7.2.2, *Margins of operation*. The δ was defined in Sect. 7.2.1.2, *Interaction potential of the half-presence sensor*, and the ϵ was defined in Sect. 7.2.2, *Gate made by the potential energy barriers*.

at the "stalled" condition, $\mu_{\mathsf{F},h} - \mu_{\mathsf{F},\ell} = \mu_{\mathsf{L},h} - \mu_{\mathsf{L},\ell}$, the efficiency of free-energy conversion drops to 0.

Prediction About "Mutants"

If one of the allosteric couplings is immobilized by some reason, the coupled transport does not function. See Fig. 8.12. In the figure, the gate of L particle is fixed independent of the detection of F particle. However, this "mutant" still controls the passive diffusion of L particle $(\mathsf{L}, h) \to (\mathsf{L}, \ell)$ through the density of F particles in the reservoir (F, h) [24].[27]

Bidirectional Control with Other Coupling Rules

We mentioned four different ways to design the bidirectional control (Sect. 8.2.2.3), of which one is denoted in Fig. 8.9. The diagrammatic representation like Fig. 8.10 allows us to find how each design works. It turns out that two out of the four designs couple the passive diffusion of F particles to that of L particles. We therefore ignore these two. The fourth rule is complementary to Fig. 8.9 in that the detection of particles of one species is linked to the accessibility of particles of the other species from its *dilute* reservoir. The resulting diagram is an inverted image of Fig. 8.10 with the inversion center being at the origin. The process of particles' migration is then [24]

$$(1_{\mathsf{F}}, 0_{\mathsf{L}}) \to (1_{\mathsf{F}}, 1_{\mathsf{L}}) \to (0_{\mathsf{F}}, 1_{\mathsf{L}}) \to (0_{\mathsf{F}}, 0_{\mathsf{L}}) \to (1_{\mathsf{F}}, 0_{\mathsf{L}}). \tag{8.27}$$

This design, therefore, works also as free-energy transducer to pump up the L particles.[28]

In Fig. 8.13 the original rule (A) and the above rule (B) are compared from the viewpoint of the *timing* of the particles' migration. In (A) the reservoirs (F, h) and

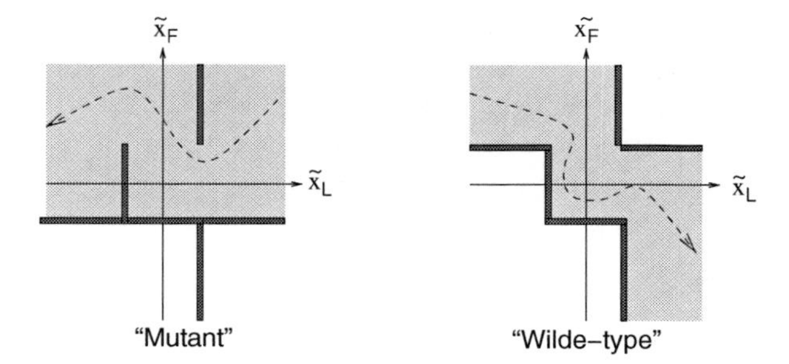

Fig. 8.12 A "mutant" of Fig. 8.10, where one of the allosteric couplings is immobilized

---

[27] L particle cannot go through the mutant transducer from the right (L, h) reservoir to the left (L, ℓ) reservoir without the assistance of the to-and-fro transitions of F particle: $0_{\mathsf{F}} \to 1_{\mathsf{F}} \to 0_{\mathsf{F}}$.

[28] This dichotomy is invariant under the exchange of the roles of Fuel and Load: the diagram of Fig. 8.10 is symmetric with respect to the two axes.

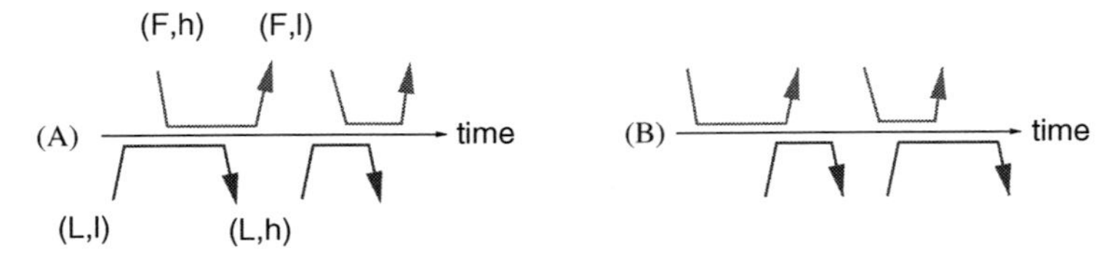

Fig. 8.13 (**A**) Typical process of particles' migration under the rules of Fig. 8.9. (**B**) The case under the complementary rules

(L, h) exchange occupancy, like the exchange of binding (see Sect. 7.2.1.4) but with delay. In (B) it is the reservoirs (L, ℓ) and (F, ℓ) that do a similar exchange. Biological motors seem to prefer type (A) (see the next section).

Bidirectional Control in Other Systems

Putting aside the energetic aspect, the logical structure of bidirectional control of autonomous systems is found also in other domains. For example, in some class of thermal ratchet models the position of a particle is detected and the potential profile is switched accordingly [25]. In Fig. 8.14 the potential energy is switched between the two profiles $U_1(x)$ and $U_2(x)$ when $x(t)$ enters the proximity of the bottom of the valleys (e.g., x_1 and x_2 in the figure). Ratchet models from the viewpoint of control have not been studied thoroughly [26–29]. As another example, the public payphone operates with two directions of control: the first control checks the credit (sensor) and admits the conversations (gate), and the other control detects the communication (sensor) and debits the credit (gate).

Information-Theoretic Approach

The above model of free-energy transducer uses two agents of communication. If the gate of an F particle is directly correlated to the sensor of this particle, for example, the system realizes feedback loop for an L particle, that is, the detection of this particle influences the accessibility of this particle. For feedback control, the information-theoretic analysis has been done [30, 31]. The generalization to the bidirectional control might be done.

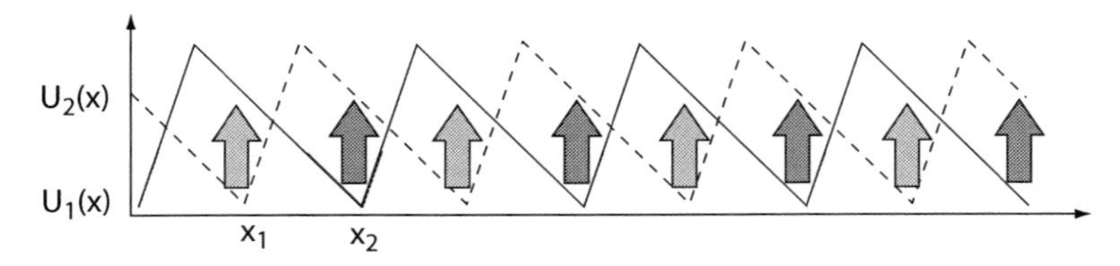

Fig. 8.14 Ratchet model driven by position-dependent switching of potentials. When $x(t)$ is detected within the zones of thick lines (around x_1, x_2, etc.) it is more probable for the actual potential function to be switched to the other potential function

8.2.3 *Structure–Function Relationship of Molecular Machines Is in the Research Domain of Stochastic Energetics*

The study of biomolecular machines started from the macroscopic level in physiology and then in biochemistry. More recently, the atomic structure of those molecular machines has been largely uncovered by X-ray crystallography on by NMR. Also, single-molecule experiments have been developed since the late twentieth century. Full molecular dynamic studies including quantum aspects are also developing.

The more the spatiotemporal resolution is improved, the more complex characteristics of the molecular machines are found. Through these discoveries, simple models made of spring and binding block, or of combustion engine and lever arm, etc., are obliged to be refined, modified, or sometimes abandoned. A single protein may accept different coarse-grained models according to different working conditions and regulations.

But at the same time, structural analysis has revealed that some local atomic structures are highly universal among a superfamily of biomolecules. For example, the notion of the "common ancestor" [32] of myosin, kinesin, and G-proteins (see [9]) stemmed from such observations.

A gap of our understanding is between this local structural universality and the global functional diversity among the members of a superfamily. For example, the nucleotide pocket of these molecular motors resembles each other very closely. Nevertheless, the relative timing between the ATP hydrolysis cycle and the filament binding/unbinding cycle is known to be very different between a conventional myosin head [33, 34] and kinesin [9].[29]

Another question is how these biomolecular machines can work under strong thermal fluctuations. Recent single-molecule experiments have started to uncover how individual molecular machines behave in nanometer and microsecond ranges. On the scales of thermal fluctuations many notions used in macroscopic world need to be reconsidered like the laws of motion, force, stability, chemical potential, heat, etc. For example, structural biology papers use the notions of gate, sensor, switch, etc., to consider the causal sequence of molecular events. However, the bidirectional control described above distinguishes between *two types of causalities*: one is the (quasi) instantaneous linkage of events through the allosteric coupling between a sensor and a gate, and the other involves the event-waiting states whose subsequent transition depends on the thermal fluctuations.

In summary, the study of intramolecular processes of biomolecular machines should clarify the structure–function relations of the machines based on the physical concepts adapted for the fluctuating world. One goal is to establish a common

---

[29] The myosin head detaches from the actin filament upon binding with an ATP molecule, while kinesin head with ATP binds to microtubule filament rather strongly. The isomerization – conformational changes associated with intramolecular hydrolysis reaction – takes place when the myosin head interacts with an actin filament only weakly, while the isomerization of a kinesin head occurs when the head is ready to detach from the microtubule.

language to describe the dynamics of different biomolecular structures. This will serve to compare, design, and predict the varieties of biomolecular functions.

As a trial in this direction the structures *and* functions of myosin and kinesin have been compared [10]. The structural data and single-molecule experimental data of these motor molecules are projected onto the above mentioned model of bidirectional control. At least for myosin, this analogy works well, and different modules in the single head of conventional myosin molecule are attributed to either one of the two allosteric degrees of freedom. This is consistent with the recent claim [35] that the structural data of the myosin with different nucleotide analogues require at least two independent degrees of freedom. At the active site where the ATP hydrolysis takes place, it is the chemical bond between β- and γ-phosphates ($\beta\gamma$–Pi bond) that is detected as an F particle in both myosin and kinesin. A surprise is that, despite the large differences in the timing of the ATP hydrolysis cycle and the filament binding/unbinding cycle (cf. footnote 29), the internal cycle of these molecules seems to be the same and approximately adapted to the schema (8.26) with an appropriate assignments of the sensors and gates for individual molecules.[30] Despite the symmetry between (8.26) and (8.27) as logical constructions, Fig. 8.13 shows that these motors, or their "common ancestor," have chosen to release the L particle after the binding of ATP (cf. Sect. 7.2.1.4). An interpretation for this broken symmetry is that, for the schema (8.26), it is easy to stay in strong binding with filaments when the ATP concentration is very low.

The approach of stochastic energetics will help us to understand how the constituting elements in the fluctuating world organize various functions as well as mutual constraints.

References

1. K. Sekimoto, F. Takagi, T. Hondou, Phys. Rev. E **62**, 7759 (2000)
2. R.P. Feynman, R.B. Leighton, M. Sands, *The Feynman Lectures on Physics – vol.1* (Addison Wesley, Reading, Massachusetts, 1963), §46.1§§–46.9
3. M. Büttiker, Zeitschrift der Physik B **68**, 161 (1987)
4. R. Landauer, J. Stat. Phys. **53,** 233 (1988)
5. L.D. Landau, E.M. Lifshitz, *Mechanics (Course of Theoretical Physics, Volume 1)*, 3rd edn. (Reed Educational and Professional Publishing Ltd, Oxford, 2002)
6. Y.G. Sinai, *Probability Theory: An Introductory Course* (Springer, New York, 1992)
7. R.D. Vale, F. Oosawa, Adv. Biomys. **26**, 97 (1990)
8. M. Nishiyama, H. Higuchi, T. Yanagida, Nat. Cell Biol. **4**, 790 (2002)
9. B. Alberts et al., *Essential Cell Biology*, 3rd edn. (Garland Pub. Inc., New York & London, 2009)
10. K. Sekimoto, C. R. Physique **8**, 650 (2007)
11. E. Eisenberg, T.L. Hill, Science **227**, 999 (1985)

[30] For example, the arrival of an L particle from the dilute reservoir is detected by the myosin at the lever arm and its comoving units, while for the kinesin the target of detection is not the tubulin but seems to be internalized.

12. B. Alberts et al., *Essential Cell Biology: An Introduction to the Molecular Biology of the Cell* (§ 12.)(Garland Pub. Inc., New York, 1998)
13. J. Abramson et al., Science **301**, 610 (2003)
14. Y.C. Kim, M. Wilkström, G. Hummer, PNAS **104**, 2169 (2007)
15. P.G. Bergmann, J.L. Lebowitz, PR **99,** 578 (1955)
16. T.L. Hill, Free Energy Transduction and Biochemical Cycle Kinetics, Dummy ed. (Springer-Verlag, 1989)
17. T. Shibata, K. Sekimoto, J. Phys. Soc. Jpn. **69,** 2455 (2000)
18. G.I. Bell, Science **200**, 618 (1987)
19. A.B. Kolomeisky, B. Widom, J. Stat. Phys. **93**, 633 (1998)
20. A.B. Kolomeisky, M.E. Fisher, Biophys. J. **84**, 1642 (2003)
21. C. Tranford, Annu. Rev. Biochem. **52,** 379 (1983)
22. P.D. Coureux, H.L. Sweeney, A. Houdusse, EMBO J. **23**, 4527 (2004)
23. M.A. Geeves, K.C. Holmes, Adv. Protein Chem. **71**, 161 (2005, chap. V)
24. K. Sekimoto, Physica D **205**, 242 (2005)
25. M. Porto, M. Urabakh, J. Klafter, Phys. Rev. Lett. **85,** 491 (2000)
26. H. Sakaguchi, J. Phys. Soc. Jpn. **67,** 709 (1998)
27. I. Derényi, R.D. Astumian, Phys. Rev. E **59**, R6219 (1999)
28. F. Jülicher, J. Prost, Phys. Rev. Lett. **75**, 2618 (1995)
29. K. Sasaki, R. Kanada, S. Amari, J. Phys. Soc. Jpn. **76**, 023003 (2007)
30. H. Touchette, S. Lloyd, Phys. Rev. Lett. **84**, 1156 (2000)
31. H. Touchette, S. Lloyd, Physica A **331**, 140 (2004)
32. F.J. Kull, R.D. Vale, R.J. Fletterick, J. Muscle Res. Cell Motil. **97,** 877 (1998)
33. R.W. Lymn, E.W. Taylor, Biochemistry **10**, 4617 (1971)
34. D.R. Trentham, J.F. Eccleston, C.R.Q. Bagshaw, Rev. Biophys. **9**, 217 (1976)
35. K.C. Holmes, R.R. Schröder, H.L. Sweeney, and A. Houdusse, Phil. Trans. R. Soc. B **359,** 1819 (2004)

Appendix A

A.1 Appendix to Chap. 1

A.1.1 Examples of Probability Distribution: Appendix to Sect. 1.1.2.1

Example 1: Binomial distribution. Suppose we have the total of N balls, of which M balls are red. We choose n balls among them at random. We assume that a ball, once chosen, will not be returned to the original ensemble. The random variable $\hat{k}$ here is the number of the red balls among the n chosen balls. We can show that

$$Prob[\hat{k} = k] \equiv P(k, n|M, N) = \frac{{}_M C_k \, {}_{N-M} C_{n-k}}{{}_N C_n},$$

where ${}_n C_k \equiv n!/(k!(n-k)!)$ is the binomial coefficient. (If both $0 \le n \le N$ and $0 \le k \le M$ are not satisfied, we can define $P(k, n|M, N) \equiv 0$.)

If we take the limits of $N \to \infty$ and $M \to \infty$ in the way that $\lim_{N\to\infty}(N/M) = p$, we have $P(k, n|M, N) \to f(k; n, p)$,[1] where

$$f(k; n, p) \equiv {}_n C_k \, p^k (1-p)^{n-k}, \qquad (0 \le k \le n\colon \text{integer}, \quad p > 0\colon \text{real}),$$

is called the binomial distribution. The parameter p is the probability of finding a red ball upon each choice. The combinatorial factor ${}_n C_k \, p^k$ comes from the possible different ways of picking up red balls among n consecutive choices.

Example 2: Poisson distribution. Now we take a different type of limits, $N \to \infty$, $M \to \infty$, and $n \to \infty$ in the way that $M/N \to 0$ and $n/N \to 0$ but $(nM/N) \to \lambda$. Then we have $P(k, n|M, N) \to f(k; \lambda)$,[2] where

---

[1] A useful asymptotic formula is $N!/(N-n)! \simeq N^n$ for $N/n \to \infty$.

[2] We will use Stirling's formula of $n!$, i.e., $n! \simeq n^n e^{-n}\sqrt{2\pi n}$ (valid up to 1% relative error for $n \ge 8$), and also use the expansion, $\log(1+\epsilon) \simeq \epsilon - \epsilon^2/2$, valid for small ϵ.

Sekimoto, K.: *Appendix*. Lect. Notes Phys. **799**, 281–320 (2010)
DOI 10.1007/978-3-642-05411-2

$$f(k;\lambda) \equiv e^{-\lambda}\frac{\lambda^k}{k!}, \qquad (0 \le k < \infty : \text{integer}, \quad \lambda > 0 : \text{real}),$$

is called the Poisson distribution. The parameter λ is the average number of red balls ($\langle\hat{k}\rangle$) found among a vast number (n) of chosen balls. The probability $f(k;\lambda)$ is maximum at around $k = \lambda$ but has tails above and below this average number.

A.1.2 A Particular Aspect of Gaussian Distribution: Fluctuation–Response: Appendix to Sect. 1.1.2.1

The Gaussian distribution is particular first because it contains only two parameters, e.g., the average and the variance, and second it has a quadratic form in the exponential. The second property gives rise to a general relation between the fluctuation of a random variable and the response of its average against the change in the parameters of the Gaussian distribution [1]. The similar relation has been well-known for general probability distribution related to thermal equilibrium, but the quadratic exponential nature of the Gaussian distribution provides this relationship irrespective of the origin of the Gaussian distribution, either equilibrium or nonequilibrium.

Suppose that a random variable $\hat{x}$ obeys the Gaussian distribution, with the probability density, $P(x;\alpha) \equiv \langle\delta(x - \hat{x})\rangle_\alpha$, where α is a parameter and $\langle\mathcal{O}(\hat{x})\rangle_\alpha = \int \mathcal{O}(x)P(x;\alpha)dx$. The average $\langle\hat{x}\rangle_\alpha$ and the variance $\sigma_a^2 \equiv \langle[\hat{x} - \langle\hat{x}\rangle_\alpha]^2\rangle_\alpha$ are, therefore, functions of α.

Because of the quadratic exponential character of Gaussian distribution, the effect of the change of parameter, $\alpha \to \alpha + \Delta\alpha$, on the probability distribution must take the form, $P(x;\alpha + \Delta\alpha) = P(x;\alpha)\, e^{uy+vy^2}/\langle e^{uy+vy^2}\rangle_\alpha$, where $y \equiv x - \langle\hat{x}\rangle_\alpha$ and the coefficients u and v depend generally on α and $\Delta\alpha$.[3] [1] showed that the effect of small change of $\Delta\alpha$ on the average, $\langle\hat{x}\rangle_\alpha$, has a simple relationship with the variance or the fluctuation, $\langle[\hat{x} - \langle\hat{x}\rangle_\alpha]^2\rangle_\alpha$. In fact

$$\begin{aligned}
\langle\hat{x}\rangle_{\alpha+\Delta\alpha} &= \frac{\langle\hat{x}e^{uy+vy^2}\rangle_\alpha}{\langle e^{uy+vy^2}\rangle_\alpha} \\
&= \langle\hat{x}\rangle_\alpha + \frac{\langle y + u^{(1)}y^2\Delta\alpha + v^{(1)}y^3\Delta\alpha + \mathcal{O}((\Delta\alpha)^2)\rangle_\alpha}{\langle 1 + u^{(1)}y\Delta\alpha + v^{(1)}y^2\Delta\alpha + \mathcal{O}((\Delta\alpha)^2)\rangle_\alpha} \\
&= \langle\hat{x}\rangle_\alpha + u^{(1)}\langle y^2\rangle_\alpha\Delta\alpha + \mathcal{O}((\Delta a)^2),
\end{aligned} \tag{A.1}$$

where $u^{(1)} \equiv \partial u/\partial(\Delta\alpha)|_{\Delta\alpha=0}$ and $v^{(1)} \equiv \partial v/\partial(\Delta\alpha)|_{\Delta\alpha=0}$. Therefore, the linear response coefficient of $\langle\hat{x}\rangle_\alpha$ with respect to $\Delta\alpha$ writes

[3] In $\langle\,\rangle_\alpha$ we should write $\hat{y} \equiv \hat{x} - \langle\hat{x}\rangle_\alpha$ rather than y, but we abuse the latter for simplicity of notation.

$$\lim_{\Delta\alpha \to 0} \frac{\langle \hat{x} \rangle_{a+\Delta\alpha} - \langle \hat{x} \rangle_{\alpha}}{\Delta\alpha} = u^{(1)} \langle [\hat{x} - \langle \hat{x} \rangle_{\alpha}]^2 \rangle_{\alpha}. \tag{A.2}$$

If u depends approximately linearly on $\Delta\alpha$ (e.g., $u \propto \Delta\alpha$), then $u^{(1)}$ is approximately constant. Then the linear response obeys approximately the conventional fluctuation–response relation. This conclusion holds whether or not the steady Gaussian distribution $P(x, a)$ corresponds to an equilibrium state.

The equilibrium fluctuation–response relation is immediately derived using the canonical distribution. If α is a small external force and x is a small deviation from the force-free equilibrium value, the distribution is

$$P_{\text{eq}}(x; \alpha) \propto \exp\left[-\frac{x^2}{2\langle x^2 \rangle_{\alpha=0}} + \frac{\alpha x}{k_{\text{B}} T}\right] \propto \exp\left[-\frac{(x - \langle x^2 \rangle_{\alpha=0} \alpha / k_{\text{B}} T)^2}{2\langle x^2 \rangle_{\alpha=0}}\right].$$

Therefore, we have

$$\frac{\partial \langle x \rangle_{\alpha}}{\partial \alpha} = \frac{\langle x^2 \rangle_{\alpha=0}}{k_{\text{B}} T}. \tag{A.3}$$

The comparison of (A.2) with (A.3) shows that the inverse of $u^{(1)}$ plays the role of an "effective temperature." This fictitious temperature can be measured as a true temperature if we attach to the system a small monitoring system ("thermometer") isolated from the thermal environment [2].

A.1.3 Sketch of Derivation of (1.10), (1.11), and (1.12): Appendix to Sect. 1.1.2.3

Let $P_N(A_N)$ be the probability distribution function of $\hat{A}_N$, or $P_N(A_N) = \langle \delta(A_N - \hat{A}_N) \rangle$. We introduce the characteristic function for $\hat{A}_N$, i.e., $\Phi_{A_N}(\phi) \equiv \langle e^{i\phi A_N} \rangle = \int_{-\infty}^{\infty} e^{iA_N\phi} P_N(A_N) dA_N$. The independence of $\hat{a}_i$ yields $\Phi_{A_N}(\phi) = [\Phi_a(\phi/N)]^N$, where $\Phi_a(\phi)$ is the characteristic function of $p(a)$, i.e., $\Phi_a(\phi) \equiv \int_{-\infty}^{\infty} e^{ia\phi} p(a) da$. By the definition of $\Phi_a(\phi)$, we have $\Phi_a(0) = 1$ (normalization), $\tilde{p}'(0) = i\langle a \rangle = ia^*$ (a^*: mean value), $\tilde{p}''(0) = -\langle a^2 \rangle$ (minus second moment), if these quantities exist for the distribution $p(a)$.[4] We then have

$$\left[\Phi_a\left(\frac{\phi}{N}\right)\right]^N \simeq \left[\Phi_a(0) + \tilde{p}'(0)\frac{\phi}{N} + \mathcal{O}\left(\frac{\phi^2}{N^2}\right)\right]^N$$
$$\to e^{ia^*\phi}, \qquad (N \to \infty). \tag{A.4}$$

[4] cf. It is not always the case; for the Lorenzian distribution, $p(a) = (\sigma/\pi)(a^2 + \sigma^2)^{-1}$, the second moment is divergent.

Since $e^{ia^*\phi}$ is the Fourier transform of $\delta(A_N - a^*)$, the last limiting property implies that $P_N(A_N) \to \delta(A_N - a^*)$ is the distribution.[5] This is called the (weak) law of large numbers. A stronger law has also been studied, but we will not discuss it here.

In order to study convergence more precisely, we focus on the narrowing peak of $P_N(A_N)$. In this purpose we introduce, instead of the empirical average, $\hat{A}_N$, the following "standardized deviation":

$$\hat{W} \equiv \frac{\hat{A}_N - a^*}{\frac{\sigma}{\sqrt{N}}} = \frac{\sum_{i=1}^N \hat{a}_i - Na^*}{\sqrt{N\sigma^2}}, \tag{A.5}$$

where $\sigma^2 = \langle (a - a^*)^2 \rangle$ is the variance of $\hat{a}_i$. We introduce the characteristic function for $\hat{W}$ as $\tilde{Q}_N(\psi) = \left\langle e^{i\psi\hat{W}} \right\rangle$. By the independence of $\hat{a}_i$, we have $\tilde{Q}_N(\psi) = \left[\left\langle e^{\frac{i\psi(a_i - a^*)}{\sqrt{N\sigma^2}}} \right\rangle\right]^N$. By developing the exponential function in terms of ψ,[6] we have $\tilde{Q}_N(\psi) \simeq \left[1 - \frac{\psi^2}{2N} + \mathcal{O}\left(\frac{\psi^2}{N^2}\right)\right]^N \to e^{-\frac{\psi^2}{2}}$ for $N \to \infty$. As $e^{-\frac{\psi^2}{2}}$ is the Fourier transformation of the standard normal distribution, $e^{w^2/2}/\sqrt{2\pi}$, the above result implies the following detailed but universal behavior of the sharp peak (de Moivre–Laplace theorem):

$$\lim_{N\to\infty} \mathrm{Prob}\left[c_1\sqrt{\frac{\sigma^2}{N}} < \hat{A}_N - a^* < c_2\sqrt{\frac{\sigma^2}{N}}\right] = \int_{c_1}^{c_2} \frac{e^{-\frac{w^2}{2}}}{\sqrt{2\pi}} dw. \tag{A.6}$$

Here c_1 and c_2 ($c_1 \le c_2$) should be of order of $N^0 (= 1)$. Therefore, within the peak region, the probability distribution for $\hat{w}$ approaches $\frac{e^{-\frac{w^2}{2}}}{\sqrt{2\pi}}$.[7] This is called the *central limit theorem.*

The deviations $\hat{A}_N - a^*$ beyond $\mathcal{O}(N^{-\frac{1}{2}})$, where we cannot apply the central limit theorem, belong to the subject of the large deviation theory. The *Cramér's theorem* tells how the probability of the large deviations behaves: (i) the probability $P_{\mathrm{d}}(x)$ of the large deviation, $x = \hat{A}_N - a^*$, obeys the law, [8]

$$P_{\mathrm{d}}(x) \sim e^{-NI(x)}, \qquad (N \to \infty), \tag{A.7}$$

---

[5] It is called the convergence in law.

[6] We assume that $\langle a^3 \rangle$ exists.

[7] In the statistical inference theory, this result is also used to assess the reliability of the empirical estimate of the mean value from $\hat{A}_N$. We expect, for large N, $\mathrm{Prob}\left[\hat{A}_N + c_1\sqrt{\frac{\sigma^2}{N}} < a^* < \hat{A}_N + c_2\sqrt{\frac{\sigma^2}{N}}\right] \simeq \int_{c_1}^{c_2} \frac{e^{-\frac{W^2}{2}}}{\sqrt{2\pi}} dW$.

[8] $X(N) \sim Y(N)$ means $\lim_{N\to\infty}(X(N)/Y(N))$ =const.

with $I(0) = 0$, and (ii) $I(x) = \sup_{t\in\mathbb{R}}[x\,t - \Lambda(t)]$. Here $\Lambda(t)$ is the cumulant generating function for $\hat{A}_N$, defined by $e^{\Lambda(t)} \equiv \langle e^{tA_N}\rangle = \Phi_{A_N}(-it)$, and we assume that $\Lambda(t)$ remains finite for all $t \in \mathbb{R}$ finite. (i) tells that except for the macroscopically expected value $\hat{A}_N = a^*$, the probability to realize any value of $\hat{A}_N$ is exponentially small for large N. The properties (i) and (ii) of $\hat{A}_N$ together are known as the large deviation principle (LDP).

A.1.4 Derivation of (1.23): Appendix to Sect. 1.1.3.2

Using (1.19), the probability density of the displacement, $\mathcal{P}(X,t) = \langle \delta X - [\hat{x}(t) - \hat{x}(0)]\rangle$, is

$$\mathcal{P}(X,t) = \left\langle \delta\left(X - \int_0^t \hat{\xi}(s)ds\right)\right\rangle. \tag{A.8}$$

By (A.8) the characteristic function for $\hat{X}$ is given as

$$\int_{-\infty}^{\infty} e^{iqX}\left\langle \delta\left(X - \int_0^t \hat{\xi}(s)ds\right)\right\rangle dX = \left\langle \exp\left[iq\int_0^t \hat{\xi}(s)ds\right]\right\rangle. \tag{A.9}$$

From (1.8), (1.9), and (1.18) together with (1.22), we find that the right-hand side of (A.9) is e^{-Dq^2t}. On the other hand, by (1.23) the characteristic function for $\hat{X}$ also yields e^{-Dq^2t}. Since the characteristic function contains (almost) the same informations as the probability density, we find the equality between $\mathcal{P}(X,t) = \langle\delta(X - [\hat{x}(t) - \hat{x}(0)])\rangle$ and (1.23).

A.1.5 Langevin Equation Obtained by the Method of Projection Operators: Appendix to Sect. 1.2.1.5

The method of projection operators is summarized without going into details, and the description is rather formal.

The gross variables (i.e., slowly varying physical observables of our interest), $\boldsymbol{A}(\boldsymbol{x},\boldsymbol{p})$, are functions of the phase space point, $\{\boldsymbol{x},\boldsymbol{p}\}$.[9] From classical Hamiltonian mechanics, we know that the time evolution of $\boldsymbol{A}(\boldsymbol{x},\boldsymbol{p})$ is governed by the Liouville equation, which is linear in $\boldsymbol{A}(\boldsymbol{x},\boldsymbol{p})$:[10]

---

[9] Except for very simple cases like the one studied in the previous section Sect. 1.2.1.4, the observable quantities of fluctuating system in general depend on many degrees of freedom. For example, the density of gas particles found in a specified small but macroscopic volume could be defined in terms of all the particles' coordinates.

[10] When we study directly the evolution of $\boldsymbol{A}(\boldsymbol{x},\boldsymbol{p})$, the arguments $\{\boldsymbol{x},\boldsymbol{p}\}$ denote the phase space point at the initial time $t = 0$, that is, $\{\boldsymbol{x}(0),\boldsymbol{p}(0)\} = \{\boldsymbol{x},\boldsymbol{p}\}$. The function $\boldsymbol{A}(\boldsymbol{x},\boldsymbol{p})$ at time t

$$\dot{A} = i\mathcal{L}A. \tag{A.10}$$

Here the linear operator, called the Liouville operator, $\mathcal{L}$, is defined using the Poisson bracket[11] $\{\,,\,\}$ and the Hamiltonian H as $\mathcal{L}A \equiv \{A, H\}$.

We will denote $\mathcal{H}$ the functional space, in which the projection operators act.

The essence of the method of projection operator is to represent the evolution of the observable A, after having (linearly) projected it into a subspace, $\mathcal{H}^{\parallel}(\subset \mathcal{H})$, spanned by a set of "slow variables" at the initial time, which are represented by an orthogonal set of the unitary vectors $\{e_j\}$ in $\mathcal{H}$ [3]. The action of the projection operator, $\mathcal{P}$, can be represented as $\mathcal{P}X = \sum_j e_j(e_j, X)$ under the properly defined scalar product, (X, Y), in the functional space $\mathcal{H}$. The temperature of the environment composed of the nonslow variables appears in the definition of the scalar product or, more precisely, in its "weighing function."[12] The formal procedure to obtain the projected evolution equation of A starts by decomposing the Liouville equation, $\dot{A} = i\mathcal{L}A$, into the two equations for $\mathcal{P}A$ and $(1-\mathcal{P})A$:

$$\frac{\partial}{\partial t}\mathcal{P}A = i\mathcal{P}\mathcal{L}[\mathcal{P}A + (1-\mathcal{P})A],$$

$$\frac{\partial}{\partial t}(1-\mathcal{P})A = i(1-\mathcal{P})\mathcal{L}[\mathcal{P}A + (1-\mathcal{P})A].$$

Then we solve formally the second equation for $(1-\mathcal{P})A$ while treating $\mathcal{P}A$ as given function of time. This procedure is analogous to (1.37). We substitute the expression for $(1-\mathcal{P})A$ thus obtained again into the above equations, where we also use the fact that $A = e^{i\mathcal{L}t}A|_{t=0}$ is the formal solution of the Liouville equation $\dot{A} = i\mathcal{L}A$ at time t (the observable at the initial time, $t = 0$, is denoted by $A|_{t=0}$). The result of the substitution is summarized as the following operator identity [4]:

$$\frac{d}{dt}e^{it\mathcal{L}} = e^{it\mathcal{L}}i\mathcal{P}\mathcal{L} + \int_0^t ds e^{i(t-s)\mathcal{L}}i\mathcal{P}\mathcal{L}e^{is(1-\mathcal{P})\mathcal{L}}i(1-\mathcal{P})\mathcal{L} + e^{it(1-\mathcal{P})\mathcal{L}}i(1-\mathcal{P})\mathcal{L}. \tag{A.11}$$

So-called Mori formula [3] is readily obtained by applying each side of the above equation to $A|_{t=0}$. For the details of the methods of projection operator, the readers should consult a good pedagogical review [5].

A difficulty met at this point was that the resulting equation of motion is always linear in the slow variable, A. The Mori formula [3] is thus reminiscent of the linear

---

connects this phase point to the value of A at that time. This viewpoint is parallel to the Lagrange picture in hydrodynamics or the Heisenberg picture in quantum mechanics. (As a complementary picture, if we defined $A(x, p)$ as a *fixed* function of $\{x, p\}$, the temporal change of the value of A would be represented as $A(x(t), p(t))$.)

[11] $\{A, B\} \equiv \sum_i \left[\frac{\partial A}{\partial x_i}\frac{\partial B}{\partial p_i} - \frac{\partial A}{\partial p_i}\frac{\partial B}{\partial x_i}\right]$, where i is the index of all the microscopic degrees of freedom.

[12] The scalar product of A and B in $\mathcal{H}$ is defined by $(A, B) \equiv k_B T \int_0^{(k_B T)^{-1}} \mathrm{Tr}\left[\rho_{eq} e^{\lambda H} A^\dagger e^{-\lambda H} B\right] d\lambda$, where $\rho_{eq} = Z^{-1} e^{-H/k_B T}$ is the equilibrium distribution or the density operator at temperature T.

Langevin equations in which the potential U is harmonic in $\boldsymbol{A}$. In order to derive Langevin equations having nonquadratic potential U with respect to $\boldsymbol{A}$, we must notice the fact that any function of $\boldsymbol{A}$ is a slow variable if $\boldsymbol{A}$ does. For example, $\boldsymbol{A}^3$ is totally dependent on $\boldsymbol{A}$ as a function of physical quantities, and the value of $\boldsymbol{A}^3$ is completely determined if that of $\boldsymbol{A}$ is known. However, as a vector in the functional space on $\{\boldsymbol{x}, \boldsymbol{p}\}$, the two functions $\boldsymbol{A}(\boldsymbol{x}, \boldsymbol{p})$ and $\boldsymbol{A}(\boldsymbol{x}, \boldsymbol{p})^3$ are independent.[13] Therefore, when we define a functional subspace of slow variables, we must include all functions of $\boldsymbol{A}$ as slow variables. A convenient and complete choice of such functions is to adopt the parameter family, $\{\delta(\boldsymbol{A} - \boldsymbol{a})\}$, with $\boldsymbol{a}$ running over a pertinent parameter space. For example, an arbitrarily chosen function of $\boldsymbol{A}$, say $f(\boldsymbol{A})$, is expressed as a form of linear combination: $f(\boldsymbol{A}) = \int f(\boldsymbol{a})\delta(\boldsymbol{A} - \boldsymbol{a})d\boldsymbol{a}$.

In 1973 Kawasaki finally derived a general form of the Langevin equation from microscopic Newtonian mechanics [4].[14] He essentially applied the above identity (A.11) to $\delta(\boldsymbol{A} - \boldsymbol{a})$. The result is (1.42) in the text, that is, $\frac{d}{dt}A_i = v_i(\mathbf{A}) - \sum_j \frac{L^0_{ij}}{T}\frac{\partial U_{\mathrm{eq}}(\mathbf{A})}{\partial A_j^*} + f_i(t)$, where the three terms on the right-hand side correspond to, respectively, the three terms on the right-hand side of (A.11).

A.1.6 The Distinction Between Different Types of Calculus: Appendix to Sect. 1.2.2.2

Sometimes we encounter in the literatures the expressions where the distinction between the Itô and the Stratonovich-type calculus is not explicit:

Example 1. Let us consider $\eta_n\sqrt{\Delta t}$, where the (discrete) Gaussian stochastic process η_n is characterized by $\langle\eta_n\rangle = 0$ and $\langle\eta_n\eta_m\rangle = b(x(t_n))\delta_{nm}$, with some nonnegative-valued function $b(x)$. It implies that $x(t_n)$ in $b()$ does not depend on the history of $\eta_{n'}$ ($n' = 1, \ldots, n$), that is, nonanticipating with respect to η_n. Therefore, the limit $\Delta t \to 0$ of the above process should be interpreted as Itô type, $\sqrt{b(x_t)}\cdot dB_t$.

Example 2. If a Langevin equation describes the dynamics in d-dimensional space or manifold,[15] the multiplicative noise term is often written as dW_t^μ ($\mu = 1, \ldots, d$) such that $\langle dW_t^\mu\rangle = 0$ and $\langle dW_t^\mu dW_t^\nu\rangle = g^{\mu\nu}(x_t)dt$, where $\langle\cdot\rangle$ denotes the average about the process between the time t and the $t + dt$, and $g^{\mu\nu}(x)$ is the (μ, ν)-component of the metric tensor $\mathsf{g}(x)$ at the point x. We can represent dW_t^μ in the form of $b(x_t)\cdot dB_t$ if we use the orthogonal matrix $\mathsf{O}(x)$ which diagonalizes the metric tensor, $\mathsf{O}(x)\mathsf{g}(x)\mathsf{O}^{\mathrm{t}}(x) = \Lambda(x)$. In fact the transformation

[13] It means that we need $a = b = 0$ if $ax + bx^3 = 0$ as a function of x. We note also that, if $\hat{y}$ is a random variable of zero average, the correlation coefficient between $\hat{y}$ and $\hat{y}^2$ is not identically equal to 1: $\langle\hat{y}\,\hat{y}^3\rangle/\sqrt{\langle\hat{y}\rangle^2\,\langle\hat{y}^3\rangle^2} \not\equiv 1$.

[14] It was Zwanzig who first proposed the equation with memory effect. [6].

[15] In Appendix to Chap. 4, we describe the energetics of the Langevin equation on a manifold (Sect. A.4.7.3).

$d\tilde{W}_t^\alpha \equiv \sum_{\mu=1}^{d} O_\mu^\alpha(x_t) \cdot dW_t^\mu$ (in Itô's sense) eliminates the cross-correlation, and we may write $d\tilde{W}_t^\alpha = \sqrt{\Lambda^\alpha(x_t)} \cdot dB_t^\alpha$, where $\Lambda^\alpha(x)$ is the αth diagonal component of the positive diagonal matrix $\Lambda(x)$ and B_t^α is the αth component, in this local frame, of d-dimensional Wiener process. Since the principal directions of the symmetric matrix $\mathrm{g}(x)$ depend on x, the suffix α at different points of x has no simple relations.

A.1.7 Conversion of Itô's Lemma (1.58) into Stratonovich Form: Appendix to Sect. 1.2.2.3

The sum of the second and third terms on the right-hand side of (1.59) yields up to $\sim dt$[16]

$$\begin{aligned}\left(f'(x_t) + f''(x_t)\frac{dx_t}{2}\right) dx_t &= f'(x_t + \frac{dx_t}{2})dx_t \\ &= \frac{f'(x_{t+dt}) + f'(x_t)}{2} dx_t \\ &= f'(x_t) \circ dx_t. \end{aligned} \tag{A.12}$$

Therefore, (1.59) simply leads to

$$df(x_t) = f'(x_t) \circ dx_t. \tag{A.13}$$

Therefore, the Stratonovich-type expression for (1.58) becomes

$$df(x_t) = a(x_t, t) f'(x_t)dt + [b(x_t, t) f'(x_t)] \circ dB_t.$$

As is seen from (A.13), formulas of the calculus of Stratonovich type take apparently the form of real analysis (integration by parts, integration by substitution, etc.). This is a merit of the calculus of Stratonovich type as compared with the Itô type.

These two types of calculus are, however, a matter of choice. The conversion from one type to the other is always possible and is given by

$$f(x_t) \circ dB_t = f(x_t) \cdot dB_t + \frac{b(x_t, t)}{2} f'(x_t)dt. \tag{A.14}$$

To understand this, we can rewrite the left-hand side as

$$\begin{aligned} f(x_t) \circ dB_t &= (1/2)(f(x_{t+dt}) + f(x_t))dB_t \\ &= [f(x_t) + \frac{1}{2}(f(x_{t+dt}) - f(x_t))]dB_t \end{aligned} \tag{A.15}$$

[16] Since $dx_t \sim \sqrt{dt}$, we keep only up to the $\sim (dx_t)^2$ terms in accordance with (1.52).

and substitute into $f(x_{t+dt}) - f(x_t)(= df(x_t))$ the formula (1.58). For $f(x_t) = b(x_t, t)$, the formula (A.14) can be easily generalized to give $b(x_t, t) \circ dB_t = b(x_t, t) \cdot dB_t + \frac{b(x_t,t)}{2}\frac{\partial b(x_t,t)}{\partial x_t}dt$. Then we arrive at (1.60).

A.1.8 Derivation of Fokker–Planck Equation (1.73) and Kramers Equation (1.74): Appendix to Sect. 1.2.3.1

We apply the Itô's lemma, (1.58), to $\delta(X - x_t)$. We obtain[17]

$$\begin{aligned}
&d\delta(X - x_t)\\
&= \left\{-a(x_t)\delta'(X - x_t) + \frac{b(x_t)^2}{2}\delta''(X - x_t)\right\} dt - b(x_t)\delta'(X - x_t) \cdot dB_t\\
&= \left\{[-a(X)\delta(X - x_t)]' + \frac{1}{2}[b(X)^2\delta(X - x_t)]''\right\} dt - b(x_t)\delta'(X - x_t) \cdot dB_t.
\end{aligned}$$

Taking the path average $\langle\ \rangle$ of each term, the terms including dB_t vanish by the rule of the Itô-type calculus. We then have

$$d\mathcal{P}(X, t) = \left\{[-a(X)\mathcal{P}(X, t)]' + \frac{1}{2}[b(X)^2\mathcal{P}(X, t)]''\right\} dt.$$

With this result and (1.72), we arrive at (1.73) in the text.

In the case with inertia, we apply the Itô's lemma to $\delta(X - x_t)\delta(P - p_t)$. We find

$$\begin{aligned}
&d[\delta(X - x_t)\delta(P - p_t)]\\
&= \delta'(X - x_t)\delta(P - p_t)\left(-\frac{p_t}{m}\right) dt\\
&+\delta(X - x_t)\delta'(P - p_t)\left[\left(\gamma\frac{p_t}{m} + \left.\frac{\partial U}{\partial x}\right|_{x=x_t}\right) dt + \sqrt{2\gamma k_B T} \cdot dB_t\right]\\
&+\frac{1}{2}\delta(X - x_t)\delta''(P - p_t)(2\gamma k_B T)dt. \qquad (A.16)
\end{aligned}$$

By taking the path average $\langle\ \rangle$ of each term, we arrive at (1.74) in the text. As noted in the footnote below (1.61), we need not to distinguish between $\circ dB_t$ and $\cdot dB_t$ in this case even if γ and/or T depend on x_t.

[17] Note that $\frac{d^n}{dy^n}[\phi(y)\delta(y - x)] = \frac{d^n}{dy^n}[\phi(x)\delta(y - x)] = \phi(x)\frac{d^n}{dy^n}\delta(y - x)$.

A.1.9 Jacobian to Transform ξ() into x(): Appendix to Sect. 1.3.1.1

Below is a heuristic explanation for the term $\partial^2 U/\partial x^2$ in (1.84). We discretize the Langevin equation in time up to $\mathcal{O}(x_{n+1} - x_n)$: $\gamma(x_{n+1} - x_n) = -\frac{1}{2}[U'(x_{n+1} + U'(x_n)]\Delta t + \sqrt{2\gamma k_B T}(B_{n+1} - B_n)$, where B_n stands for the Wiener process at time $t = n\Delta t$ and $x_n = x(n\Delta t)$. This trapezoidal rule, reminiscent of the Stratonovich interpretation, assures the correct limit $\Delta t \to 0$ in $P[x]$ of (1.84). (It is not the case with the other choices like the simple Euler scheme where $U'(x_n)$ is used instead of $\frac{1}{2}[U'(x_{n+1} + U'(x_n)]$, see the later Sect. 4.1.2.5.) Then the above discretized form can be rewritten as follows: $\gamma(x_{n+1} - x_n)[1 + \frac{\Delta t}{2\gamma}U''(x_n)] = \sqrt{2\gamma k_B T}(B_{n+1} - B_n) - U'(x_n)\Delta t$. This relation gives a linear transformation of $\{(B_{n+1} - B_n)\}$ into $\{(x_{n+1} - x_n)\}$ of $\mathcal{O}(\sqrt{\Delta t})$. The Jacobian then consists of the multiplication of $\partial(B_{n+1} - B_n)/\partial(x_{n+1} - x_n) = c\left[1 + \frac{\Delta t}{2\gamma}U''(x_n)\right]$, with c being constant through all the time-steps. ($U'(x_n)\Delta t$ is nonanticipating and does not depend on $x_{n+1} - x_n$.) Thus in the limit of $\Delta t \to 0$ we have $\prod_{n=0}^{(t/\Delta t)}[1 + \frac{\Delta t}{2\gamma}U''(x_n)] \to \exp(\frac{1}{2\gamma}\int_0^t U''(x)ds)$.

A.1.10 Derivation of Fluctuation–Dissipation (FD) Relation: Appendix to Sect. 1.3.1.2

Let $\boldsymbol{X}$ be the set of variables which evolves as a Markov process,

$$\frac{d\boldsymbol{X}}{dt} = \mathcal{V}(X(t), \xi(t)).$$

This can be a Langevin equation, where $\boldsymbol{X} = (x, p)$ or x; a Hamiltonian equation, where $\boldsymbol{X}$ represents all the positions and momenta and we ignore $\xi(t)$; or any other Markovian evolution equation. We adjust the additive constants in $\boldsymbol{X}$ such that its equilibrium average without external force vanishes.

The above equation (without external force) can be formally solved as an initial value problem. We denote the solution as follows:

$$\boldsymbol{X}(t) = \mathcal{X}(\xi(); \boldsymbol{X}_0) \qquad (t \leq t_0), \tag{A.17}$$

where $\boldsymbol{X}_0 \equiv \boldsymbol{X}(t_0)$ is the initial value of $\boldsymbol{X}$ at the time t_0.

We denote by $C_{X,X}(t - t_0) \equiv \langle \boldsymbol{X}(t)\boldsymbol{X}(t_0)\rangle_{\mathrm{eq}}$ the *canonical equilibrium correlation function* of $\boldsymbol{X}(t)$ with the initial variable $\boldsymbol{X}(t_0)$. This correlation function is given by the average of the above equation over the stochasticity of the process $\xi()$ and also over the initial equilibrium distribution, $P^{\mathrm{eq}}(\boldsymbol{X}_0; T)$

$$\langle \boldsymbol{X}(t)\boldsymbol{X}(t_0)\rangle_{\mathrm{eq}} = \int \langle \mathcal{X}(\xi(); \boldsymbol{X}_0)\rangle^{\xi()}\boldsymbol{X}_0 P^{\mathrm{eq}}(\boldsymbol{X}_0; T)\, d\boldsymbol{X}_0, \tag{A.18}$$

where $\langle\,\rangle^{\xi()}$ in the integrand denotes the average taken over the random thermal noise $\xi()$ but at a fixed initial value of $\boldsymbol{X}$.

In the linear response theory, the *relaxation function* of $\boldsymbol{X}(t)$ against a force on this variable is defined through the evolution of $\langle \boldsymbol{X}(t)\rangle$ given that a weak constant external force $\boldsymbol{h}$ on $\boldsymbol{X}$ has been applied until $t = t_0$ and is switched off at $t = t_0$. Because of the exponential Boltzmann factor form of the canonical distribution, the equilibrium density in the presence of the weak force takes the form

$$e^{\frac{\boldsymbol{X}_0\cdot\boldsymbol{h}}{k_BT}}P^{\rm eq}(\boldsymbol{X}_0;T) \simeq \left(1+\frac{\boldsymbol{X}_0\cdot\boldsymbol{h}}{k_BT}\right)P^{\rm eq}(\boldsymbol{X}_0;T)$$

up to the first order of $\boldsymbol{h}$. Therefore, the average of $\boldsymbol{X}(t)$ is given by

$$\begin{aligned}\langle \boldsymbol{X}(t)\rangle &= \int \langle \mathcal{X}(\xi(); \boldsymbol{X}_0)\rangle^{\xi()}\left(1+\frac{\boldsymbol{X}_0\cdot\boldsymbol{h}}{k_BT}\right)P^{\rm eq}(\boldsymbol{X}_0;T)\,d\boldsymbol{X}_0\,\boldsymbol{h}\\ &= \frac{1}{k_BT}C_{X,X}(t-t_0)\cdot\boldsymbol{h}.\end{aligned} \tag{A.19}$$

On the right-hand side, the coefficient of $\boldsymbol{h}$ is called the (linear) relaxation function, $R_{X,X}(t-t_0)$.

In conclusion, we have the following formula, which is essentially the fluctuation–dissipation relation:

$$C_{X,X}(t-t_0) = k_BT[R_{X,X}(t-t_0)+R_{X,X}(t_0-t)],$$

where we took into account the causality $R_{X,X}(t) = 0$ if $t < 0$. The generalization to the quantum case requires the noncommutativity of $\boldsymbol{X}(t)$ and $\boldsymbol{X}_0$. The generalization to the fluctuations of "currents" needs some derivation and integration [7], but we do not go into those details for simplicity's sake. From the above derivation, it is clear why this relation does not depend on the particular evolution model. We notice that neither the concrete form of $P^{\rm eq}(\boldsymbol{X}_0;T)$ nor the detailed balance conditions have been used except for the fact that the force perturbs $P^{\rm eq}(\boldsymbol{X}_0;T)$ by a factor $\left(1+\frac{\boldsymbol{X}_0\cdot\boldsymbol{h}}{k_BT}\right)$. This suggests the generalizability of the above relationship to the nonequilibrium steady states.

A.1.11 Derivation of (1.93) and (1.94): Appendix to Sect. 1.3.2.1

The global strategy is to (1) integrate the starting SDE (1.31) for the time step Δt ($\gg m/\gamma$), (2) eliminate the less dominant terms in Δt, and finally (3) take the formal limit of $\Delta t \to 0$. In this way the final SDE does not explicitly include the time resolution, Δt.

Step (1): We integrate formally the first equation of (1.31) between t and $t+\Delta t$, regarding $x(t)$ and $a(t)$ therein as known functions of time:

$$p(t')=\int_0^{t'} e^{-\frac{t'-s}{\tau_p}}\left[\sqrt{2\gamma k_{\mathrm{B}}T(x(s))}\xi(s)-\frac{\partial U(x(s),a(s))}{\partial x}\right]ds+e^{-\frac{t'}{\tau_p}}p(0).$$

We substitute this expression for the $p(t')$ in the time integral of the second equation in (1.31) over $[t, t+\Delta t]$,

$$\gamma[x(t+\Delta t)-x(t)]=(\gamma/m)\int_t^{t+\Delta t} p(t')dt'. \tag{A.20}$$

Step (2): The estimation of the resulting double integral requires some care: we use the following identity:

$$\int_t^{t+\Delta t} dt'\int_0^{t'} ds f(s,t')=\int_0^t ds\int_t^{t+\Delta t} dt' f(s,t')+\int_t^{t+\Delta t} ds\int_s^{t+\Delta t} dt' f(s,t'). \tag{A.21}$$

Also we use the development of the functions of $x(s)$ in powers of $[x(s)-x(t)]$, like

$$\sqrt{T(x(s))}=\sqrt{T(x(t))}+\left.\frac{d\sqrt{T(x)}}{dx}\right|_{x=x(t)}[x(s)-x(t)]+\cdots. \tag{A.22}$$

For $[x(s)-x(t)]$ in (A.22) we substitute (A.20) iteratively, replacing $t+\Delta t$ by s, so that the result is correct up to the order $\mathcal{O}(\Delta t)$.[18] Bringing all these techniques together and using the formulas such as (1.55), we arrive at the following result:[19]

$$\begin{aligned}\gamma[x(t+\Delta t)-x(t)] &= \sqrt{2\gamma k_{\mathrm{B}}T(x(t))}[B_{t+\Delta t}-B_t] \\ &+k_{\mathrm{B}}T'(x(t))\{[B_{t+\Delta t}-B_t]^2-\Delta t\}-\frac{\partial U(x(t),a(t))}{\partial x}\Delta t \\ &+o(\Delta t).\end{aligned} \tag{A.23}$$

The first and the second lines on the right-hand side of (A.23) are, respectively, of the order $\mathcal{O}((\Delta t)^{1/2})$ and $\mathcal{O}(\Delta t)$. $[B_{t+\Delta t}-B_t]^2$ is known to obey the so-called χ^2-distribution of one degree of freedom. Those B_t's in the first and the second line of (A.23) represent the same realization (path).

Step (3): In the limit of $\Delta t\to 0$, the term $\{[B_{t+\Delta t}-B_t]^2-\Delta t\}$ vanishes because of the law $dB_t^2=dt$ (see (1.52)). We then arrive at (1.93) and (1.94). When we solve numerically these equations, the formula (A.23) is useful.

This form is essentially identical to so-called Milstein scheme [8] to solve the SDE of the form (1.93). It means that the coarse-grained form (A.23) is not a result peculiar to the elimination of the inertia effect.

---

[18] Note that $a(s)-a(t)$ is treated as $\mathcal{O}(\Delta t)$.

[19] $o(\Delta t)$ denotes the terms higher order than Δt, i.e., $\lim_{\Delta t\to 0}[o(\Delta t)/\Delta t]=0$.

A.1.12 Derivation of (1.108): Appendix to Sect. 1.3.3.2

The reasoning is based on the previous result (1.104): As τ_{op} is the duration time of the protocol, this time can be arbitrarily larger than τ_{erg}. On the other hand, if τ_{op} is large enough, there is a time scale Δt such that $\Delta t \gg \tau_{erg}$ but at the same time the functions $\Phi(\,, a(t))$ and $\mathcal{P}^{eq}(\,, a(t); T)$ remain almost constant in the time domain $[t, t + \Delta t]$. Then we can regard the time integral on the right-hand side of (1.107) as the sum of integrals over $[n\Delta t, (n + 1)\Delta t]$. In each of the integrals,

$$\hat{I}_n \equiv \int_{n\Delta t}^{(n+1)\Delta t} \Phi(\hat{x}(t), a)\frac{da}{dt}dt,$$

$a(t)$ is effectively constant (by definition of Δt) while $\Phi(\hat{x}(t), a)$ can be rewritten as $\int_{-\infty}^{+\infty} \Phi(X, a)\delta(X - \hat{x}(t))dX$. Therefore

$$\begin{aligned}\mathcal{I}_n &\simeq \int_{-\infty}^{+\infty} \Phi(X, a)\left[\frac{1}{\Delta t}\int_{n\Delta t}^{(n+1)\Delta t} \delta(X - \hat{x}(t))dt\right]dX\, \tilde{a}'\left(\frac{n\Delta t}{\tau_{op}}\right)\frac{\Delta t}{\tau_{op}} \\ &\simeq \langle\Phi(\cdot, a)\rangle_{eq}[a((n + 1)\Delta t) - a(n\Delta t)]. \end{aligned} \tag{A.24}$$

Adding up these integral of segments, we find the result (1.108) in the text. If a has $n(> 1)$ components, the product $\Phi\, da$ is understood as the scalar product.

A.1.13 Derivation of the Mean First Passage Time: Appendix to Sect. 1.3.3.3 [20]

A.1.13.1 The Fokker–Planck Equation with Absorbing Boundary Condition

When a Brownian particle obeys the Langevin equation, $0 = -\frac{\partial U(x,a)}{\partial x} - \gamma\frac{dx}{dt} + \xi(t)$, and if we follow the particle only up to its first passage of the boundary of Ω, we can suppose that the particle is absorbed perfectly by the boundary.

In the language of the associated Fokker–Planck equation,

$$\frac{\partial P}{\partial t} = \mathcal{L}P, \tag{A.25}$$

this absorbing boundary imposes the Dirichlet boundary condition on the probability density, $P(x, t) \equiv \langle\delta(x - \hat{x}(t))\rangle$:

$$P(x, t) = 0, \quad x \in \partial\Omega, \tag{A.26}$$

where $\partial\Omega$ denotes the boundary of Ω.

[20] The text follows the derivation in [9].

If (A.25) is solved under this boundary condition and the concentrated initial condition, $P(x,0) = \delta(x - x_0)$, $(x_0 \in \Omega)$, the total probability to find the particle in Ω, i.e., $\int_\Omega P(x,t)dx$, decreases in time for $t > 0$.

A.1.13.2 Green's Function of the Problem

Using the above solution $P(x,t)$, we define

$$G(x|x_0) = -\int_0^{+\infty} P(x,t)dt. \tag{A.27}$$

$G(x|x_0)$ satisfies $\mathcal{L}G(x|x_0) = \delta(x - x_0)$ $(x, x_0 \in \Omega)$ and the Dirichlet condition on x, i.e., $G(x|x_0) = 0, \quad x \in \partial\Omega$. This can be verified by directly operating $\mathcal{L}$ to the right-hand side and using the Fokker–Planck equation. The function $G(x|x_0)$ is called Green's function of the operator $\mathcal{L}$.

A.1.13.3 Integral of $\psi(\hat{x})$ up to the First Passage Time

Let us regard the FPT $\hat{\tau}_\Omega$ as random variable. Each realization of $\hat{x}(\,)$ gives a value of $\hat{\tau}_\Omega$. For any function $\psi(x)$ defined on Ω, the following equality holds:

$$\int_0^{\hat{\tau}_\Omega} \psi(\hat{x}_t)dt = \int_0^{+\infty}\left[\int_\Omega \psi(x)\delta(x - \hat{x}_t)dx\right]dt, \tag{A.28}$$

because the integral in [ ] is $\psi(\hat{x})$ while $\hat{x} \in \Omega$ but is 0 after the absorption at $t = \hat{\tau}_\Omega$.

Next we take the path average of (A.28) with a fixed initial condition, $\hat{x}(0) = x_0$. Using (A.27) and $P(x,t) = \langle \delta(x - \hat{x}(t)) \rangle$, the result is as follows:

$$\left\langle \int_0^{\hat{\tau}_\Omega} \psi(\hat{x}_t)dt \right\rangle = -\int_\Omega \psi(x)G(x|x_0)dx. \tag{A.29}$$

A.1.13.4 A Form of the Feynman–Kac Formula and MFPT

We define $\phi(x)$ as the solution of $\psi(x) = \mathcal{L}^*\phi(x)$ with the Dirichlet boundary condition, $\phi(x) = 0$ for $x \in \partial\Omega$. Then we have

$$\int_\Omega [\mathcal{L}^*\phi(x)]G(x|x_0)dx = \int_\Omega \phi(x)[\mathcal{L}G(x|x_0)]dx = \phi(x_0), \tag{A.30}$$

where the definition of adjoint operator and the property $\mathcal{L}G(x|x_0) = \delta(x - x_0)$ were used. This result is called (a version of) Feynman–Kac formula. Finally putting $\psi(x) = -1$ in (A.29), we obtain

$$\langle \hat{\tau}_\Omega \rangle = \phi(x_0). \tag{A.31}$$

A.1.13.5 Addenda: Potential Theory

Equation (A.27) suggests a way to calculate $G(x|x_0)$. We solve the above Langevin equation with the initial condition $x(0) = x_0$ and measure the residence time in every element dx in Ω. The path average of such residence time is $dx \int_0^{+\infty} P(x,t)dt$. If $U(x,a) = 0$, the equation $\mathcal{L}G = \delta(x - x_0)$ is the Laplace equation, $\Delta G = \delta(x - x_0)$ with a point source. Then the above method gives the solution of Laplace equation with a Dirichlet boundary condition from the observation of free Brownian motion. This type of analysis is called potential theory.

A.2 Appendix to Chap. 2

A.2.1 Maxwell Relation in the Fundamental Relation Assures the Existence of Thermodynamic Function: Appendix to Sect. 2.1.3.2

We show that, if the Maxwell-type relation, $\frac{\partial a}{\partial y} = \frac{\partial b}{\partial x}$, is satisfied in the equation of differentials, $dz = a(x,y)dx + b(x,y)dy$, then we can reconstitute a monovalent function $z = z(x,y)$.

It is sufficient to show this fact on a small segment, $\Box ABCD$, with $A = (0,0)$, $B = (\Delta x, 0)$, $C = (\Delta x, \Delta y)$, and $D = (0, \Delta y)$. We show that the integral of $dz = a(x,y)dx + b(x,y)dy$ along the path $A \to B \to C$, denoted by $\Delta z(ABC)$, and the integral along the path $A \to D \to C$, denoted by $\Delta z(ADC)$, give the same value up to $\mathcal{O}(\Delta x \Delta y)$. Using the illustration in Fig. A.1, we find

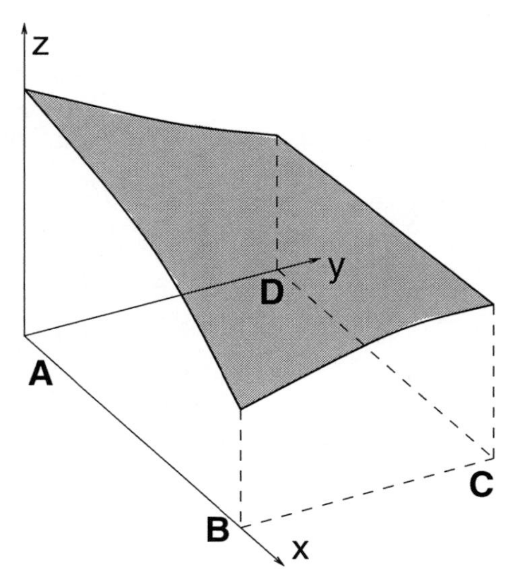

Fig. A.1 Integrals of $dz = a(x,y)dx + b(x,y)dy$ along the path $A \to B \to C$ and along the path $A \to D \to C$

$$\begin{aligned}
&\Delta z(ADC) - \Delta z(ABC) \\
&= [\Delta z(DC) - \Delta z(AB)] - [\Delta z(BC) - \Delta z(AD)] \\
&\simeq [a(\frac{1}{2}\Delta x, \Delta y) - a(\frac{1}{2}\Delta x, 0)]\Delta x - [b(\Delta x, \frac{1}{2}\Delta y) - b(0, \frac{1}{2}\Delta y)]\Delta y \\
&\simeq [\frac{\partial a}{\partial y}\Delta y]\Delta x - [\frac{\partial b}{\partial x}\Delta x]\Delta y. \qquad \text{(A.32)}
\end{aligned}$$

Then the relation $\frac{\partial a}{\partial y} = \frac{\partial b}{\partial x}$ assures the existence of a function $z(x, y)$.

A.2.2 *Invariance of Thermodynamic Relations and the Choice of Reference Energy and Entropy: Appendix to Sect. 2.1.3.4*

In Newtonian mechanics, the choice of a nonaccelerating frame (inertial frame) of reference is arbitrary, that is, the form of Newton equation of motion is invariant under change of the frame (the principle of relativity of Galilee). In classical (vs. quantum) macroscopic thermodynamics, the choice of zero points of energy and entropy is arbitrary (to be stated more precisely below). We expect, therefore, that the thermodynamic relations discussed in the text are invariant under the change of these zero points.

Zero volume, $V = 0$, and zero particles, $N = 0$, have direct physical meanings. However, the zero point of the energy or the entropy *per particle* is a matter of choice in the classical (*not* quantum) thermodynamics. The following redefinitions of the energy and entropy of a system with N particles should, therefore, have no physical consequences [10]:

$$E \mapsto \tilde{E} = E + e^* N, \quad S \mapsto \tilde{S} = S + s^* N, \qquad \text{(A.33)}$$

where e^* and s^* are arbitrary constants. Evidently the transformations of (A.33) are consistent with (2.4).

The less evident consequences of the above transformations (A.33) are the changes in other complete thermodynamic functions derived through the Legendre transformation [21]:

$$\begin{aligned}
F \equiv E - TS \; &\mapsto \tilde{F} = F + (e^* - Ts^*)N \\
G \equiv E - TS + pV \; &\mapsto \tilde{G} = G + (e^* - Ts^*)N. \qquad \text{(A.34)}
\end{aligned}$$

The last formula defines how the chemical potential, μ, should transform: With the relation $G = \mu N$, we have

$$\mu \mapsto \tilde{\mu} = \mu + (e^* - Ts^*). \qquad \text{(A.35)}$$

[21] From here up to the end of this section we will take the example of a single-component gas system: The generalization of the proofs is not difficult.

The following two examples demonstrate the invariance of the basic thermodynamic relations (cf. Sect. 2.1.3.1).

(i) Fundamental relation:

$$dE = TdS - pdV + \mu dN. \tag{A.36}$$

We rewrite this relation in terms of $\tilde{E}$, $\tilde{S}$, and $\tilde{\mu}$, using (A.33) and (A.35). The result is $d\tilde{E} = Td\tilde{S} - pdV + \tilde{\mu}dN$, which is identical form to the original one.

(ii) Thermodynamic derivative:

$$\frac{\mu}{T} = \frac{\partial S(E, V, N)}{\partial N}. \tag{A.37}$$

We notice the change of the arguments of the entropy: $\tilde{S}(\tilde{E}, V, N) = S(E, V, N) + s^*N = S(\tilde{E} - e^*N, V, N) + s^*N$. The partial derivative of this expression with respect to N then yields

$$\frac{\partial \tilde{S}}{\partial N} = \frac{\partial}{\partial N}[S(\tilde{E} - e^*N, V, N) + s^*N] = (-e^*)\frac{\partial S}{\partial E} + \frac{\mu}{T} + s^*. \tag{A.38}$$

In using $\partial S/\partial E = 1/T$ and (A.35), the right-hand side becomes $(-e^*)/T + \mu/T + s^* = \tilde{\mu}/T$.

Remark: When we deal with the multicomponent systems with chemical reactions, we must respect the constraints among the e^*'s and s^*'s. Where the quantum effects appear, the arbitrariness of the additive constants is more restricted. When several chemical substances participate in a chemical reaction, the third law of thermodynamics (Sect. 2.1.2) about the uniqueness of the ground state at $T = 0$ induces the relations among the specific quantities of those substances. See, for example, Chap. 9 of [11].

A.3 Appendix to Chap. 3

A.3.1 Derivation of (3.27): Appendix to Sect. 3.3.1.4

The inequality (3.27), that is, $D(\boldsymbol{P} \parallel \boldsymbol{Q}) \geq D(\mathsf{K}\boldsymbol{P} \parallel \mathsf{K}\boldsymbol{Q})$, is a special case of the "log sum inequality," (A.40) below.[22] The starting point is Jensen's inequality,

[22] The demonstration below follows [12]. Y. Oono pointed this out to me.

which holds for any strictly convex function $f(t)$,[23]

$$\sum_{i=1}^{n} \alpha_i f(t_i) \geq f\left(\sum_{i=1}^{n} \alpha_i t_i\right), \tag{A.39}$$

where $\alpha_i \geq 0$, $\sum_{i=1}^{n} \alpha_i = 1$. By substituting $f(t) = t \ln t$, $\alpha_i = b_i / \sum_{j=1}^{n} b_j$, and $t_i = a_i/b_i$ in (A.39) with $a_i > 0$ and $b_i > 0$, we have the log sum inequality:

$$\sum_{j=1}^{n} a_j \ln \frac{a_j}{b_j} \geq \left(\sum_{j=1}^{n} a_j\right) \ln \frac{\sum_{k=1}^{n} a_k}{\sum_{l=1}^{n} b_l}. \tag{A.40}$$

We then further substitute $a_j = a_j^{(i)} \equiv K_{ij} P_j$ and $b_j = b_j^{(i)} \equiv K_{ij} Q_j$ in (A.40) and take the sum over i of each side of the inequality. The result is

$$\sum_{i=1}^{n} \sum_{j=1}^{n} K_{ij} P_j \ln \frac{P_j}{Q_j} \geq \sum_{i=1}^{n} \left(\sum_{j=1}^{n} K_{ij} P_j\right) \ln \frac{\sum_{k=1}^{n} K_{ik} P_k}{\sum_{l=1}^{n} K_{il} Q_l}. \tag{A.41}$$

Using $\sum_{i=1}^{n} K_{ij} = 1$ on the left-hand side, we arrive at (3.27).

A.3.2 Derivation of (3.72) and (3.73): Appendix to Sect. 3.3.3.4

The simplifying approximations we use are:

(1) That the formation of the complex ES is a Markov process, so that the probability $P_v(t_v)$ that the enzyme remains vacant, E, for the period up to t_v is[24]

$$P_v(t_v) = e^{-t_v/T_v}.$$

We then have the average and the variance: $\langle t_v \rangle = T_v$, and $\langle {t_v}^2 \rangle - \langle t_v \rangle^2 = {T_v}^2$.

(2) The transition rate ${T_v}^{-1}$ is proportional to the concentration of the substrate (around the active site of the enzyme). It is, therefore, ${T_v}^{-1} = \kappa[S]$ as written above.

(3) That the termination of the complex ES is a Markov process *and* is independent of its fate, either E+S or E+P. The probability $P_r(t_r)$ that the complex ES exists for the period t_r is

$$P_r(t_r) = e^{-t_r/T_r}.$$

We then have the average and the variance: $\langle t_r \rangle = T_r$, and $\langle {t_r}^2 \rangle - \langle t_r \rangle^2 = {T_r}^2$.

---

[23] The strictly convex function is defined such that $\frac{1}{2}(f(x) + f(y)) \geq f(\frac{x+y}{2})$.

[24] To obtain this, one should solve $dP_v(t_v) = -{T_v}^{-1} dt_v$ with the initial condition, $P_v(0) = 1$. Similar argument applies to $P_r(t_r)$ below.

(4) That the probability by which the complex ES dissociates into E+P, rather than into E+S, is independent of the past history, so that the probability $P_P(n)$ that the product P is formed in the nth formation of the complex ES is

$$P_P(n) = (1-q)q^n.$$

We then have the average and the variance: $\langle n \rangle = (1-q)^{-1}$ and $\langle n^2 \rangle - \langle n \rangle^2 = (1-q)^{-2}q$.

With the assumption made above, the sum in (3.71), that is, $\sum_{k=1}^{n}(t_r^{(k)} + t_v^{(k)})$ consists of independent random numbers, $t_r^{(k)}$, $t_v^{(k)}$, and n. Therefore, we have the mean value of t_P and that of $t_P{}^2$ as

$$\langle t_P \rangle = \sum_{k=1}^{n} P_P(n) \left\langle \sum_{k=1}^{n} (t_r^{(k)} + t_v^{(k)}) \right\rangle = \langle n \rangle [\langle t_v \rangle + \langle t_r \rangle], \tag{A.42}$$

$$\begin{aligned} \langle t_P{}^2 \rangle &= \sum_{k=1}^{n} P_P(n) \left\langle \left[\sum_{k=1}^{n} (t_r^{(k)} + t_v^{(k)}) \right]^2 \right\rangle \\ &= \langle n^2 \rangle [\langle t_v \rangle + \langle t_r \rangle]^2 + \langle n \rangle [\langle t_v{}^2 \rangle - \langle t_v \rangle^2 + \langle t_r{}^2 \rangle - \langle t_r \rangle^2]. \end{aligned} \tag{A.43}$$

By substituting the averages and variances of t_v, t_r, and n, into (A.42) and (A.43), we finally obtain the two concise equations (3.72) and (3.73) in the text.

A.4 Appendix to Chap. 4

A.4.1 Derivation of (4.13): Appendix to Sect. 4.1.2.2

We prepare an SDE of Stratonovich type which corresponds to (4.1):

$$dp = -\gamma \frac{p}{m} dt - \frac{\partial U}{\partial x} dt + \sqrt{2\gamma k_B T}\, dB_t.$$

We multiply each term of this equation by (p/m). Here the product of Stratonovich type with p_t and that of Itô type are related via $\circ p_t \Leftrightarrow \cdot(p_t + dp_t/2)$. In the $dp_t/2$ we can reiterate the above SDE. Finally, using $(dB_t)^2 = dt$, we arrive at (4.13), i.e.,

$$dE = -\frac{2\gamma}{m}\left(\frac{p^2}{2m} - \frac{k_B T}{2}\right) dt + d'W + \sqrt{2\gamma k_B T}\, \frac{p}{m} \cdot dB_t.$$

In order to recover the energy balance equation (4.7) from (4.13), we apply the identity, $d(p^2/2m) = (p/m) \circ dp$ in (4.13). Then we rewrite dp using the SDE (1.61). Finally we rewrite $p \circ dB_t$ using (A.14), where x should be reread as p, and $b(p_t, t)$ should be reread as $\sqrt{2\gamma k_B T}$.

A.4.2 Error in the Euler Scheme: Appendix to Sect. 4.1.2.5

When we integrate (4.14) from t to $t+h$, the result may differ from (4.15) only due to the error in the term $dU(\tilde{x}_t)/dx$. If we expand this in term in powers of $(\theta - \frac{1}{2})$, we have

$$\frac{dU(\tilde{x}_t)}{dx} = \left.\frac{dU(\tilde{x}_t)}{dx}\right|_{\theta=\frac{1}{2}} + \left(\theta - \frac{1}{2}\right) \left.\frac{d^2U(\tilde{x}_t)}{dx^2}\right|_{\theta=\frac{1}{2}} (x_{t+h} - x_t) + \cdots . \quad \text{(A.44)}$$

This is also the expansion in powers of $(x_{t+h} - x_t) = \mathcal{O}(h^{0.5})$.[25] Therefore, the error in (4.14) is $\mathcal{O}(h^{1.5})$. The approximate solution $x(t)$ is basically the sum of (4.14), and the cumulated error in $x(t)$ is, therefore, $\mathcal{O}(h^{1.5}) \times N = \mathcal{O}(h^{0.5})$.

About the energetics, we calculate $\Delta r = \Delta U - \Delta Q$ defined in the main text. With the discretization of the solution (4.14), the heat (4.16) is $\Delta Q = \frac{dU(\tilde{x}_t)}{dx}(x_{t+h} - x_t)$. We expand $U(x_t)$ and $U(x_{t+h})$ in ΔU, as well as $U(\tilde{x}_t)$, around $\bar{x}_t \equiv (x_t + x_{t+h})/2$ up to second order. We then have

$$\begin{aligned} \Delta r &= \left(\theta - \frac{1}{2}\right) \left.\frac{d^2U(\tilde{x}_t)}{dx^2}\right|_{\theta=\frac{1}{2}} (x_{t+h} - x_t)^2 + o(h) \\ &\simeq \left(\theta - \frac{1}{2}\right) \frac{d^2U(\bar{x}_t)}{dx^2} \frac{2k_BT}{\gamma} h + o(h), \end{aligned} \quad \text{(A.45)}$$

where we have used $(x_{t+h} - x_t)^2 \sim h$ to go to the second line, and $o(h)$ is such that $\lim_{h\to 0} o(h)/h = 0$. Therefore, the cumulated error should become $N\mathcal{O}(h) = \mathcal{O}(1)$ unless we choose $\theta = \frac{1}{2}$.

A.4.3 Derivation of (4.27): Appendix to Sect. 4.1.3.2

Substitution of (4.2) into $d'Q = (-\gamma dx/dt + \xi(t)) \circ dx$ gives $d'Q = (\partial U/\partial x) \circ /dx$. We then substitute for dx the SDE corresponding to (4.2), i.e., $dx = -\gamma^{-1}(\partial U/\partial x)dt + \sqrt{2k_BT/\gamma}\, dB_t$. The product $(\partial U/\partial x) \circ dB_t$ can be converted into the Itô-type representation using the Itô's lemma:[26]

$$\frac{\partial U}{\partial x} \circ dB_t = \frac{\partial U}{\partial x} \cdot dB_t + \frac{1}{2}\frac{\partial^2 U}{\partial x^2}\sqrt{2k_BT/\gamma}\, dt. \quad \text{(A.46)}$$

---

[25] In a very short time, x_t undergoes free Brownian motion, and therefore, $(x_{t+h} - x_t) \simeq \frac{1}{\gamma} w_{t,t+h} = \mathcal{O}(h^{0.5})$.

[26] cf. (1.65).

Taking everything together, we find a less obvious formula for $d'Q$ as compared with (4.25):

$$d'Q = -\frac{1}{\gamma}\left[\left(\frac{\partial U}{\partial x}\right)^2 - k_B T \frac{\partial^2 U}{\partial x^2}\right]dt + \frac{\partial U}{\partial x} \cdot \sqrt{2k_B T/\gamma}\, dB_t. \tag{A.47}$$

Because $\frac{\partial U}{\partial x}$ in the rightmost term is nonanticipating with respect to dB_t, we obtain the result (4.27) in the text.

A.4.4 Derivation of (4.36): Appendix to Sect. 4.2.1.1

Following exactly the same procedure as in Sect. 4.1.3.1 for a single thermal environment, we can show the following expression for $d'Q_i$:

$$d'Q_i = -\frac{2\gamma_i}{m_i}\left(\frac{p_i^2}{2m_i} - \frac{k_B T_i}{2}\right) dt + \sqrt{2\gamma k_B T_i}\, \frac{p_i}{m_i} \cdot dB_{i,t}, \tag{A.48}$$

where the Wiener processes $\{B_{i,t}\}$ are mutually independent and satisfy $(dB_{i,t})^2 = dt$.

If the effect of inertia is negligible, we will use, instead of (4.32), the following Langevin equation:

$$0 = -\gamma_i \frac{dx_i}{dt} - \frac{\partial U}{\partial x_i} + \sqrt{2\gamma_i k_B T_i}\, \theta_i(t), \tag{A.49}$$

and the energy balance (4.35) holds as before except that the energy E should be read as U. For the ith heat, $d'Q_i = \left(-\gamma \frac{dx_i}{dt} + \xi_i(t)\right) \circ dx_i$, we can rewrite as follows:

$$d'Q_i = -\frac{1}{\gamma_i}\left[\left(\frac{\partial U}{\partial x_i}\right)^2 - k_B T_i \frac{\partial^2 U}{\partial x_i^2}\right]dt + \frac{\partial U}{\partial x_i} \cdot \sqrt{2k_B T_i/\gamma_i}\, dB_{i,t}. \tag{A.50}$$

Given these formula of $d'Q_i$, we rewrite the terms $\propto dt$, the terms which are *not* nonanticipating, using probability fluxes and probability density. We introduce the following identities:

$$\frac{\partial E}{\partial X_i} J_{i,x} + \frac{\partial E}{\partial P_i} J_{i,p} = -\frac{2\gamma_i}{m_i}\left[\frac{P_i^2}{2m_i} - \frac{k_B T_i}{2}\right]\mathcal{P} - \frac{\partial}{\partial P_i}\left[\gamma_i k_B T_i \frac{P_i}{m_i}\mathcal{P}\right], \tag{A.51}$$

$$\frac{\partial U}{\partial X_i} J_{i,x} = -\frac{1}{\gamma_i}\left[\left(\frac{\partial U}{\partial x_i}\right)^2 - k_B T_i \frac{\partial^2 U}{\partial x_i^2}\right]\mathcal{P} - \frac{k_B T_i}{\gamma_i}\frac{\partial}{\partial X_i}\left[\frac{\partial U}{\partial X_i}\mathcal{P}\right], \tag{A.52}$$

Combining (A.48) with (A.51) and also (A.50) with (A.52), we arrive at (4.36) in the text.

A.4.5 Derivation of (4.55): Appendix to Sect. 4.2.1.2

Hereafter we use the unit of $K = k_{\mathrm{B}} = 1$. The calculation of $\mathcal{D}$ is somewhat lengthy. We use the relations

$$\mu \circ (\mu - \xi) = (\mu - T) - \mu \cdot \xi, \quad \mu \circ (\mu + \xi') = (\mu - T') + \mu \cdot \xi',$$

which yield

$$\gamma d'Q = -(\mu^2 - T)dt + \mu \cdot \sqrt{2\gamma T} dB_t, \quad \gamma' d'Q' = -(\mu^2 - T')dt - \mu \cdot \sqrt{2\gamma' T'} dB'_t.$$

If we ignore the initial term, the explicit solution of $\mu(t)$ is

$$\mu(t) = \int_0^t e^{-a(t-s)} (\sqrt{2D} dB_s - \sqrt{2D'} dB'_s).$$

In the double integral to calculate the left-hand side of (4.54), we can replace $dB_{u_1} dB_{u_2}$ by $\delta(u_1 - u_2) du_1 du_2$. We also use the formula about the average of the product of four Gaussian variables:

$$\langle ABCD \rangle = \langle AB \rangle \langle CD \rangle + \langle AC \rangle \langle BD \rangle + \langle AD \rangle \langle BC \rangle.$$

We then have the following formulas:

$$\langle \mu(u_1) \mu(u_2) \rangle = (\gamma' T + \gamma T')/(\gamma + \gamma') e^{-a|u_1 - u_2|},$$
$$\langle \mu(u_1) \cdot \sqrt{2\gamma T} dB_{u_1} \rangle = 2T\theta(u_1 - u_2) e^{-a(u_1 - u_2)} du_2,$$

where $a \equiv 1/\gamma + 1/\gamma'$. Putting everything together in the double integral of (4.54), we arrive at (4.55).

A.4.6 Definition of the Energy of Open System: Appendix to Sect. 4.2.3.3

1. In the presence of interparticle interactions, the single-particle energy $U_1(x_1, a)$ for a particle at x_1 is defined as the energy of this particle when it is well isolated from all the other $N - 1$ particles.
2. The two-particle interaction energy, $U_2(x_1, x_2)$, between one particle at x_1 and the other one at x_2 is defined so that the energy of these particles is $U_1(x_1, a)+$

$U_1(x_2, a)+ U_2(x_1, x_2)$ when these two particles are well isolated from all the other $N-2$ particles.

3. In the same manner the p-particle interaction energy $U_p(x_1, \ldots, x_p)$ $(p \geq 3)$ for a cluster of p particles is defined as the energy to be added to all the contribution of $U_1, \ldots, U_{p-1}$, when this cluster of p particles is well isolated from the $(N-p)$ other particles. By definition, $U_p(x_1, \ldots, x_p)$ $(p \geq 2)$ vanishes unless all these positions are close to each other.
4. The characteristic function for the p-particle cluster, $\theta_\Omega^{(p)}(x_1, \ldots, x_p)$, $(p \geq 1)$, can be written in terms of $\theta_\Omega(x)$ as follows:

$$\theta_\Omega^{(p)}(x_1, \ldots, x_p) \equiv 1 - \Pi_{j=1}^p [1 - \theta_\Omega(x_j)]. \tag{A.53}$$

5. Using $U_p(x_1, \ldots, x_p)$ and $\theta_\Omega^{(p)}(x_1, \ldots, x_p)$, the energy of the open system, E, is written as (4.65) in the text.

A.4.7 Application of Energy Balance to Other Forms of Langevin Equation: Appendix to Sect. 4.3

A.4.7.1 Fluctuating Hydrodynamics

Summary of the Framework of Fluctuating Hydrodynamics

The dynamics of the fluctuating hydrodynamics is given by a Langevin equation for the velocity field [13]. The evolution of the mass density ρ and the velocity field v is

$$\frac{\partial \rho}{\partial t} + \nabla \cdot (\rho v) = 0, \tag{A.54}$$

$$\rho\left(\frac{\partial v}{\partial t} + v \cdot \nabla v\right) = \rho \boldsymbol{f}_{\text{ext}} - \nabla(p\mathbf{1} - \sigma'), \tag{A.55}$$

where p is the hydrostatic pressure and $\boldsymbol{f}_{\text{ext}}$ is the distant force per mass exerted by an external system. The symmetric tensor σ' represents the deviatoric (traceless) stress, with $(-p\mathbf{1}+\sigma')$ being the stress.[27] This σ' is [13] (α and β are the Cartesian indexes)

$$\sigma'_{\alpha\beta} = 2\eta\left(\frac{\partial v_\alpha}{\partial x_\beta} + \frac{\partial v_\beta}{\partial x_\alpha} - \frac{2}{3}\delta_{\alpha\beta}\sum_\lambda \frac{\partial v_\lambda}{\partial x_\lambda}\right) + s_{\alpha\beta}, \tag{A.56}$$

where η is the shear viscosity of the fluid. For simplicity of demonstration we have ignored the second viscosity against dilatation/contraction. The last term $s_{\alpha\beta}$

[27] That is, $(p\mathbf{1} - \sigma')$ is the conductive momentum flux.

is the deviatric thermal random stress, obeying white Gaussian statistics with the following moments:

$$\langle s_{\alpha\beta}(x,t)\rangle = 0. \tag{A.57}$$

$$\langle s_{\alpha\beta}(x,t)s_{\mu\nu}(x',t')\rangle = 2\eta k_{\mathrm{B}}T\delta(x-x')\delta(t-t')\left[\delta_{\alpha\mu}\delta_{\beta\nu}+\delta_{\alpha\nu}\delta_{\beta\mu}-\frac{2}{3}\delta_{\alpha\beta}\delta_{\mu\nu}\right]. \tag{A.58}$$

We denote by $V(t)$ a connected volume which moves with the mass of the fluid. The equation of continuity (A.54) is, simply,[28]

$$\frac{d}{dt}\int_{V(t)}\rho\, dV = 0. \tag{A.59}$$

(A.55) is[29]

$$\frac{d}{dt}\int_{V(t)}\rho v\, dV = \int_{V(t)}\rho \boldsymbol{f}_{\mathrm{ext}}\, dV - \oint_{\partial V(t)}(p\mathbf{1}-\sigma')\cdot\hat{n}\, dS. \tag{A.60}$$

The surface integral on the right-hand side cancels with its neighbors because the outward unit normal vector $\hat{n}$ of the neighboring volume is oriented opposite to that of $V(t)$.

We introduce further simplifying assumptions that the heat capacity and the thermal conductivity of the environment are large enough that the temperature T is kept constant. We define $\tilde{u}$ by $d\tilde{u} = -p\,d\left(\frac{1}{\rho}\right)$. This represents the change of the Helmholtz free energy in the isothermal process. By taking the scalar product of each term of (A.55) with the velocity v in the Stratonovich sense and also using (A.54), the balance of the *total* energy of the fluid is [30]

$$\frac{\partial}{\partial t}\rho\left(\frac{v^2}{2}+\tilde{u}\right)+\nabla\cdot\left[\rho\left(\frac{v^2}{2}+\tilde{u}\right)v+(p\mathbf{1}-\sigma')\cdot v\right] = -\sigma':(\nabla v)+\rho\boldsymbol{f}_{\mathrm{ext}}\cdot v. \tag{A.61}$$

[28] The derivation uses the kinematical identity, $\frac{d}{dt}\int_{V(t)}\mathcal{A}dV = \int_{V(t)}\frac{\partial\mathcal{A}}{\partial t}dV + \oint_{\partial V(t)}\mathcal{A}\, v\cdot\hat{n}\, dS$, and the integral theorem of Gauss, $\oint_{\partial V}\mathcal{B}\hat{n}\, dS = \int_V \nabla\mathcal{B}\, dV$. $\partial V(t)$, is the surface of the volume $V(t)$, and $\hat{\boldsymbol{n}}$ is the unit outward normal vector on this surface.

[29] The derivation uses the kinematic identity mentioned before with (A.54), and also rewriting, $v\nabla\cdot(\rho v) = -\nabla\cdot(\rho v v)+\rho v\cdot\nabla v$.

[30] S.I. Sasa pointed out an early mistake of the author about the treatment of the terms including σ'. The symbol "$\cdot$" hereafter is not for Itô calculus, but for the scalar product. The symbol ":" of $\mathcal{A}:\mathcal{B}$ for symmetric any rank-2 tensors $\mathcal{A}$ and $\mathcal{B}$ means $\sum_\alpha\sum_\beta \mathcal{A}_{\alpha\beta}\mathcal{B}_{\alpha\beta}$. The above derivation uses $(v\nabla):\sigma' = \nabla v:\sigma' - \sigma':\nabla v$.

Hereafter, all the ∇v in $\sigma' : (\nabla v)$ can be replaced by its symmetric deviatric part[31] due to the deviatric and symmetric property of σ'. The integral form of (A.61) is

$$\frac{d}{dt}\int_{V(t)} \rho\left(\frac{v^2}{2}+\tilde{u}\right)dV = -\int_{V(t)} \sigma' : (\nabla v)dV + \int_{V(t)} \rho \boldsymbol{f}_{\text{ext}} \cdot v dV - \oint_{\partial V(t)} (p\mathbf{1}-\sigma') : (v\hat{n})dA. \tag{A.62}$$

As in (A.60) the surface integral in the second line on the right-hand side cancels with its neighbors.

Stochastic Energetics

The object is to rewrite (A.62) so that the structure of the energy balance, $dE = d'Q + d'W$, is visible.

For the fluctuating fluid system, the system is "fluid particle," or macroscopically small connected element of mass that contains a large number of fluid molecules. The local state of this system is characterized by the density field ρ and the velocity field v. The (local) thermal environment of the system consists, therefore, of the degrees of freedom in the fluid *less* the fields ρ and v.

When there is no mechanical support within the fluid, the momentum conservation, or the Galilee invariance, prohibits an isolated spontaneous force as a vector. The random thermal force, therefore, takes the form of the force dipole (a symmetric tensor). This is the entity of $s_{\alpha\beta}$. The $s_{\alpha\beta}$ arises as a result of coincidental spatiotemporal local coherence in the molecular motions.

$\boldsymbol{\sigma}'$ of (A.56) is the force dipole on the system from the environment. The two terms on the right-hand side of (A.56), therefore, play the roles of $(-\gamma\frac{dx}{dt})$ and $\xi(t)$ of the Brownian motion.

Let us regard $V(t)$ as the volume occupied by a system, i.e., a fluid particle. Equation (A.62) then describes the energy balance for the system. As mentioned above, the surface integral in the second line on the right-hand side cancels with its neighbors. The remaining terms look similar to the form of the energy balance, $dE = d'Q + d'W$.

We should, however, remember the second remark about the work (Sect. 4.1.2.2): If $\boldsymbol{f}_{\text{ext}}$ is constant, for example, the work by the external system is 0, while $\boldsymbol{f}_{\text{ext}} \cdot v$ in (A.62) does not vanish. We will then rewrite (A.62) in the case that the external force has a potential, i.e.,

$$\boldsymbol{f}_{\text{ext}} = -\nabla \Phi_{\text{ext}}(x, a), \tag{A.63}$$

with a being the control parameter. Concomitantly we define the new energy that includes Φ_{ext}: $u \equiv \tilde{u} + \Phi_{\text{ext}} = -\int p\, d\left(\frac{1}{\rho}\right) + \Phi_{\text{ext}}$. We rewrite the force term using the following identity:[32]

---

[31] $\frac{1}{2}[\nabla v + (\nabla v)^t - \frac{2}{3}(\nabla \cdot v)\mathbf{1}]$.

[32] The derivation uses the aforementioned kinetic identity and (A.54).

$$\int_{V(t)} \rho \boldsymbol{f}_{\mathrm{ext}} \cdot v \, dV = -\frac{d}{dt} \int_{V(t)} \rho \, \Phi_{\mathrm{ext}} \, dV + \int_{V(t)} \rho \frac{\partial \, \Phi_{\mathrm{ext}}}{\partial a} \frac{da}{dt} dV.$$

We finally have

$$\begin{aligned} d\left[\int_{V(t)} \rho \left(\frac{v^2}{2} + u \right) dV \right] &= -dt \int_{V(t)} \sigma' : (\nabla v) dV + da \int_{V(t)} \rho \frac{\partial u}{\partial a} dV \\ &\quad - dt \oint_{\partial V(t)} (p\mathbf{1} - \sigma') : (v\hat{n}) dA \\ &= dt \int_{V(t)} (\nabla \cdot \sigma') \cdot v dV + da \int_{V(t)} \rho \frac{\partial u}{\partial a} dV \\ &\quad - dt \oint_{\partial V(t)} p v \cdot \hat{n} dA. \end{aligned} \tag{A.64}$$

This formula shows the balance of energy: $dE = d'Q + d'W +$ (surface term). In the second equality, the surface term transmits only the mechanical force, not the force due to the thermal environment. Note that $\nabla \cdot \boldsymbol{\sigma}'$ is the force on the fluid element.

In case a part of the boundary $\partial V(t)$ is the surface of a "bead," the equation is more convenient, because its surface term is the work by the *total* force on the bead. The isotropic form of the stress (A.56) may be modified on the surface of the bead, but its importance on the total force on the bead is not yet clear.[33]

A.4.7.2 Suspension of Hard Spheres

The fluctuating motion of suspended particle has been studied since long time [14, 15]. Recently, a systematic derivation of the Langevin equation of the suspended particles was carried out [16, 17]. It is done by noting that the spatiotemporal scales of movement of suspended particles are much bigger than those of solvent molecules. For monodisperse suspension particles of mass m and radius a, their positions $\{x_i(t)\}$ ($i, j, \ldots$ distinguish the particles) obey the following equation [16, 17]:

$$m \frac{d^2 x_i}{dt^2} = -\frac{\partial U(\{x_j\}, a)}{\partial x_i} + M_i(t), \tag{A.65}$$

$$M_i(t) = -\zeta_0 \frac{dx_i}{dt} - \sum_{j(\neq i)} \mathcal{G}(x_i - x_j) \cdot M_j + R_i(t), \tag{A.66}$$

where $\zeta_0 = 6\pi \eta a$ with η being the viscosity of the solvent and $\mathcal{G}$ represents the hydrodynamic interaction (of Oseen and dipole types):

[33] Discussion with M. Schindler at ESPCI on this issue is acknowledged.

$$\mathcal{G}(x) = \frac{3}{4}\frac{a}{\|x\|}\left[\mathbf{1} + \frac{x}{\|x\|}\frac{x}{\|x\|}\right] + \frac{1}{2}\left(\frac{a}{\|x\|}\right)^3\left[\mathbf{1} - 3\frac{x}{\|x\|}\frac{x}{\|x\|}\right]. \tag{A.67}$$

R_i is the Gaussian Markov random force on the ith particle, with zero mean and

$$\langle R_i(t)R_j(t')\rangle = \begin{cases} 2\zeta_0 k_{\mathrm{B}} T\delta(t-t')\mathbf{1} & (i=j) \\ 2\zeta_0 k_{\mathrm{B}} T\delta(t-t')\mathcal{G}(x_i - x_j) & (i \neq j) \end{cases}. \tag{A.68}$$

In stochastic energetics, we regard the solvent as the thermal environment. Since the forces M_i are exerted by the solvent, the heat can be defined. Let us denote by dx_i the true displacement of ith sphere obeying (A.65) and (A.66). By multiplying, in the Stratonovich sense, each term of (A.66) by $dx_i(t)$ and summing over the particles (the index i), we have $dE = d'Q + d'W$, where

$$\begin{aligned} E &= \sum_i \frac{m}{2}\left(\frac{dx}{dt}\right)^2 + U(\{x_j\}, a), \\ d'Q &= \sum_i M_i \cdot dx_i, \\ d'W &= \frac{\partial U(\{x_j\}, a)}{\partial a} da. \end{aligned} \tag{A.69}$$

Here all the kinetic energy of the solvent is ignored.

A.4.7.3 Langevin Equation on Manifolds

It often occurs that the stochastic variable, $\hat{x}(t)$, is of more than one component and is defined on a curved space. For example, $\hat{x}(t)$ can represent the position of a membrane protein on a spherical lipid bilayer membrane. Or, $\hat{x}(t)$ can represent a unit vector that undergoes the 3D rotational Brownian motion. Also $\hat{x}(t)$ can be the variable parameterizing complicated conformations of the protein. One way to represent the fluctuating motion in a curved space is to regard it as the motion within an Euclid space of higher dimensionality, subjected under constraining/penalty potential energy. A formulation using the Lagrange multiplier is found in [18]. Another way is to use the covariant, or coordinate-independent, formulation of Langevin equation [19]. The latter is also useful for flat space with curved coordinate system. Below we discuss the latter way of description.

When we study stochastic processes in a curved space and/or by using non-Cartesian coordinates, we need to use the Riemannian metric tensor, g. This tensor gives the infinitesimal distance ds through the formula, $ds^2 = \sum_{\mu,\nu} g_{\mu\nu} dx^\mu dx^\nu$.[34]

[34] A concise and clear introduction to the metric geometry is found in a book of Dirac [20]. Below, we will also use g as the determinant of $g_{\mu\nu}$ and $g^{\mu\nu}$ as the inverse matrix of $g_{\mu\nu}$.

The covariant Langevin equation is [19] [35]

$$\gamma(x^\mu_{m+1} - x^\mu_m) = \left[-g^{\mu\nu} \frac{\partial U(x_m, a)}{\partial x^\nu} + \frac{k_B T}{\sqrt{g(x_m)}} \frac{\partial(\sqrt{g} g^{\mu\nu})}{\partial x^\nu} \right] \Delta t + \eta^\mu_m \sqrt{\Delta t}, \quad \text{(A.70)}$$

$$\langle \eta^\mu_m \rangle = 0, \quad \langle \eta^\mu_m \eta^\nu_n \rangle = 2\gamma k_B T g^{\mu\nu}(x_n) \delta_{mn}, \quad \text{(A.71)}$$

where time is discretized and represented by the indices m and n, etc. As for the multiplicative noise η_n, the limit of $\Delta t \to 0$ should be interpreted as of the Itô type. That is, the random force η_n takes place between t_n and t_{n+1} while x_n is the value at t_n. (See, Examples 1 and 2 of Appendix Sect. A.1.6).

The above SDE is equivalent to the following Fokker–Planck equation for the scalar density $\rho = \sqrt{g} P$ with P being the probability density:

$$\frac{\partial \rho}{\partial t} = \frac{1}{\gamma} \frac{\partial}{\partial x^\mu} \sqrt{g} g^{\mu\nu} \left\{ \frac{\partial U}{\partial x^\nu} \frac{\rho}{\sqrt{g}} + k_B T \frac{\partial}{\partial x^\nu} \left(\frac{\rho}{\sqrt{g}} \right) \right\}. \quad \text{(A.72)}$$

For a being fixed, the canonical equilibrium distribution $P \propto e^{-U/k_B T}$ is the stationary solution of the above Fokker–Planck equation.

We can obtain the energy balance, $\Delta U = \Delta' W + \Delta' Q$, by operating $(\Delta t)^{-1} g_{\mu\sigma}$ $(x^\sigma_{m+1} - x^\sigma_m)$ on each term of (A.70)[36]:

$$\Delta U = U(x_{m+1}, a_{m+1}) - U(x_m, a_m)$$

$$\Delta' W = \frac{1}{2} \left[\frac{\partial U(x_{m+1}, a_{m+1})}{\partial a} + \frac{\partial U(x_m, a_m)}{\partial a} \right] [a_{m+1} - a_m]$$

$$\begin{aligned} \Delta' Q &= \left[-\gamma \frac{x^\mu_{m+1} - x^\mu_m}{\Delta t} + \frac{k_B T}{\sqrt{g(x_m)}} \frac{\partial(\sqrt{g} g^{\mu\nu})}{\partial x^\nu} + \frac{\eta^\mu_m}{\sqrt{\Delta t}} \right] g_{\mu\sigma}(x^\sigma_{m+1} - x^\sigma_m) \\ &= \frac{1}{2} \left[\frac{\partial U(x_{m+1}, a_{m+1})}{\partial x^\mu} + \frac{\partial U(x_m, a_m)}{\partial x^\mu} \right] (x^\mu_{m+1} - x^\mu_m). \end{aligned} \quad \text{(A.73)}$$

By direct expansion around $(\bar{x}, \bar{a}) \equiv (\frac{1}{2}(x_m + x_{m+1}), \frac{1}{2}(a_m + a_{m+1}))$, we can verify that error in the energy balance, $\Delta U = \Delta' W + \Delta' Q$, is $\mathcal{O}(\Delta t^{3/2})$. Recall that the error must be at most $\mathcal{O}(\Delta t^{3/2})$ to assure the correct energetics (Sect. 4.1.2.5).

---

[35] Hereafter we use Einstein summation convention for the Greek indices appearing pairwise; $A^\mu B_\mu$ implies $\sum_\mu A^\mu B_\mu$.

[36] On the curved space (Riemannian manifold), the first-order scheme for Brownian motion is given in [21, 22], but we will need the order of $\Delta t^{1.5}$ accuracy. [cf. A.B. Cruzeiro, C. Alves, Monte-Carlo simulation of stochastic differential systems — a geometrical approach (preprint).]

Therefore, the above result assures the convergence with $\Delta t \rightarrow 0$ to the correct energy balance over a finite time interval.

A.4.8 General Growth Process: Insertion of Functions as well as Variables: Appendix to Sect. 4.3

For illustration purpose, let us introduce a toy model that imitates (very roughly) the growth of actin gel. See Fig. A.2:

1. The system consists of Hookean springs (wiggly lines) on the substrates (base line), which are vertically joined by beads (open circles).
2. The external force $a(t)$ is applied on the uppermost end of the system (filled circle).
3. The insertion and removal of the springs occur only at the substrate ($x = 0$).
4. The positions of the beads are denoted by $\{x_1, \ldots, x_n\}$ from top to bottom, where n is also a stochastic variable. When there are n springs, $x_{n+1} \equiv 0$.
5. The natural length ℓ_i and the elastic stiffness k_i of a Hookean spring are determined *when* the spring is inserted. They depend on the force on the lowermost spring before the insertion (state-to-function information).

While $\{x_i\}$ specify the state of the gel, $\{\ell_i\}$ and $\{k_i\}$ are the parameters of the gel's function as an elastic body. Upon the creation of $(n + 1)$th spring, the state variable x_{n+1}, and other positions, are determined after the functional parameters, (k_{n+1}, ℓ_{n+1}), are specified (function-to-state information).

Through the loop of (state-to-function information) and (function-to-state information), the growth of an open system carries the memory of its past state.[37]

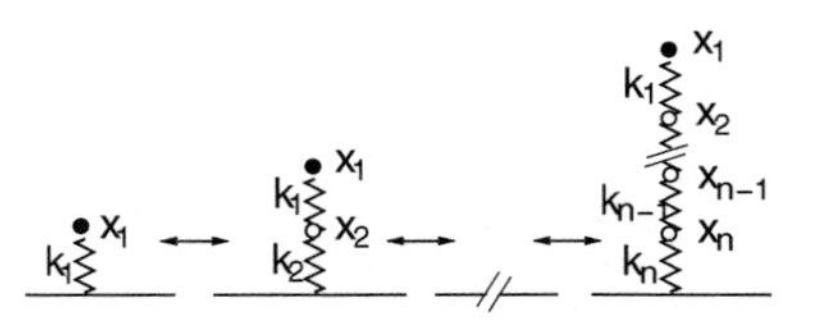

Fig. A.2 A simplified model of gel growth from the substrate. From the *left* to the *right*, there are $n = 1$, 2, and n springs incorporated into the gel. When the $(n + 1)$th spring is inserted, not only the variable x_{n+1} for a new joint appears, but also a new energy function, $(k_{n+1}/2)(x_{n+1} - \ell_{n+1})^2$, with new parameters k_{n+1} and ℓ_{n+1} appear

[37] The usage of the words "state" and "function" may not be proper since the state influences the function also. Lambda calculus [23] describes object/subject duality.

A.5 Appendix to Chap. 5

A.5.1 Statistical Mechanical Derivation of (5.19): Appendix to Sect. 5.2.1.4

A.5.1.1 Super-Entire System

We first imagine a closed super-entire system of the volume Ω_{sup} and the number of particles N_{sup}. We define its Helmholtz free energy F_{sup} through

$$e^{-F_{\text{sup}}/k_{\text{B}}T} = \int_{\Omega_{\text{sup}}} e^{-E_{\text{sup}}/k_{\text{B}}T} d^{N_{\text{sup}}}x. \tag{A.74}$$

Since there is no thermodynamic operation to change the parameters Ω_{sup} and N_{sup}, we can arbitrarily choose the additive constant in its free energy,

A.5.1.2 Open Entire System

Then we separate from the super-entire system an "entire system" of the macroscopic volume $\Omega_{\text{tot}}(\ll \Omega_{\text{sup}})$. Assuming a weak coupling between Ω_{tot} and $(\Omega_{\text{sup}}\backslash \Omega_{\text{tot}})$, we have the following relation between F_{sup} and F_{tot}:

$$e^{-F_{\text{sup}}/k_{\text{B}}T} = \sum_{N_{\text{tot}}=0}^{N_{\text{sup}}} \frac{N_{\text{sup}}!\, e^{-F_{\text{tot}}/k_{\text{B}}T}}{(N_{\text{sup}} - N_{\text{tot}})!} \int_{(\Omega_{\text{sup}}\backslash\Omega_{\text{tot}})} e^{-E^{c}_{N_{\text{sup}}-N_{\text{tot}}}/k_{\text{B}}T} d^{N_{\text{sup}}-N_{\text{tot}}}x, \tag{A.75}$$

where F_{tot} is the Helmholtz free energy of the entire system, defined by

$$e^{-F_{\text{tot}}/k_{\text{B}}T} = \frac{1}{N_{\text{tot}}!} \int_{\Omega_{\text{tot}}} e^{-E_{\text{tot}}/k_{\text{B}}T} d^{N_{\text{tot}}}x, \tag{A.76}$$

We approximate the integral (A.75) assuming ideal gas $\int_{(\Omega_{\text{sup}}\backslash\Omega_{\text{tot}})} e^{-E_{N_{\text{sup}}-N_{\text{tot}}}/k_{\text{B}}T} d^{N_{\text{sup}}-N_{\text{tot}}}x = (\|\Omega_{\text{sup}}\backslash\Omega_{\text{tot}}\| A)^{N_{\text{sup}}-N_{\text{tot}}}$, where A does not depend on $\|\Omega_{\text{sup}}\backslash\Omega_{\text{tot}}\|$ or the number of particles. Using the approximation, $N_{\text{sup}}!/(N_{\text{sup}} - N_{\text{tot}})! \simeq {N_{\text{sup}}}^{N_{\text{tot}}}$, we have

$$e^{-F_{\text{sup}}/k_{\text{B}}T} \simeq \sum_{N_{\text{tot}}=0}^{N_{\text{sup}}} e^{-F_{\text{tot}}/k_{\text{B}}T} \left(\frac{N_{\text{sup}}}{\|\Omega_{\text{sup}}\backslash\Omega_{\text{tot}}\| A}\right)^{N_{\text{tot}}} \times (\|\Omega_{\text{sup}}\backslash\Omega_{\text{tot}}\| A)^{N_{\text{sup}}}.$$

Identifying $\left(\frac{N_{\text{tot}}}{\|\Omega_{\text{sup}}\backslash\Omega_{\text{tot}}\| A}\right) = e^{\mu/k_{\text{B}}T}$, we obtain

$$e^{-F_{\text{sup}}/k_{\text{B}}T} \simeq \left[\sum_{N_{\text{tot}}=0}^{N_{\text{sup}}} e^{-[F_{\text{tot}}-\mu N_{\text{tot}}]/k_{\text{B}}T}\right] \times e^{(\text{indep. } N_{\text{tot}})}.$$

Although the last exponent is not extensive with respect to $N_{\rm sup}$ or to $\Omega_{\rm sup}$, there is no causality because our operations will not change these parameters. We thus justified $(N_{\rm tot}!)^{-1}$ in the definition (A.76).

A.5.1.3 Open (Small) System

We suppose that $\Omega_{\rm tot}$ consists of a system Ω and its "environment" Ω^c. Again we assume that the interaction between the "environment" and the system is negligible. Equation (A.76) can then be rewritten as

$$e^{-F_{\rm tot}/k_{\rm B}T} = \sum_{n=0}^{N_{\rm tot}} \frac{1}{(N_{\rm tot}-n)!} e^{-F^{(n)}/k_{\rm B}T} \int_{(\Omega^c)} e^{-E^c_{N_{\rm tot}-n}/k_{\rm B}T} d^{N_{\rm tot}-n}x,$$

where $F^{(n)}$ has been defined in (5.22) in the main text. Repeating the same calculation as above, we obtain

$$e^{-F_{\rm tot}/k_{\rm B}T} \simeq e^{-\mu N_{\rm tot}/k_{\rm B}T} \sum_{n=0}^{N_{\rm tot}} e^{-(F^{(n)}-\mu n)/k_{\rm B}T}.$$

By transferring the global factor on the right-hand side to the left-hand side, we finally obtain

$$\exp\left[-\frac{F_{\rm tot}-\mu N_{\rm tot}}{k_{\rm B}T}\right] \simeq \sum_{n=0}^{N_{\rm tot}} e^{-(F^{(n)}-\mu n)/k_{\rm B}T}. \tag{A.77}$$

From this expression, (5.19) with the definition (5.21) is derived by taking the limit of $N_{\rm tot} \propto \|\Omega_{\rm tot}\| \to \infty$. The equilibrium probability P_n to find n particles in the open system is found from (5.21), i.e.,

$$P_n = e^{(J-F^{(n)}+\mu n)/k_{\rm B}T}. \tag{A.78}$$

A.5.2 *Quasistatic Transport of Particle: Appendix to Sect. 5.2.3*

The analysis of the transport is in some aspect analogous to that of the "Berry phase" [24] in quantum physics. Since $J[P^{(\rm eq)}] = 0$ as mentioned in the text, we need to take into account the first-order correction in the time derivative, $\dot{a}(t)$, to this probability:

$$P(x,t) = P^{(\rm eq)}(x,a(t)) + \sum_\alpha \varphi_\alpha(x,a(t))\,\dot{a}_\alpha(t) + \mathcal{O}(|\dot{a}|^2), \tag{A.79}$$

where $\dot{a} = da/dt$, and $\varphi(x, a) = \{\varphi_\alpha\}$ is a function having the same number of components as $a \equiv \{a_\alpha\}$. The sum is taken for all the components of these quantities. The probability current $J[P]$, which is a linear functional of P, is then written as

$$J[P] = \sum_\alpha J[\varphi_\alpha(x, a(t))]\dot{a}_\alpha(t) + \mathcal{O}(|\dot{a}|^2). \tag{A.80}$$

The average number of particles, $\mathcal{N}$, that cross a fixed spatial point (e.g., at x_0) to the right per cycle of operation is given by the time integral of $J[P]_{x=x_0}$. From (A.80) this time integral can be written as an integral over $\tilde{a}$ along the closed trajectory on the parameter space [25]:[38]

$$\mathcal{N} = \oint \sum_\alpha J[\varphi_\alpha(x_0, \tilde{a})]\, d\tilde{a}_\alpha \qquad \text{(quasistatic limit)}. \tag{A.81}$$

$\{J[\varphi_\alpha(x_0, \hat{a})]\}$ is not of the gradient form with respect to the parameter.[39] Its line integral along a closed loop generally, therefore, does not vanish. The result for $\mathcal{N}$ is, therefore, generally nonzero. Although the integral of (A.81) looks purely static, with no dependence on $\dot{a}$, this effect is kinetic: $J[P]$ includes the kinetic coefficient γ.

A.5.3 Derivation of (5.37): Appendix to Sect. 5.3.1

We will show (5.37) in several steps.

A.5.3.1 Perturbative Treatment of Fokker–Planck Equation and Irreversible Work

For slow variation of $a(t)$, the probability distribution $P(x, t)$ obeying the Fokker–Planck equation[40] is almost the canonical equilibrium probability, $P^{(\mathrm{eq})}(x, a(t)) = e^{\beta(F(a,\beta)-U(x,a))}$ ($\beta = 1/k_B T$), with a small correction proportional to $da(t)/dt = \mathcal{O}((\tau_{\mathrm{op}})^{-1})$.[41] (See, (A.79) of Sect. 5.2.3.4.) If we write this correction as P_1 ($= \mathcal{O}((\tau_{\mathrm{op}})^{-1})$), the first-order perturbation of the Fokker–Planck equation, $\frac{\partial P}{\partial t} = \mathcal{L}(a(t))P$, with $P(x, t) = P^{(\mathrm{eq})}(x, a(t)) + P_1(x, t)$ is as follows:

---

[38] We use $\dot{a}(t)dt = d\tilde{a}(s)/ds \cdot ds = d\tilde{a}$.

[39] That is, the sum called the 1-form, $\sum_\alpha J[\varphi_\alpha(x_0, \hat{a})]da_\alpha$, cannot be written as $d\Phi(a)$ with some function Φ, called exact 1-form.

[40] Equation (1.73) with γ and T being constants.

[41] With $s \equiv t/\tau_{\mathrm{op}}$, we have $da(t)/dt = (\tau_{\mathrm{op}})^{-1} d\tilde{a}(s)/ds$.

$$\frac{\partial P^{(\text{eq})}}{\partial t} = \mathcal{L} P_1. \tag{A.82}$$

The left-hand side of (A.82) is $\mathcal{O}((\tau_{\text{op}})^{-1})$. From the explicit form of $P^{(\text{eq})}$, this term satisfies

$$\frac{\partial P^{(\text{eq})}(x, a(t))}{\partial t} = -\beta \frac{da}{dt} P^{(\text{eq})}(x, a(t)) \left(\frac{\partial U}{\partial a} - \left\langle \frac{\partial U}{\partial a} \right\rangle \right). \tag{A.83}$$

P_1 then depends on the time only through $a(t)$ and its time derivative.

The average work W up to $\mathcal{O}((\tau_{\text{op}})^{-1})$ is given as $\langle W \rangle = \int \left[\int \frac{\partial U}{\partial a} (P^{(\text{eq})} + P_1) dx \right] da$. The average irreversible work $\langle W_{\text{irr}} \rangle$ is then

$$\langle W_{\text{irr}} \rangle = \int_{t=0}^{t=\tau_{\text{op}}} \left[\int \frac{\partial U}{\partial a} P_1 dx \right] da(t) + \mathcal{O}((\tau_{\text{op}})^{-2}). \tag{A.84}$$

A.5.3.2 Green Function, $g(x, x'; a)$

We define the function $g(x, x'; a)$ by the following equation:

$$\mathcal{L}(a)[P^{(\text{eq})}(x, a) g(x, x'; a)] = \delta(x - x'), \tag{A.85}$$

where $\mathcal{L}(a)$ operates on the x variable to the right. g depends on β and a through this equation. We require that g vanishes at the infinite boundaries.

Using the explicit form of the Gibbs distribution, and also the explicit form of the Fokker–Planck operator, $\mathcal{L}(a)$, we can rewrite (A.85) as (5.40) in the text.

A.5.3.3 Solution for P_1

The equation for P_1, (A.82), can be solved using the Green function g: letting $P_1(x, t) = P^{(\text{eq})}(x, a(t)) \pi(x, a(t))$, then π is given as the superposition of $g(x, x', a)$ the weight $\partial P^{(\text{eq})}(x', a)/\partial t$ up to an additive constant. That is,

$$P_1 = P^{(\text{eq})} \left\{ g^* \frac{\partial P^{(\text{eq})}}{\partial t} + \chi \right\}, \tag{A.86}$$

where we have introduced the symbol * to denote $g^* f(x) \equiv \int g(x, x') f(x') dx'$. The additive constant, χ, is determined by recalling the normalization condition of the probability, $\int P dx = 1$. To $\mathcal{O}((\tau_{\text{op}})^{-1})$, this condition imposes $\int P_1 dx = 0$. With (A.86) the last condition yields

$$\chi = -\int P^{(\text{eq})} \left(g^* \frac{\partial P^{(\text{eq})}}{\partial t} \right) dx \equiv -\left\langle g^* \frac{\partial P^{(\text{eq})}}{\partial t} \right\rangle_{\text{eq}}. \tag{A.87}$$

A.5.3.4 The Average Irreversible Work

Substituting P_{l} thus obtained into (A.84) the average irreversible work $\langle W_{\mathrm{irr}}\rangle$ up to $\mathcal{O}((\tau_{\mathrm{op}})^{-1})$ is given as follows:

$$\langle W\rangle_{\mathrm{irr}} = \int_0^{\tau_{\mathrm{op}}}\left[\int \frac{\partial U}{\partial a}P^{(\mathrm{eq})}\left\{g^*\frac{\partial P^{(\mathrm{eq})}}{\partial t} - \left\langle g^*\frac{\partial P^{(\mathrm{eq})}}{\partial t}\right\rangle_{\mathrm{eq}}\right\}\right]\frac{da}{dt}dt$$

$$= \int_0^{\tau_{\mathrm{op}}}\left[\left\langle\frac{\partial U}{\partial a}\left(g^*\frac{\partial P^{(\mathrm{eq})}}{\partial t}\right)\right\rangle_{\mathrm{eq}} - \left\langle\frac{\partial U}{\partial a}\right\rangle_{\mathrm{eq}}\left\langle g^*\frac{\partial P^{(\mathrm{eq})}}{\partial t}\right\rangle_{\mathrm{eq}}\right]\frac{da}{dt}dt$$

$$+\mathcal{O}((\tau_{\mathrm{op}})^{-2}). \tag{A.88}$$

Substituting the expression (A.83) for $\partial P^{(\mathrm{eq})}/\partial t$, we have, after some rearrangement, the symmetric form (5.37) in the text.

A.5.4 Derivation of the Fluctuation Theorem (FT): Appendix to Sect. 5.4.2

The derivation below is adapted from [26].

A.5.4.1 Transition Rate and Path Probability

The probability that the system does not undergo any transition from a particular state i through the period $t' < s < t''$ is given as [27]

$$\mathsf{M}^{(i)}_{[t',t'']} \equiv \exp\left[-\int_{t'}^{t''}\sum_{j(\neq i)} w_{i\to j}(a(s))ds\right]. \tag{A.89}$$

We will consider a particular realization of the process, which starts from the state i_0 at time $\tau = 0$ and jumps at time t_α from the state $i_{\alpha-1}$ to the state i_α $(1 \le \alpha \le n)$ until $\tau = t$, where $t_\alpha < t_{\alpha+1}$.[42] We denote by $\mathrm{Prob}[i(\tau), a(\tau)]$ the *path probability* for the realization of such history $i(\tau)$, with the condition that the process starts with the canonical equilibrium state corresponding to the given initial parameter, $P^{\mathrm{eq}}_{i_0}(a(0))$. The path probability is constructed as follows:

$$\mathrm{Prob}[i(\tau), a(\tau)] = P^{\mathrm{eq}}_{i_0}(a(0))\mathsf{M}^{(i_0)}_{[0,t_1]}w_{i_0\to i_1}(a(t_1))\mathsf{M}^{(i_1)}_{[t_1,t_2]}w_{i_1\to i_2}(a(t_2)) \times$$
$$\cdots \times w_{i_{n-1}\to i_n}(a(t_n))\mathsf{M}^{(i_n)}_{[t_n,t]}. \tag{A.90}$$

[42] Hereafter in this section, the τ in $i(\tau)$ and $a(\tau)$ implies being a generic time and $i(\tau)$ means a particular realization or a history.

A.5.4.2 Time-Reversed Protocol and Time-Reversed Process

Under the time-reversed protocol $\tilde{a}(\tau) \equiv a(t-\tau)$ for $0 < \tau < t$, we consider the time-reversed process, $\tilde{i}(\tau) \equiv i(t-\tau)$. What corresponds to $\mathsf{M}^{(i)}_{[t',t'']}$ for $\tilde{a}(\tau)$, which we denote by $\tilde{\mathsf{M}}^{(i)}_{[t',t'']}$, satisfies the identity

$$\tilde{\mathsf{M}}^{(i)}_{[t',t'']} = \mathsf{M}^{(i)}_{[t-t'',t-t']}. \tag{A.91}$$

The path probability for the realization of the time-reversed process $\tilde{i}(\tau)$ under the time reversed protocol $\tilde{a}(\tau)$, in condition that the process starts with the canonical equilibrium state corresponding to the given (reversed) initial parameter, $P^{\mathrm{eq}}_{i_n}(a(t))$, is then given by

$$\begin{aligned}\mathrm{Prob}[\tilde{i}(\tau), \tilde{a}(\tau)] = P^{\mathrm{eq}}_{i_n}(a(t))\mathsf{M}^{(i_n)}_{[t_n,t]} w_{i_n \to i_{n-1}}(a(t_n))\mathsf{M}^{(i_{n-1})}_{[t_{n-1},t_n]} \times \\ \ldots \times w_{i_2 \to i_1}(a(t_2))\mathsf{M}^{(i_1)}_{[t_1,t_2]} w_{i_1 \to i_0}(a(t_1)))\mathsf{M}^{(i_0)}_{[0,t_1]}.\end{aligned} \tag{A.92}$$

A.5.4.3 Ratio of Path Probabilities of a Trajectory and its Time Reversal

If we take the ratio, $\mathrm{Prob}[\tilde{i}(\tau), \tilde{a}(\tau)]/\mathrm{Prob}[i(\tau), a(\tau)]$, we find that all the factors $\mathsf{M}^{(i)}_{\tau',\tau''}$ cancel out. The result is

$$\frac{\mathrm{Prob}[\tilde{i}(\tau), \tilde{a}(\tau)]}{\mathrm{Prob}[i(\tau), a(\tau)]} = \frac{w_{i_1 \to i_0}(a(t_1))w_{i_2 \to i_1}(a(t_2)) \cdots w_{i_n \to i_{n-1}}(a(t_n))P^{\mathrm{eq}}(i_n; a(t))}{P^{\mathrm{eq}}_{i_0}(a(0))w_{i_0 \to i_1}(a(t_1))w_{i_1 \to i_2}(a(t_2)) \cdots w_{i_{n-1} \to i_n}(a(t_n))}. \tag{A.93}$$

Then we can use the detailed balance condition (5.62) to eliminate all the transition rates:

$$\frac{\mathrm{Prob}[\tilde{i}(\tau), \tilde{a}(\tau)]}{\mathrm{Prob}[i(\tau), a(\tau)]} = \frac{P^{\mathrm{eq}}_{i_0}(a(t_1))P^{\mathrm{eq}}_{i_1}(a(t_2)) \cdots P^{\mathrm{eq}}_{i_{n-1}}(a(t_n))P^{\mathrm{eq}}_{i_n}(a(t))}{P^{\mathrm{eq}}_{i_0}(a(0))P^{\mathrm{eq}}_{i_1}(a(t_1))P^{\mathrm{eq}}_{i_2}(a(t_2)) \cdots P^{\mathrm{eq}}_{i_n}(a(t_n))}. \tag{A.94}$$

A.5.4.4 A lemma

We introduce $R[i(\tau), a(\tau)]$ through the definition,[43]

$$e^{-R[i(\tau),a(\tau)]} \equiv \frac{\mathrm{Prob}[\tilde{i}(\tau), \tilde{a}(\tau)]}{\mathrm{Prob}[i(\tau), a(\tau)]}. \tag{A.95}$$

This quantity is a functional of the history, $i(\tau)$, under a given protocol of the external parameter, $a(\tau)$, *if* we know how the right-hand side depends on $i(\tau)$ and $a(\tau)$. At least we know the property,

$$R[\tilde{i}(\tau), \tilde{a}(\tau)] = -R[i(\tau), a(\tau)], \tag{A.96}$$

---

[43] [31]. See also [30].

which holds immediately from the definition (A.95). The probability distribution of $R[i(\tau), a(\tau)]$ can be deduced through the average using the path probability, $\text{Prob}[i(\tau), a(\tau)]$:

$$P_R(r) \equiv \langle \delta(r - R[i(\tau), a(\tau)]) \rangle, \tag{A.97}$$

where the average of $\mathcal{O}[i(\tau)]$ over the ensemble of processes is defined by

$$\langle \mathcal{O}[i(\tau)] \rangle \equiv \sum_{\{i(\tau)\}} \mathcal{O}[i(\tau)] \text{Prob}[i(\tau), a(\tau)]. \tag{A.98}$$

The following equality holds [28, 29] (see also [30]):

$$\begin{aligned}
e^{-r} P_R(r) &= \langle e^{-R[i(\tau),a(\tau)]} \delta(r - R[i(\tau), a(\tau)]) \rangle \\
&= \left\langle \frac{\text{Prob}[\tilde{i}(\tau), \tilde{a}(\tau)]}{\text{Prob}[i(\tau), a(\tau)]} \delta(r - R[i(\tau), a(\tau)]) \right\rangle \\
&= \sum_{\{i(\tau)\}} \text{Prob}[\tilde{i}(\tau), \tilde{a}(\tau)] \delta(r - R[i(\tau), a(\tau)]) \\
&= \sum_{\{i(\tau)\}} \text{Prob}[\tilde{i}(\tau), \tilde{a}(\tau)] \delta(r + R[\tilde{i}(\tau), \tilde{a}(\tau)]) \\
&= P_R(-r),
\end{aligned} \tag{A.99}$$

where (A.96) is used for the second last equality.

Using $P_i^{\text{eq}}(a) = \exp[\beta(F(a,\beta) - E_i(a,\beta))]$, the right-hand side of (A.94) is equal to $\exp[\beta(\Delta F - W)]$, where $\Delta F = F(a(t), \beta) - F(a(0), \beta)$, and W is given by (4.20). Therefore, $R[i(\tau), a(\tau)] = \beta(W - \Delta F) \equiv \beta W_{\text{irr}}$ is (β times) the irreversible work. Then the Lemma $e^{-r} P_R(r) = P_R(-r)$ gives a law of the probability density of the irreversible work (5.63) in the text.

A.6 Appendix to Chap. 6

(There is no appendix to Chap. 6)

A.7 Appendix to Chap. 7

A.7.1 Derivation of (7.4): Appendix to Sect. 7.1.1.4

The average work $\langle W \rangle_{[a^*, a_{\text{f}}]}$ after the crossover of timescales, i.e., during $a^* < a(t) \le a_{\text{f}}$, is

$$\langle W \rangle_{[a^*,a_f]} = \int_{a^*}^{a_f} \left\{ \int_{x\in\Omega} \mathcal{P}(x,t) \frac{\partial U(x,a)}{\partial a} dx \right\} da. \tag{A.100}$$

Since the potential barrier is raised beyond $U(0, a^*)$, we can expect that $\mathcal{P}(x, t)$ inside the barrier region Ω is smaller than $\mathcal{P}^{\rm eq}(x, a^*; T)$.[44] Then we can have an upper bound of $\langle W \rangle_{[a^*,a_f]}$:

$$\begin{aligned} \langle W \rangle_{[a^*,a_f]} &\leq \int_{a^*}^{a_f} \left\{ \int_{x\in\Omega} \mathcal{P}^{\rm eq}(x,a^*;T) \frac{\partial U(x,a)}{\partial a} dx \right\} da \\ &\leq [U(0,a_f) - U(0,a^*)] \int_{x\in\Omega} \mathcal{P}^{\rm eq}(x,a^*;T) dx, \end{aligned} \tag{A.101}$$

where we have used the inequality, $U(0, a_f) - U(0, a^*) \geq U(x, a_f) - U(x, a^*)$. We can then roughly estimate the integral above, i.e., the probability to find the particle within Ω, to be $\simeq c(\delta/L)\, e^{-U(0,a^*)/k_BT}$ with c being a numerical constant. Therefore, we find the result (7.4) in the text.

A.7.2 Simple Model of Aging and Plastic Flow: Appendix to Sect. 7.1.1.6

There are very interesting cases where $\tau_{\rm sys}$ adjusts itself to approach and remains in the proximity of $\tau_{\rm op}$. Such cases are observed when a system contains the *internal* feedback mechanism from the fluctuating variables (like x in (7.7)) to the control parameters (like a or b, ibid). The case where $\tau_{\rm sys}$ approaches $\tau_{\rm op}$ from *below*,

$$\tau_{\rm sys} \ll \tau_{\rm op} \qquad \longrightarrow \qquad \tau_{\rm sys} \simeq \tau_{\rm op}, \tag{A.102}$$

will be referred as aging,[45] while the case where $\tau_{\rm sys}$ approaches $\tau_{\rm op}$ from *above*,

$$\tau_{\rm sys} \gg \tau_{\rm op} \qquad \longrightarrow \qquad \tau_{\rm sys} \simeq \tau_{\rm op}, \tag{A.103}$$

will be referred as plastic flow.

A.7.2.1 Aging

Figure A.3 shows a typical situation of what we call aging. For each double well, a "particle" (a black dot) is attached to a spring under tension. The tension favors the thermally assisted transition from the left well to the right well. If the distance

---

[44] We admit that it is not rigorous: The exception will occur if the tail of the barrier is too much extended, and the particle excluded from the barrier top becomes stagnant in that tail region.

[45] Aging can be defined in different ways depending on the context. Here we adopt the version of P. G. de Gennes (Lecture at the Collège de France, 2002).

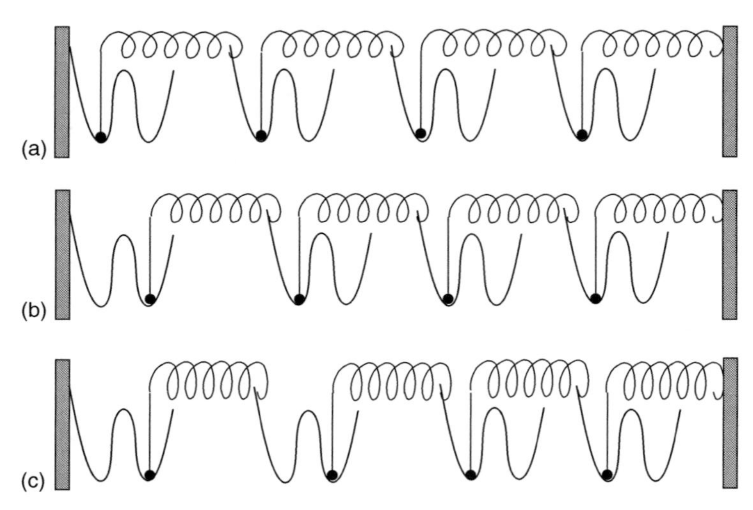

Fig. A.3 A schematic model of aging: Both the masses (*black dots*) and the double-well potential are mobile. The time proceeds from (**a**) to (**c**). The relaxation time τ_{sys} increases with the (observation) time t

between the two walls is fixed, the transition in one of the double wells leads to the diminution of the tension on the springs. Thus the system remains in (b) for a time longer than in (a) until the further transition to, for example, the state (c). In this manner, as τ_{op} proceeds with the time of observation, the transition is observed when $\tau_{\mathrm{sys}} \simeq \tau_{\mathrm{op}}$ is attained. We, therefore, have $\tau_{\mathrm{sys}} = t$. As consequence, the tension of the spring, to which τ_{sys} is exponentially related, decreases logarithmically in time.

A.7.2.2 Plastic Flow

Figure A.4 shows a typical situation of what we call plastic flow. Unlike aging, the distance between the walls is increased slowly in time. The increase in the tension on the springs causes the diminution of the relaxation time, τ_{sys}. Once the transition occurs after a time $\sim \tau_{\mathrm{sys}}$, the tension on the springs is relaxed partially, leading to the transient increase of τ_{sys}.

In the standard view, aging is the nonequilibrium state in which an "age" evolves as state parameter, while plastic flow is the nonequilibrium state in which the elastic displacements are continually updated. Here we view the latter phenomenon from the same viewpoint as the former. The dynamics of timescales gives us a common viewpoint to discuss phenomena of slow dynamics. The dynamics of heterogeneously distributed relaxation times exhibits another aspects, such as the internal stresses and memory effect [32–34].

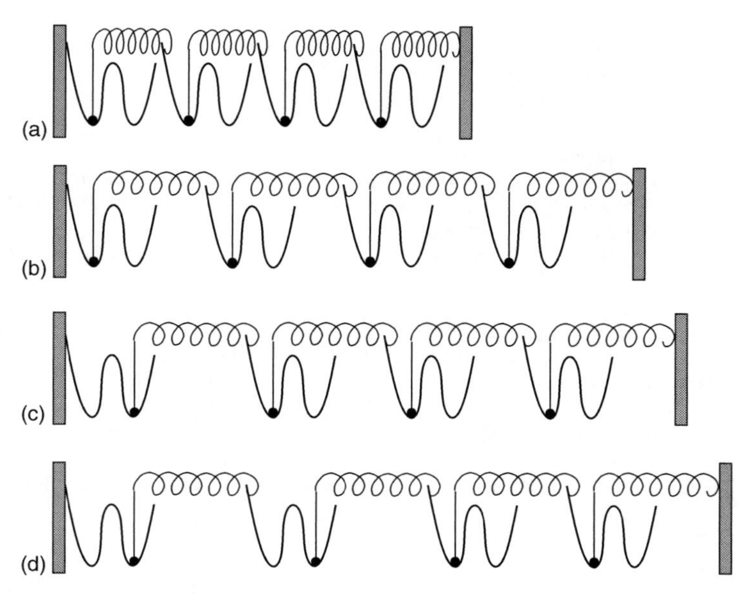

Fig. A.4 A schematic model of plastic flow: The time proceeds from (**a**) to (**d**)

A.8 Appendix to Chap. 8

(There is no appendix to Chap. 8)

References

1. K. Sato, Y. Ito, T. Yomo, K. Kaneko, PNAS **100**, 14086 (2003)
2. L.F. Cugliandolo, J. Kurchan, L. Peliti, Phys. Rev. E **55**, 3898 (1997)
3. H. Mori, Prog. Theor. Phys. **33**, 423 (1965)
4. K. Kawasaki, J. Phys. A **6**, 1289 (1973)
5. S. Nordholm, R. Zwanzig, J. Stat. Phys. **13**, 347 (1975)
6. R. Zwanzig, Phys. Rev. **124**, 983 (1961)
7. R. Kubo, M. Toda, N. Hashitsume, *Statistical Physics II: Nonequilibrium Statistical Mechanics*, 2nd edn. (Springer-Verlag, Berlin, 1991)
8. G. Milstein, *Numerical Integration of Stochastic Differential Equations* (Kluwer Academic, Dordrecht, 1995)
9. H. Qian, J. Math. Chem. **27**, 219 (2000)
10. T. Shibata, K. Sekimoto, J. Phys. Soc. Jpn. **69**, 2455 (2000)
11. I. Prigogine, R. Defay (translated by D.H. Everett), *Chemical Thermodynamics* (Longmans, Green & Co., London, 1954)
12. T.M. Cover, J.A. Thomas, *Elements of Information Theory* (John Wiley & Sons, Inc., New York, 1991)
13. L.D. Landau, E.M. Lifshitz, *Fluid Mechanics (Course of Theoretical Physics, Volume 6)*, 2nd edn. (Reed Educational and Professional Publishing Ltd, Oxford, 2000)
14. E.J. Hinch, J. Fluid Mech. **72**, 499 (1975)

15. P.N. Pusey, R.J.A. Tough, J. Phys. A: Math. Gen. **15**, 1291 (1982)
16. M. Tokuyama, I. Oppenheim, Phys. Rev. E **50**, R16 (1994)
17. M. Tokuyama, I. Oppenheim, Physica A **216**, 85 (1995)
18. R. Mochizuki, Prog. Theor. Phys. **88**, L1233 (1992)
19. G.G. Batrouni, H. Kawai, P. Rossi, J. Math. Phys. **27**, 1646 (1986)
20. P.A.M. Dirac, *General Theory of Relativity* (John Wiley & Sons, Inc., New York, 1975)
21. A. Cruzeiro, P. Malliavin, A. Thalmaier, C. R. Acad. Sci. Paris, Ser. I **338**, 481 (2004)
22. A. Cruzeiro, P. Malliavin, Stochastic Process. Appl. **116**, 1088 (2006)
23. H.P. Barendregt, *The Lambda Calculus (Studies in Logic and the foundation of mathematics, vol. 103)*, Revised edn. (Elsevier, Amsterdam, 1984)
24. M.V. Berry, Proc. Roy. Soc. London, A **392**, 45 (1984)
25. J.M.R. Parrondo, Phys. Rev. E **57**, 7297 (1998)
26. U. Seifert, J. Phys. A: Math. Gen. **37**, L517 (2004)
27. D. Gillespie, J. Comput. Phys. **22**, 403 (1976)
28. J.L. Lebowitz, H. Spohn, J. Stat. Phys. **95**, 333 (1999)
29. G.E. Crooks, Phys. Rev. E **61**, 2361 (2000)
30. C. Maes, *On the Origin and the Use of fluctuation Relations for the Entropy*: Poincaré Seminar 2003, ed. by J. Dalibard, B. Duplantier, V. Rivasseau (Birkhäuser Verlag, Basel, 2004), pp. 145–191
31. G.E. Crooks, J. Stat. Phys. **90**, 1481 (1998)
32. T. Ooshida, K. Sekimoto, Phys. Rev. Lett. **95**, 108301 (2005)
33. Y. Miyamoto, K. Fukao, H. Yamao, K. Sekimoto, Phys. Rev. Lett. **88**, 225504 (2002)
34. K. Sekimoto, in *Chemomechanical Instabilities in Responsive Materials (NATO Science for Peace and Security Series A: Chemistry and Biology)*, ed. by P. Borckmans, et al., (Springer, New York, 2009)

Index